THE NETWORK RELIABILITY OF TRANSPORT

PROCEEDINGS OF THE 1ST INTERNATIONAL SYMPOSIUM ON TRANSPORTATION NETWORK RELIABILITY (INSTR)

Related books

BELL (ed.)	Transportation Networks: Recent Methodological Advances
DAGANZO	Fundamentals of Transportation and Traffic Operations
GRIFFITHS (ed.)	Mathematics in Transport Planning and Control
HENSHER (ed.)	Travel Behaviour Research: The Leading Edge
HENSHER & BUTTON (eds.)	Handbooks in Transport Series
LAM & BELL (eds.)	Advanced Modeling for Transit Operations and Service Planning
MAHMASSANI (ed.)	In Perpetual Motion: Travel Behaviour Research Opportunities and Application Challenges
TANIGUCHI	City Logistics
TAYLOR (ed.)	Transportation and Traffic Theory in the 21st Century
TILANUS (ed.)	Information Systems in Logistics and Transportation

Related Journals

Transport Policy;
Editor: M. Ben-Akiva

International Journal of Transport Management:
Editor: M. McDonald

Journal of Transport Geography
Editor: R.D. Knowles

Transportation Research A : Policy and Practice
Editor: F.A. Haight

Transportation Research B: Methodological
Editor: F.A. Haight

Transportation Research C: Emerging Technologies
Editor: S. Ritchie

Transportation Research D: Transport and Environment
Editor: K. Button

Transportation Research E: Logistics and Transportation Review
Editor: W. Talley

transport connect

THE NETWORK RELIABILITY OF TRANSPORT

PROCEEDINGS OF THE 1ST INTERNATIONAL SYMPOSIUM ON TRANSPORTATION NETWORK RELIABILITY (INSTR)

EDITED BY

MICHAEL G.H. BELL
Imperial College London, UK

YASUNORI IIDA
Kyoto University, Japan

2003

PERGAMON
An Imprint of Elsevier Science

Amsterdam – Boston – London – New York – Oxford – Paris
San Diego – San Francisco – Singapore – Sydney - Tokyo

ELSEVIER SCIENCE Ltd
The Boulevard, Langford Lane
Kidlington, Oxford OX5 1GB, UK

First edition 2003

Library of Congress Cataloging in Publication Data
A catalog record from the Library of Congress has been applied for.

British Library Cataloguing in Publication Data
A catalogue record from the British Library has been applied for.

ISBN: 0 080 44109 2

♾ The paper used in this publication meets the requirements of ANSI/NISO Z39.48-1992 (Permanence of Paper).
Printed in The Netherlands.

Preface

The reliability of transportation networks has become an increasingly important issue as sustained economic growth and improvements to the quality of life around the world lead to increases in the value of time. Consequently, schedules and routes need to be able to accommodate the unexpected, like accidents, disasters or traffic flow fluctuations, with as little loss in operational efficiency as possible. Sources of unreliability include variation of demand and supply. People in the 21st century will desire a more stable transportation system with less travel time uncertainty. It is widely expected that network reliability analysis will play a more important role in the planning, design and management of transportation facilities and networks in the future.

The First International Symposium on Transport Network Reliability (INSTR) was held at Kyoto International Community House, Kyoto, Japan on 31st, July and 1st, August in 2001. The aim of the symposium was to bring together researchers and professionals interested in transportation network reliability to discuss both recent research topics and future directions in this expanding research field. Fifty-five persons participated and thirty-eight papers were presented from all over the world.

This book, The Network Reliability of Transport, is an outcome of the symposium, consisting of twenty-four selected papers. It covers various aspects of transport network reliability, such as definitions and methodological developments for reliability indices, behavioural analysis under uncertainty, evaluation methods for the disaster resistance of transport networks, and simulation / observation of travel time reliability. We believe that this book successfully encapsulates current understanding of transport network reliability and will become a useful reference for future research activities.

Yasunori Iida
Chairperson, Organising Committee of the first INSTR
Professor, Kyoto University, Japan

Michael G. H. Bell
Chairperson, International Scientific Committee of the first INSTR
Professor, Imperial College London, United Kingdom

September, 2002

Contributors

Takamasa Akiyama	Department of Civil Engineering, Gifu University, Gifu, Japan
Zarko Andjic	Gabites Porter Consultants, Christchurch, New Zealand
Yasuo Asakura	Graduate School of Science and Technology, Kobe University, Kobe, Japan
Katsuhiko Asaoka	Osaka Port Transport System Co., Ltd., Osaka, Japan
Michael G H Bell	Centre for Transport Studies, Imperial College London, UK, London, UK
Katja Berdica	Royal Institute of Technology, Stockholm, Sweden
Chris Cassir	Institute for Transport Research, DLR Berlin, Germany
Anthony Chen	Department of Civil and Environmental Engineering, Utah State University, USA
Erica Dalziell	vivas ltd, Strategic Risk Management Consultants, London, UK
Glen M D'Este	PPK Environment and Infrastructure, Brisbane, Australia
Satoshi Fujii	Department of Civil Engineering, Tokyo Institute of Technology, Tokyo, Japan
Eiji Hato	Civil and Environmental Eng., Ehime University, Matsuyama, Japan
Hitoshi Ieda	Department of Civil Engineering, University of Tokyo, Tokyo, Japan
Yasunori Iida	Department of Civil Engineering, Kyoto University, Kyoto, Japan
Kensuke Ikeda	Toyota Motors Ltd., Toyota, Japan
Hiroshi Inouye	Himeji Institute of Technology, Himeji, Japan
Zhaowang Ji	Department of Civil and Environmental Engineering, Utah State University, USA
Hiroyuki Kameda	Disaster Prevention Research Institute, Kyoto University, Japan
Masuo Kashiwadani	Civil and Environmental Eng., Ehime University, Matsuyama, Japan
Seiichiro Kawaratani	OMRON Corporation, Osaka, Japan
Shinya Kikuchi	Department of Civil and Environmental Engineering, University of Delaware, Delaware, USA
Ryuichi Kitamura	Department of Civil Engineering Systems, Kyoto University, Kyoto, Japan
Fumitaka Kurauchi	Department of Civil Engineering, Kyoto University, Kyoto, Japan
Tsz Hang Lam	Department of Civil Engineering, The Hong Kong University of Science and Technology, Kowloon, Hong Kong, P. R China
William H. K. Lam	Department of Civil and Structural Engineering, The Hong Kong Polytechnic University, Hong Kong, P.R China
Yung - Lung Lee	Department of Land Management and Development, Chang Jung Christian University, Taiwan, R. O. C.

David Levinson Department of Civil Engineering, University of Minnesota, USA

Yan Li Ritsumeikan Asia Pacific University, Oita, Japan

Anthony D. May Institute for Transport Studies, University of Leeds, Leeds, UK

Alan J Nicholson Department of Civil Engineering, University of Canterbury, Christchurch, New Zealand

Keiichi Ogawa Department of Environment Systems Engineering, Ritsumeikan University, Shiga, Japan

Will Recker Institute of Transportation Studies, University of California at Irvine, USA

Jan-Dirk Schmöcker Transport Operations Research Group, Newcastle University, Newcastle upon Tyne, UK

Simon P. Shepherd Institute for Transport Studies, University of Leeds, Leeds, UK

Hirofumi Shimada Ministry of Land, Infrastructure and Transport, Tokyo, Japan

Masumi Suganuma Department of Civil Engineering, Kyoto University, Kyoto, Japan

Agachai Sumalee Institute for Transport Studies, University of Leeds, Leeds, UK

Mei-Lam Tam Department of Civil and Structural Engineering, The Hong Kong Polytechnic University, Hong Kong, P.R China

Wilson H. Tang Department of Civil Engineering, The Hong Kong University of Science and Technology, Kowloon, Hong Kong, P. R China

Eiichi Taniguchi Department of Civil Engineering Systems, Kyoto University, Kyoto, Japan

Michael A P Taylor Transport Systems Centre, University of South Australia, Adelaide, Australia

Hiroshi Tsukaguchi Department of Environment Systems Engineering, Ritsumeikan University, Shiga, Japan

Nobuhiro Uno Department of Civil Engineering, Kyoto University, Kyoto, Japan

Hiroshi Wakabayashi Faculty of Urban Science, Meijo University, Kani, Japan

Tadashi Yamada Department of Social and Environmental Engineering, Hiroshima University, Higashi-Hiroshima, Japan

Hai Yang Department of Civil Engineering, The Hong Kong University of Science and Technology, Kowloon, Hong Kong, P. R China

Kuang - Yih Yeh Department of Urban Planning, National Cheng Kung University, Taiwan, R. O. C.

Yafeng Yin Institute of Transportation Studies, University of California, Berkeley, USA

Lei Zhang Department of Civil Engineering, University of Minnesota, USA

Henk J van Zuylen Delft University of Technology, Delft, the Netherlands

Contents

1

Assessing Transport Reliability: Malevolence and User Knowledge

Alan Nicholson, Department of Civil Engineering, University of Canterbury, Christchurch, New Zealand
Jan-Dirk Schmöcker, Transport Operations Research Group, Newcastle University, Newcastle upon Tyne, UK
Michael G H Bell, Centre for Transport Studies, Imperial College London, UK
Yasunori Iida, Department of Civil Engineering, Kyoto University, Kyoto, Japan

Abstract

The objective of this paper is to give an overview of various reliability concepts that have been developed in the last decades. The paper first summarises various indicators that have been developed in order to measure the reliability of a network and then looks at techniques to calculate these indicators. The usefulness and limitations of the different indicators is discussed. The paper suggests that there is no single perfect indicator but that the choice of indicator and technique depends on several factors, including the viewpoint of the analyst and the type and range of interventions being considered. In order to assess the impact of incidents the authors propose to distinguish between three types of intervention, namely "benevolent", "neutral" or random, and "malevolent". Also discussed is why the provision of up-to-date information to the traveller has a central role to play when trying to minimise the impact of an incident.

Introduction

Transport reliability has been of interest for many years. For instance, Garrison (1960) used graph theory concepts to assess the connectivity of the USA Interstate Highway system and the accessibility of major urban areas to the system. He also compared the accessibility of cities in the south-east of the USA to the Interstate Highway System, with their accessibility to the railroad system. There is a variety of measures of reliability, Chapman (1976) identifying sixteen possible measures of bus reliability.

In basic terms, a facility is 'reliable' if the expectations of users are almost always met, and the reliability increases as the frequency and/or consequence of failing to meet user expectations decrease. Expectations can vary considerably between users, from 'unrealistically high' to 'quite modest'. They can also vary within users (i.e. can vary with time), with the spread of the 'just-in-time' philosophy indicating an increase in the expectations of users of the transport system.

Transport networks are subject to two types of phenomena that may affect reliability:

(1) variations in the demand for transport services; and
(2) variations in the supply of transport services.

The latter has been primary focus of the research to date, with most studies investigating the effect of service disruptions (e.g. roads being closed or public transport services being cancelled) on the performance of the network.

In recent years there has been an increasing interest in the health, financial and environmental risks facing society, and this has led to the development of techniques and procedures for risk evaluation and management. A transport system user faces a number of risks, including the risk that his expectations will not be met (i.e. they may not complete their trip safely and/or in good time). Risk evaluation and management techniques involve considering both the probability and consequence of adverse events. For transport systems, there are numerous models for helping to identify the consequence of system degradation (e.g. increased travel times, distances and costs), but there is little experience in the identification of the probability of disruption.

One approach is to consider the worst-case scenario, where a network is under attack from a malevolent agency, which is seeking to cause the greatest possible adverse impact. Alternatively, one might consider an event where the intention is to minimise the adverse effect (e.g. the temporary partial closure of a traffic lane at a building site to facilitate construction). In this case there is no malevolence involved; indeed, the effort to minimise the adverse effect suggests the opposite of malevolence, namely benevolence. The actual situation might involve random degradations and thus lie somewhere between these two extreme cases. The case of random degradations can be termed 'neutral', in that neither malevolence nor benevolence are involved.

The failure probability for the component whose failure will have the greatest adverse impact will lie somewhere between zero and unity (the probabilities of a 'benevolent' and malevolent attack on that component, respectively). The difficulty is to estimate where in this range the probability lies. Similarly, it is difficult to estimate the probability of degradation for the other components.

Malevolence can affect both the probability and consequence of disruption. The more malevolent the persons involved in an attack, the more determined they are to succeed in causing disruption (i.e. the greater the probability of degradation) and the more determined they are to maximise the consequence of the degradation. An intervention can be fully characterised in terms of its location (i.e. which link), the level of degradation (i.e. whether the link is completely or partly closed, whether the public transport service has stopped or headways are more variable), and the duration of the degradation (i.e. whether long or short). As the level of malevolence increases, the greater will be the effort to degrade 'critical' components, and to maximise the level and duration of the degradation (i.e. to maximise the severity of the degradation).

The importance of a link for network reliability depends upon both its ability to withstand degradation (e.g. a bridge may be strengthened to enable it to survive the largest reasonably foreseeable earthquake) and the consequences of degradation (e.g. the extra user costs if a bridge fails, plus the costs of any repairs). The level of degradation would be characterised by the physical damage to the link (i.e. the effect on the link characteristics), but the effect of the degradation (i.e. travel disruption) is much broader, and includes the extra user costs.

The effect may well be mitigated (i.e. the consequence of degradation will be alleviated) through users adapting to the disruption. The response can take various forms, including:

(1) cancelling their trip (i.e. a change in trip generation);
(2) postponing their trip (i.e. a change in the temporal distribution of trips);
(3) choosing another destination (i.e. a change in the spatial distribution of trips);
(4) travelling by another mode (i.e. a change mode split);
(5) choosing a different route (i.e. a change in the traffic assignment).

When estimating the expected costs of disruption, allowing for no user response and all possible responses gives the worst-case and best-case scenarios respectively, given that an event has occurred.

Users need to have some information about the nature of any degradation in order to respond. If the user has a high level of information (i.e. information is provided well in advance, and route guidance is available once the trip has commenced), the range of available options is greater and the consequence of the degradation is reduced. If there is a low level of information (i.e. there is no advance warning or route guidance), the only option is to re-route once the user arrives at the location of the degradation or encounters the congestion caused by it. If the user is not familiar with the network, the scope for minimising the effect of the disruption associated with the degradation is less. If the user has access to radio broadcasts giving general advice on conditions in the network, along with experience driving in the network, then that might be considered an 'intermediate' level of user information.

Assuming that users are rational and will use whatever information they have to minimise the consequence of any disruption by choosing the best available responses for themselves, models for assessing the reliability of a network can be categorised according to the malevolence of the intervention and the level of user knowledge. The estimate of the expected cost of disruption will vary according to the type of intervention and level of user knowledge, as shown in Table 1. As the level of driver knowledge increases, the effect of an intervention will reduce as users have greater opportunity to mitigate the effect through changing their travel behaviour. As the level of malevolence increases (i.e. the intended effect becomes increasingly adverse), the actual effect should also increase, as users will generally be unable to adjust their behaviour sufficiently to compensate fully for the increased effect of the event.

The term 'attack' applies to an intervention involving a high level of malevolence. Such events differ from natural events (e.g. earthquakes, floods, road accidents), which are generally considered to be 'random' (i.e. they involve a substantial element of randomness) and to involve no malevolent intent.

Table 1 Effect of Intervention Type and User Knowledge on Expected Cost

	Type of Intervention		
Level of Knowledge	**Benevolent**(1)	**Neutral**(2)	**Malevolent**(3)
High	Low	Low/Medium	Medium
Medium	Low/Medium	Medium	Medium/High
Low	Medium	Medium/High	High

Note: (1) aim is impact minimisation
(2) random events
(3) aim is impact maximisation

If one assumes malicious intent when none exists, then the result is likely to be an over-estimate of the effect. Conversely, if one does not allow for malicious intent when it does exist, one will tend to under-estimate the effect of disruption.

One can obtain lower and upper bounds for the disruption effect, by considering the high-knowledge with benevolent-intent and the low-knowledge with malevolent-intent combinations, respectively. It may be that the lower and upper bounds are fairly similar (i.e. the effect of disruption is not sensitive to the level of knowledge and the type of intervention). If this is not the case, it will be necessary to identify the level of knowledge and malevolence in order to obtain an accurate estimate of the effect.

Before assessing transport network reliability, one must first decide how reliability is to be measured and then what modelling technique will be used to estimate the reliability.

MEASURES OF RELIABILITY

Introduction

Wakabayashi and Iida (1992) define reliability as "the probability of a device performing its purpose adequately for the period of time intended under the operating conditions encountered". This is probably one of the most commonly accepted definitions of reliability. It is, however, very general and the word "adequately" will have very different meanings to different users. Before every journey network users might have several questions regarding the reliability of the route they plan to take (as discussed later). The questions will become more important as the likelihood of disruptions in the network increases, and the range of questions shows that reliability has many different aspects. In some cases it might be of primary interest that the journey is finished by a certain time (e.g. just in time delivery), while in other cases it is more important to assess whether a disruption might be encountered on the journey (e.g. the risk of being involved in an incident when transporting hazardous materials).

Previous research has therefore developed several measures to highlight different aspects of reliability, and Berdica (2000) reviews different reliability concepts. Here we introduce some other concepts based on more recent research. It should be noted that the measures here are all very general and can be applied to all kinds of transport network. More specific measures might be introduced especially for scheduled transit networks (Chapman, 1976).

Variance

Some studies have involved using classical statistical measures of the variability of some measure of network performance (e.g. the variance of the travel time between pairs of origins and destinations) as a measure of reliability. It is generally assumed that the higher the variance relative to the mean, the lower is the reliability. When the variance is zero, there is complete predictability and reliability is maximised, but as the variance becomes very large, the predictability and reliability both reduce. If the expected travel times for two routes are equal, but the variance of the travel time for one route (A, say) is much less than for the other (B, say), as shown in Figure 1, then risk-neutral users should choose the route with the smaller probability of the travel time exceeding their threshold value. This is however only valid if the threshold value is larger than the mean travel time (is x_2, say). This is generally the case, but it is also possible to have situations where the threshold is lower than the mean travel time; for example, if delays have already occurred and it is no longer likely that the destination can be reached in time (the threshold is x_1, say), travellers might choose the route with the higher travel time variation to improve their (low) chances of arriving on time.

While an increase in the variance relative to the mean generally means a reduction in reliability, a doubling of the variance does not necessarily mean a halving of the reliability (i.e. the relationship between the reliability and variance is not necessarily linear). While the reliability is well defined when the variance is zero or very large, the relationship is not so clear for intermediate levels of variance, when the perceived level of reliability will depend upon the circumstances, as illustrated in Figure 2.

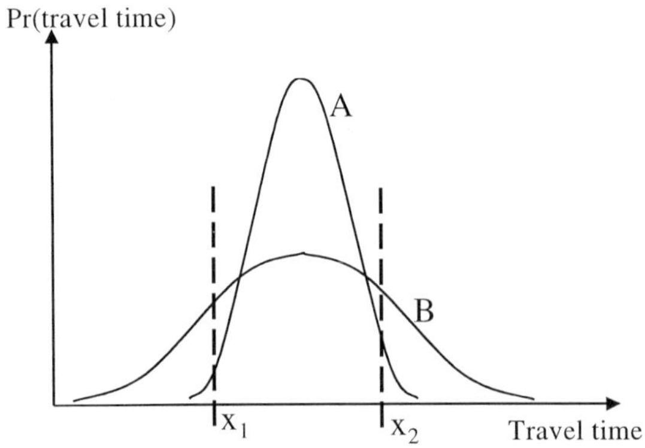

Figure 1 Two probability density functions of travel time

Two extreme situations are illustrated in Figure 2. Where users have no attractive alternative means of travel, they will be quite insensitive to low levels of variation in the performance measure and may perceive the reliability as being very high, with the perceived reliability falling slowly as the variance increases. If, however, the user has attractive alternatives, they may be very sensitive to even low levels of variation, with the perceived reliability falling quickly as the variance increase.

The variance is not a good measure when the distribution of the network performance measure (travel time, say) varies according to a very skew distribution. As noted by Bonsall (2000), when users of transport systems are confronted with an unpredictable situation, they appear to adopt strategies that are not able to be explained in terms of the variance of the performance measure. In such circumstances, it is necessary to consider the full distribution.

It seems that while the variance of travel time, say, is likely to influence reliability, the variance alone is not a good reliability measure, because a high variance does not necessarily mean that the reliability is low. In the case of a low travel time threshold, or if the travel time distribution is very skewed, conclusions about the reliability are very difficult.

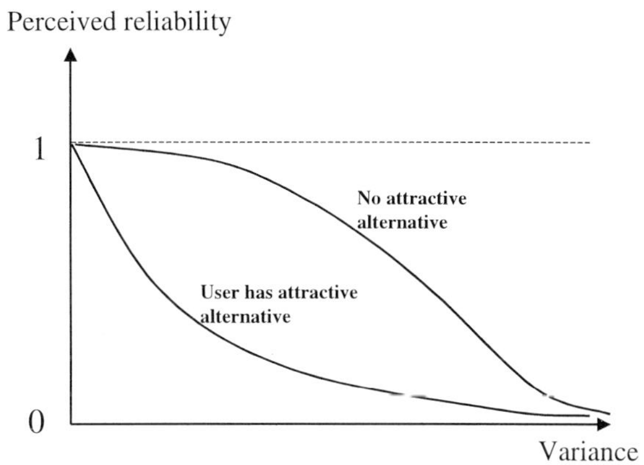

Figure 2 Perceived reliability depending on available route alternatives

Terminal Reliability

Terminal reliability is commonly defined as "the probability that nodes are connected, i.e. it is possible to reach the destination", and is the simplest, most fundamental reliability measure. It is appropriate only if links are either open or closed, and if an origin and destination are not connected, further questions regarding route or mode choice are pointless. If the network state (i.e. the condition of all links in the network) is known, the terminal reliability for each OD pair will be binary, otherwise it can be expressed as the expected value. In this case the reliability has a value between zero and one, relating to the probability that it is possible to reach the destination on any route.

The research into connectivity has a long history (e.g. Garrison, 1960). Delay due to congestion was not an important issue then, and the research was mainly motivated by the relatively sparse nature of the network; the loss of a link was likely to result in long detours being necessary. Wakabayashi and Iida (1992) and Bell and Iida (1997) also employed basic graph theory methods to calculate the terminal reliability, and Bell and Iida (2001) have recently proved that a variant of Floyd's algorithm can also be used efficiently. This later method has the advantage that path sets do not have to be established, unlike the method in Bell and Iida (1997).

Encountered Reliability

Bell and Schmöcker (2002) defined encountered reliability to be "the probability of not encountering a link degradation on the path with least (expected) costs". This measures the likelihood of users encountering a disruption on their preferred route (a matter not addressed by terminal reliability). Even though users might still be able to get to their destination by a circuitous route without experiencing a degraded link, it may be likely that they encounter degradation if all the cheaper routes are affected by the event. Encountered reliability 'labels' all users who have taken a link that is degraded (i.e. it does not include users who are delayed as a result of traffic diverting from degraded links). Unlike connectivity for terminal reliability a link 'failure' does not have to be a total link closure; the disruption can also be an event that only reduces the maximum possible link flow. Road users often encounter big delays due to passing roadside construction work, but that does not necessarily stop them from reaching their destination by this route. However, users will often try to avoid degraded links, and links which may be degraded. Therefore, the level of information to the users is important when considering encountered reliability. In addition, one has to consider how "risk averse" users are; do mean travel cost considerations outweigh unreliability and travel cost variance, or are the users so concerned to avoid disruptions that they will travel a longer, more reliable route (i.e. how much do they tend to avoid routes that have a small probability of being affected by a failure)?

If very risk averse, users perceive any small probability of difficulty as a major hindrance (just like closure). Less risk averse travellers will still travel on their preferred link (based on mean cost considerations), even though it might be degraded, instead of taking another route, which will almost certainly not be degraded, but which involves an enormous detour and therefore guarantees a big cost increase. If this is the situation, then there is a substantial difference between terminal and encountered reliability. In Figure 3, a reliable but high cost alternative exists (Route B), and therefore the terminal reliability will be high. The encountered reliability might be low, however, because of users who still prefer to travel Route A. Given that users have and follow information on the expected link failure probabilities, the more risk averse they are, the more similar are encountered reliability and terminal reliability (Bell and Schmöcker, 2002). The difference between the expected terminal reliability and encountered reliability therefore indicates how much the reliability could be increased if all re-routeing options were explored and capacity restrictions are ignored. To analyse terminal reliability the link costs do not have to be known, but they do for encountered reliability (to enable the optimal routes to be found).

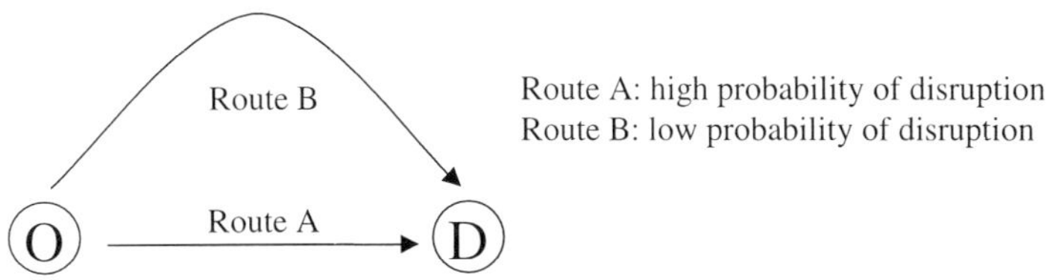

Figure 3 Example for terminal and encountered reliability

Capacity Reliability

This has been defined by Chen et al. (1999) as "the probability that the network can accommodate a specific demand level", and is another measure of the likelihood of users experiencing difficulties on their chosen path. Terminal reliability does not take into account capacity restrictions of links and therefore might overestimate the reliability. Encountered reliability does take user costs into account and assigns the traffic to the paths with lower costs. If a link gets congested, i.e. capacity is reached, the costs will increase and users, whose route choice is based on distance and travel time, will tend to avoid these links. Encountered reliability therefore implicitly takes link capacities into account, but as the risk averseness of travellers increases, the less is the importance of link capacities, because risk averseness was defined such that users who are extremely risk averse will avoid degraded links (or links which may be degraded) 'at any price'. This may include paying little attention to capacity problems on the non-degraded links.

Encountered reliability calculates the probability that users do not experience a link degradation, whereas capacity reliability estimates the number of users who are unable to travel on a link because of the capacity restraint. Chen et al. (1999) acknowledged that it is difficult to estimate the maximum flow capacity, because every link is influenced by many capacity-affecting factors, like physical and operational characteristics as well as weather conditions. Therefore they employed a stochastic approach to estimate the capacity, using defined probability distributions. They calculated the highest possible demand that could be accommodated by the network by a Monte-Carlo simulation approach. This is an extension of the reserve capacity concept for signalised intersections. The probability that the highest possible demand is greater than or equal to the actual demand is equal to the capacity reliability. This highest possible demand can be calculated for different scenarios and levels of link degradation.

Travel Time and Cost Reliability

This is commonly defined as "the probability that a trip can be successfully finished within a specified time interval (or less than a specified cost)". This measure has been developed because neither terminal nor encountered reliability identify the impact caused by network degradation. For example, is the disruption so severe that it causes a big increase in total travel time? This question can be asked for a link closure as well as for any capacity reducing disruption. In some cases a slight road capacity reduction might lead to great delays, while in other cases the abundance of re-routeing options might mean that the delays to the network users are almost zero.

Cassir (2000) defines the acceptable level of travel time as the travel time in normal conditions, plus a safety margin. Nicholson and Du (1997) distinguish two kinds of travel time reliability according to the source of uncertainty. Travel time can vary because of daily traffic variations and/or events that lead to link degradations. Especially in the latter case the information available to the users clearly influences their route choice and therefore the travel time reliability. Precise information about network disruptions will mean less impact (user delay, say), as illustrated in Table 1. Travel time reliability is thus a direct indicator of the impact on the network users.

The concept of travel time reliability can be extended to travel cost reliability (Schmöcker and Bell, 2001). In this case the acceptable level is not defined in terms of travel time but generalised travel costs, including time, distance, vehicle operating costs and public transport fares. A path is considered to be reliable if the destination can be reached with less cost than a threshold. The "acceptability threshold" is defined in much the same way as the acceptable travel time. Whereas travel time and travel cost reliability are often very similar if only re-routeing is considered, considerable differences might be observed if mode choice is included in the behavioural responses to link degradation. If the users do not change their mode of travel the cost of re-routeing is likely to be very much connected to the travel time increase. If the users, however, change to a different mode to avoid the link degradation (e.g. flying instead of driving) this might not result in a higher travel time but nonetheless lead to much higher travel costs.

It should be noted that the setting of an "acceptability threshold" for travel time and cost is of course difficult. For every journey there will be a threshold, but it will depend upon the journey purpose and the individual. However, the definition of this reliability measure is a good indicator if the travel time or cost respectively is likely to increase significantly due to the disruption.

Flow Decrement Reliability

Du and Nicholson (1997) proposed measuring reliability using the probability that the reduction in flow (as a result of supply-demand interaction) is not less than a threshold, for both OD pairs and the network. They defined network reliability in terms of the reliabilities of the sub-networks connecting individual origin-destination (OD) pairs. When a link is degraded, the cost of travel between one or more OD pairs may be affected, and the flow between those OD pairs will be affected, as a result of supply-demand interaction (Nicholson and Du, 1997). The greater the level of link degradation, the greater will be the reduction in flow between each of the affected OD pairs, in general. Du and Nicholson defined the 'flow decrement' for each sub-network as the reduction in flow between each OD pair, as a proportion of the flow when there is no degradation. That is, the sub-network flow decrement can vary between zero (when there is no flow reduction) and unity (when the flow is reduced to zero). They suggested that a 'threshold flow decrement' could be set for each sub-network, with a sub-network being considered to have failed if the estimated flow decrement exceeded the threshold value for that sub-network. The sub-network reliability was defined to be the probability that the sub-network would not fail.

The network flow decrement was equated to the flow-weighted average of the sub-network flow decrements, with the weight for a sub-network being the ratio of the flow in the sub-network to the flow in the whole network, when there is no degradation. The network reliability was defined to be the probability that the network flow decrement does not exceed a threshold value (i.e. no failure).

Du and Nicholson showed that using this approach, which is based on classical reliability analysis techniques, can lead to a network being considered to have not failed even though one or more sub-networks had failed. This is consistent with a well-known result in reliability engineering that a system may have high reliability even though some sub-systems have low reliabilities. Hence a sparse network (such as may exist in rural areas) may be considered reliable, even though residents in some small areas experience considerable unreliability in their transport connections

A variation on this measure is the "probability that the travel demand is not less than the demand satisfaction ratio" (Lam and Zhang, 2000), who considered the case of a network with varying demand. They note that the demand is not only a function of the travel time but also of demand variations due to "latent demand events". These include recurrent events like peak-period traffic, and "regular special events" (e.g. sporting events) that cause additional demand for a certain period of time. The demand satisfaction ratio is measured as the ratio between equilibrium travel demand and the "latent travel demand". The equilibrium travel demand is the demand that is satisfied (i.e. the number of users that are travelling at the current network costs). The latent travel demand is the sum of the equilibrium travel demand and the non-satisfied travel demand (i.e. the suppressed demand). They define the reliability to be the probability that the latent travel demand is satisfied, and also note that ignoring variations in travel demand can lead to error in the estimation of network reliability.

Vulnerability

Two quite different definitions of 'vulnerability' have been proposed. For Nicholson and Du (1994) and Berdica (2000), it is simply the susceptibility to incidents that result in route closures, and it increases as the probability and/or consequence of failing to meet user expectations increases. That is, vulnerability is essentially the opposite of reliability, and increases as reliability decreases. This interpretation of vulnerability encompasses both probability and consequence, and is closely related to risk (discussed below).

D'Este and Taylor (2001) define vulnerability to be the likelihood of severe adverse consequences if a small number of links (or possibly a single link) is degraded. They distinguish between 'connective vulnerability' and 'access vulnerability'. The former considers a pair of nodes and the generalised cost of travel between them; if the loss or substantial degradation of one or more network links leads to a substantial increase in that cost, then the connection between those nodes is deemed vulnerable. Access vulnerability considers a single node and the overall quality of access from that node to all other parts of the network; a node is vulnerable if the loss of substantial degradation of a small number of links results in a significant reduction in the accessibility of that node, as measured by a standard index of accessibility. It should be noted that the second definition of vulnerability ignores probability; this vulnerability is really a measure of the consequence of degradation.

D'Este and Taylor note that a network is not necessarily most vulnerable where the link flows are greatest, because alternative routes may be available nearby, allowing a new equilibrium flow pattern with little reduction in overall network performance. They suggest that vulnerable links be identified using the probabilities of a link being used for travel between OD pairs; the higher the probabilities relative to the probabilities of the alternative routes, the greater the impact of the link being degraded (i.e. the greater the link vulnerability). If there are three routes (A, B and C, say), carrying 50%, 40% and 10% of the traffic respectively, then the closure of route A is not as disruptive as it would be if the proportions were 90%, 5% and 5% respectively. In the former situation, there is a similarly attractive alternative to route A, but this is not the case in the latter situation.

This approach considers only the consequence of link degradation, and ignores the probability of link degradation. Nicholson and Du (1997) argue that links are 'weak' if the probability of degradation is high and 'important' if the consequences of degradation are high, with the 'critical' links (i.e. the links which should be treated) being those that are both important <u>and</u> weak. The danger with focusing on links that are only important <u>or</u> only weak, is that resources will be wasted on non-critical links, while critical links are deprived of the resources needed to improve them and the reliability of the network.

Risk

The 'risk' of a hazard is the product of the probability of the hazard occurring and the consequence of its occurring (i.e. the expectation of the hazard or threat). Hence, if the hazard has different levels *i*, each with its own probability p_i and cost c_i, then the expected cost is simply

$$E = \sum_i p_i c_i$$

Risk evaluation and management is a well-established process for identifying hazards, identifying their probabilities and consequences, assessing the acceptability of the risks, and taking action to address unacceptable risks (Nicholson and Dalziell, 2001).

Risk evaluation includes low and high probability hazards and minor and major consequence hazards, and is thus a more complete or comprehensive technique. It seeks to identify the critical links. Risk is consequently the most complex and demanding reliability indicator, as it requires information about the probabilities of hazards occurring and links being degraded. Such information can be difficult to obtain and may require a substantial data collection effort and/or a detailed knowledge of the processes underlying the occurrence of the hazards.

The higher the risk (i.e. the higher the expected annual cost of degradation), the lower is the reliability. That is, the reliability is not simply a probability; it is the expected cost of degradation, taking account of the probability of degradation occurring and a wide range of behavioural responses (Nicholson and Dalziell, 2001).

Discussion

The reliability measures outlined earlier are different and highlight different aspects of reliability. Some indicators might be useful for network planning purposes only, while others might also provide valuable information for individual users. Terminal reliability requires the least data but gives information about the connectivity only. The expected terminal reliability is an estimate of how likely it is that the destination can be reached. Encountered reliability takes travel cost and user information level into account, and acknowledges that some users will experience difficulties even though there are other (perhaps more expensive) routes without any difficulties. Both measures might be interesting for network planners as well as for individual users, while travel time and cost reliability are probably most interesting to users. These measures indicate directly the impact caused by the link degradation to the user. Risk is the most comprehensive measure available to the network planner, and if the risk associated with each possible hazard were known, this would give a good guide as to which links should be made more robust. The other reliability measures like capacity reliability and flow-decrement reliability can further highlight links in the network that are critical (i.e. links that are important and likely to fail).

The indicators discussed here cannot be used to analyse the effects of all behavioural responses. Whereas the impact reducing effects of route choice and mode choice are often considered, the effect of the other responses (e.g. postponing or cancelling the trip, or choosing a different destination) are generally not considered.

Tables 2 and 3 show reliability issues from the network user and network planner viewpoint, respectively. Table 2 lists questions users might ask before they start their journey, if they suspect that there could be disruption of the network. The list proposes a sequence in which the questions will be considered. It further shows which information can be gained from the different types of reliability measures. This table includes only the reliability measures of interest to individual (selfish) network users, who will be interested in reliability indicators at the OD and path level. This information is also of interest to the network planner, who might also want to know network-wide values for these reliability indicators. Table 3 includes additional questions that might be asked from a planning point of view (i.e. questions aimed at improving the current network to make it more robust).

The last two questions in Table 2 are added to emphasize some shortcomings of the measures summarised in this chapter. None of the measures, as described here, takes the duration of the link closure into account (i.e. the analysis is static and estimates of the benefits and costs of postponing trips may well be inaccurate). The measures generally do not identify the net economic benefit of a trip (trip utility minus trip costs) and therefore cannot give any information on the dis-benefit of the behavioural responses "cancelling the trip" and "choosing another destination".

The assessments in Tables 2 and 3 are indicative only. The measures focus on different aspects of reliability (e.g. either the user or planner viewpoint) and thus have their advantages and disadvantages. There are substantial differences in the information obtained from the different measures.

Table 2 Reliability aspects for network users and planners (at OD and route level)

	Variance	Terminal	Encountered	Travel time/cost
Is it still possible to reach my destination (which OD pairs are not connected)?	M	G	S	S
Is the route that I normally take likely to be closed (which routes are closed)?	M	G	S	S
If it is open, am I likely to encounter an unusual event (which users encounter such events)?	S	M	G	S
How severe are the delays on my preferred route (what delays without re-routeing)?	S	M	S	G
Is it advantageous to choose a different route or mode (delays with re-routeing)?	S	S	S	G
Should I travel later (suggest postponing travel)?	N			
Should I choose a different destination, or cancel the trip (provide travel advice)?	N			

M – minimal usefulness; S – some usefulness; G - good usefulness

Table 3 Additional reliability aspects for network planner (network level)

	Vulner-ability (1)	Flow Decrement	Capacity	Risk
How many users will not travel to their destination?	M	G	S	
Which links will be congested ?	S	S	G	
Which are the important links in the network?	G	S	S	
Which are the critical (important and weak) links?	M	S	S	G
What are the expected economic costs of hazards?	M	S	S	G

M – no usefulness; S – some usefulness; G - good usefulness

(1) The consequence-based definition of vulnerability.

Reliability Analysis Techniques

Introduction

The previous section summarised different measures of reliability. These reliability indicators might be calculated in different ways. Often a certain approach might be chosen, because of the scenario that is considered. Table 1 includes nine scenarios to differentiate between different levels of network user information and different levels of malevolence. The reliability measures found through different approaches must always be seen in the context of these scenarios (e.g. the travel time reliability might be much higher for a network user with a high level of information compared to a user who does not have any information about the possible network problems and re-routeing options).

Another difference between the techniques is whether they require that the probabilities of the adverse event are known. Deterministic approaches do not because the failure probabilities are a result of the analysis. This makes the use of deterministic models substantially easier. The disadvantage is that the calculated probabilities are only upper or lower bounds (i.e. they do not reflect 'real life'). Therefore, deterministic approaches can only be used to analyse a specific scenario of Table 1, whereas other approaches can cover many scenarios. The deterministic game theory approach, for example, always assumes high malevolence and a medium level of user knowledge.

Terminal Reliability Estimation Methods

It has already been mentioned that connectivity can be calculated by establishing path sets as done in Bell and Iida (1997). The reliability R of a series of links can be calculated as $R = \Pi_i \, r_i$ and the reliability of parallel links can be calculated as $R = 1 - \Pi_i \, (1 - r_i)$, where r_i is the expected link reliability and has a value between 0 and 1. If r_i is 1 the link is certainly operational and if r_i is 0 the link certainly fails. Every network can (theoretically) be decomposed into a system of series and parallel links and therefore the terminal reliability can be calculated. The advantage of using path sets (or cut sets) is that the expected reliability can be calculated directly. The disadvantage, however, is that it might be difficult or even impossible to identify all paths for larger networks.

Bell and Iida (2001) used a variant of Floyd's shortest path algorithm. In practice any algorithm, which calculates the costs for all possible routes between origin and destination, could be used. The task is not to find the cheapest route but to deliver 1 or 0 depending on whether at least one route is functional or not. Using a shortest path algorithm does not require the prior identification of paths, but it can only calculate a binary value for the reliability. Therefore r_i has to be 1 or 0. To find the expected reliability with this method heuristics have to be used. The link states (1 or 0) could be sampled and the evaluation repeated until convergence.

Absorbing Markov Chains

The absorbing Markov chain approach focuses specifically on estimating the encountered reliability. The basic idea of this approach is to set up transition probabilities for each possible move through the network. If no link exists between two nodes the transition probability will be zero, and if the transition is part of the shortest route to the destination the probability will be 1 (AON assignment) or a value between zero and 1 (SUE assignment). In addition to the transition to a next node, Bell and Schmöcker (2002) defined a transition probability to a so-called 'bin' for each node. This means, instead of travelling smoothly to the next node, there is also a certain probability that the user encounters a network problem (i.e. transition to the notional 'bin'). Therefore, all the Markov chain approach does is to distinguish between users who arrive at their destination without encountering a degraded link and users who do. The approach does not need any specification of the severity of the degradation (e.g. whether completely or only partly closed) and therefore cannot give information on travel time or travel cost reliability. The costs with disruption are not calculated, but users are just labelled as having reached the 'bin' (i.e. they have encountered a network problem).

Unlike the game theory approach, which estimates the worst-case failure probabilities, the failure probabilities have to be found before using the Markov chain approach. This, however, enables the modelling of different levels of "malevolence". Estimates of the actual failure probabilities, as well as worst-case failure probabilities, can both be input to the model. The transition probabilities are calculated according to the shortest routes, which depend on the link cost

$$c_{ij} = d_{ij} - \beta \ln (r_i)$$

where d_{ij} is the cost ignoring unreliability, β is the risk averseness factor, and r_i is the reliability of node i. If β is zero the link reliability is not considered, but if β is large and the user has information on the reliability r_i, then the reliability is an important factor in the user's cost estimation. Bell and Schmöcker (2002) have shown that the encountered reliability increases with increasing user risk averseness.

Game Theory Approach

The game theory approach has been mainly used to find the upper bound of the impact on a network of a possible link degradation, while assuming users adjust their routes according to the expected link failure probabilities (i.e. a medium level of information). The situation can be considered a 'minimax game', in which network users try to minimise their costs and a network 'spoiler' seeks to disrupt the network to maximise user costs. Bell (1999) allows the 'spoiler' the freedom to degrade one link in the network. The 'spoiler' chooses the link that causes the most impact and is therefore highly malevolent (or malicious). The probability of the 'spoiler' choosing each link reflects the vulnerability (or consequence of failure) of that link. The impact is calculated as the sum of the increased costs to all users. The network users are assumed to be aware of which links might be failed and to adjust their route choice according to the expected link costs c_l

$$c_l = r_l\, c_{l,\,failed\_conditions} + (1 - r_l)\, c_{l,\,unfailed\_conditions}$$

where r_l is the link failure probability.

The new network user route choice probabilities in turn might lead to a change in the network spoiler strategy. Link failure probabilities that caused the most impact before the re-routeing might not be the worst-case failure probabilities anymore. Therefore, the network spoiler updates the failure probabilities with the Method of Successive Averages. This, again, causes a network user reaction. The game will continue until the link failure probabilities and the route choice probabilities do not change significantly anymore. At this point the so-called mixed strategy Nash Equilibrium is reached. It is an equilibrium situation in the sense that neither the network spoiler nor the network user can further improve their route choice probabilities or link failure probabilities respectively. The results of the game are therefore worst case link failure probabilities which can be used to calculate the above mentioned reliability measures.

Travel time (and cost) reliability can be calculated by comparing the costs without and with failure probabilities. The theory of the game approach is described in more detail in Cassir and Bell (2000) and Bell (2000), while applications and extensions can be found in Cassir (2001), Cassir et al. (2001) and Schmöcker and Bell (2001). Bell (1999) considers the case of intermediate levels of malevolence (i.e. the decisions of the spoiler might be slightly random), using a logit model to find the link failure probabilities.

Monte Carlo Simulation

This technique is often used to identify the effects on a dependent variable, as a result of variations in those variables upon which the first variable depends, especially when the relationship between the dependent variable and the other variables is complex, and/or the variations in the other variables is not described by a mathematically-convenient distribution (e.g. are not Normally distributed). It is also used when the dependent variable depends upon many other variables, even though the relationship is fairly simple and the latter variables are

Normally distributed (say). Monte Carlo simulation involves simulating the random behaviour of the other variables, to identify the random behaviour of the dependent variable.

Monte Carlo simulation was used extensively by Dalziell and Nicholson (2001) to develop closure duration probability distributions for hazards that may cause link closure. Chen et al. (1999) also used the technique to identify the distribution of the largest OD demand multiplier that does not cause link capacity violations, given a set of random link capacities. Given the complexity of the interactions between variables describing or affecting transport system performance, and the complexity of the hazard mechanisms affecting link capacity or availability, it is likely that Monte Carlo simulation will receive increasing use in future network reliability studies.

Microsimulation

While the use of equilibrium-based models may be appropriate when assessing the effect of long-term degradations, as is generally the case with natural disasters, degradations will frequently be for short durations (e.g. lanes being blocked by truck during loading and unloading). In such circumstances, it is very unlikely that an equilibrium situation will be achieved. If one is concerned about increases in the demand for travel (e.g. before and after major sporting events), the increases are unlikely to sustained long enough for equilibrium to be achieved. For both short-term degradations and demand increases, the use of an equilibrium-based method is likely to lead to error in assessing the effect. Even if equilibrium is achieved, the effect during the period when the traffic situation is adjusting from an equilibrium situation to another cannot be accurately assessed (Berdica, 2000).

A recent study into the effects of short-term blockages of a link in an urban network indicates that microsimulation models are more sensitive to disturbances (including disturbance duration) than equilibrium models, which underestimate the effects of the short-term disturbances (Berdica et al. 2001).

CONCLUSION

Interest in transport network reliability research has significantly increased in the last years and reliability issues are likely to become even more important in the future. This is due to the increased expectations of the users of transport systems (e.g. the increased prevalence of the 'just-in-time' philosophy, which relies upon goods deliveries being very predictable). Investment in increasing the capacity of transport systems has not been keeping up with the increasing use of the systems, and there is increasingly widespread congestion of road and other transport networks (including public transport systems). Secondly, there is increasing evidence that environmental pollution is increasing the threat of natural disasters (e.g. flooding and hurricanes). Finally, transport systems are increasingly the target of malevolent acts (or attacks) by those who seek to cause severe hardship to society by disrupting transport systems, which play such an especially important role in developed societies.

This paper gives a brief overview of reliability issues, summarising and discussing several reliability measures and reliability analysis techniques. While the notion of variance underlies several of the reliability measures, the variance is not a good reliability measure because of difficulty in its interpretation. It is important that reliability measures are seen within the context of different scenarios, which can be classified according to the levels of malevolence and user knowledge. The malevolence of an intervention (or event) affects the location, level of degradation and duration of the event. The more malevolent the intervention and the lower the level of user knowledge, the higher will be the impact. Network planners have no control over the former, but may be able to reduce the impact of disruptions (i.e. improve the network reliability) by making more information available to users.

There is, however, a lack of quantitative information regarding how users will respond to such information, and how travel behaviour is affected by unreliability in the transport system, and there is clearly a need for further research on the human factors aspects. It is important to distinguish between the network user and planner viewpoints; users and planners might well have somewhat different information requirements. The former are likely to focus on their individual needs, while the latter should consider the performance of the whole network and the welfare of all users.

Some behavioural responses (especially postponing a trip, choosing a different destination, cancelling a trip, en-route travel decisions) are difficult to allow for using most reliability measures and analysis techniques. The primary focus of traffic models has been the analysis of normal, steady state conditions (viz. the emphasis on equilibrium-based methods), and they are generally not well-suited to modelling dynamic conditions, such as occur at the start and finish of a disruption. If the disruption duration is short, traffic conditions might not be even approximately steady state. There is thus a need for further research to develop traffic models appropriate for assessing the impact of short-term disruptions. Microsimulation models seem to offer considerable potential for meeting this need.

There will be a need to calibrate such models, to ensure they predict what actually happens when closures occur reasonably accurately. There is a lack of empirical data on actual traffic behaviour during disruptions, to enable calibration of a model, and there is thus a need for field studies of traffic behaviour in such circumstances. It should be noted that a model that accurately predicts traffic behaviour in normal, steady state conditions will not necessarily predict traffic behaviour during short-term closures accurately.

It seems clear, nevertheless, that it is much more difficult and laborious to estimate event probabilities reasonably accurately than it is to estimate the impact of network degradation if it occurs. The deterministic game theory approach tries to circumvent this problem by looking at worst-case scenarios. There is a need to develop techniques for estimating event probabilities, even if only approximately. One could then obtain an approximate estimate of the expected cost of degradation and ascertain the sensitivity of that estimate to variations in the approximate probabilities. One could then decide whether to proceed with a more detailed and thorough evaluation of the impacts.

REFERENCES

Bell, M.G.H. (2000). A game theory approach to measuring the performance reliability of transport networks. *Transportation Research B,* **34(6)**, 533-546.

Bell, M.G.H. and Iida, Y. (1997). *Transportation Network Analysis.* Wiley: Chichester.

Bell, M.G.H. (1999). Measuring network reliability: A game theoretic approach *Journal of Advanced Transportation*, **33(2)**, 135-146.

Bell, M.G.H. and Iida, Y. (2001). Estimating the terminal reliability of degradable transport networks. *Triennial Symposium on Transportation Analysis IV (TRISTAN),* June 13-19, *São Miguel,* Azores Islands, Portugal.

Bell, M.G.H. and Schmöcker, J.-D. (2002). Public transport network reliability: Topological effects. *Submitted to Third International Conference on Transportation and Traffic Studies (ICTTS),* Guilin, China, 2002.

Berdica, K. (2000). Analysing Vulnerability in the Road Transportation System; Putting Theory into Practice Using Sweden as an Example. *Working Paper TRITA-IP FR 00-76,* Department of Infrastructure and Planning, Royal Institute of Technology, Stockholm, Sweden.

Berdica, K., Andjic, Z. and Nicholson, A.J. (2001). Simulating road traffic interruptions: Does it matter what model we use? *Procs of 1st International Symposium on Transportation Network Reliability (INSTR)*, July/August 2001, Kyoto University.

Bonsall, P. (2000). Travellers' Response to Uncertainty. *In "Reliability of Transport Networks" (Eds. M.G.H. Bell and C. Cassir)*, Research Studies Press, Baldock (Herts.), United Kingdom, 1-10.

Cassir, C. (2001). A Flow Model for the Analysis of Transport Network Reliability. *PhD Thesis*, University of Newcastle, United Kingdom.

Cassir, C., Bell, M.G.H. and Schmöcker, J.-D. (2001) A normative assessment of transport network reliability based on game theory. *Procs of 1st International Symposium on Transportation Network Reliability (INSTR)*, July/August 2001, Kyoto University.

Cassir, C. and Bell, M.G.H. (2000). The n+m person game approach to network reliability. *In "Reliability of Transport Networks" (Eds. M.G.H. Bell and C. Cassir),* Research Studies Press, Baldock (Herts.), United Kingdom, 91–102.

Chapman, R.A. (1976). Bus Reliability: Definition and Measurement. *Research Report 18*, Transport Operations Research Group, University of Newcastle upon Tyne.

Chen, A., Yang, H., Lo, H.K. and Tang, W.H. (1999). A Capacity Related Reliability for Transportation Networks. *Journal of Advanced Transportation*, **33(2)**, 183-200.

Dalziell, E.P. and Nicholson, A.J. (2001). Risk and Impact of Natural Hazards on a Road Network. *ASCE Journal of Transportation Engineering*, **127(2)**, 159-166.

D'Este, G.M. and Taylor, M.A.P. (2001). Network Vulnerability: An Issue for Regional, National and International Strategic Transport Networks. *Procs of 1st International Symposium on Transportation Network Reliability (INSTR)*, July/August 2001, Kyoto University.

Du, Z.P. and Nicholson, A.J. (1997). Degradable Transportation Systems: Sensitivity and Reliability Analysis. *Transportation Research B*, **31(3)**, 225-237.

Garrison, W.L. (1960). Connectivity of the Interstate Highway System. *Papers and Proceedings of the Regional Science Association*, **6**, 121-137.

Lam, W.H.K. and Zhang, N. (2000). A New Concept of Travel Demand Satisfaction Reliability for Assessing Road Network Performance. *Paper presented at the Matsuyama Workshop on Transport Network Analysis*, August 2000, Matsuyama, Japan.

Nicholson, A.J. and Du, Z.P. (1994). Improving Network Reliability: A Framework. *Procs 17th Australian Road Research Board Conference*, **17(6)**, 1-17.

Nicholson, A.J. and Du, Z.P. (1997). Degradable Transportation Systems: An Integrated Equilibrium Model. *Transportation Research B*, **31(3)**, 209-223.

Schmöcker, J.-D. and M. G. H. Bell (2002). The PFE as a Tool for Robust Multi-Modal Network Planning. *Traffic Engineering & Control*, **44(3)**, 108-115.

Wakabayashi, H. and Iida, Y. (1992). Upper and lower bounds of terminal reliability of road networks: an efficient method with Boolean algebra. *Journal of Natural Disaster Science*, **14**, 29-44.

2

Network Vulnerability: An Approach to Reliability Analysis at the Level of National Strategic Transport Networks

G M D'Este, PPK Environment and Infrastructure, Brisbane, Australia
M A P Taylor, Transport Systems Centre, University of South Australia, Adelaide, Australia

Abstract

The importance of the adverse impacts of network degradation has stimulated substantial international research interest in transport network reliability, that is, the ability of degraded transport networks to cope with travel demand. Most of the recent research effort has focused on the reliability of urban passenger transport networks, in terms of the probability that the network will deliver a required standard of performance. This situation is characterised by high levels of congestion, a dense road network, and quantifiable probability of degradation of the network. Outside major urban centres, the situation is very different. The main dominant consideration in transport network infrastructure provision is accessibility - linking urban centres, providing regional coverage, and basic levels of accessibility for the non-urban community and economy. The network is sparse, congestion is not a significant issue, and access to essential community services and to markets is the major driving force underlying network development. In this context, the **vulnerability** of the network is perhaps more important than 'reliability'. This paper develops the concept of network vulnerability. It begins by reviewing the current state of research into network reliability, then proposes extensions and adaptations to the reliability concepts that are more appropriate for strategic-level multi-modal transport systems. Several alternative definitions for vulnerability are proposed. The paper also discusses the development of algorithmic and visualisation tools that may be used to identify specific 'weak spots' in a network, where failure of some part of the transport infrastructure would have the most serious effects on access to specific locations and on overall system performance. Finally, the paper describes potential applications of network vulnerability concepts, and proposes directions for further research.

INTRODUCTION

Transport networks allow convenient access to work and leisure in urban areas; they underpin the national and urban economy; and in rural and remote areas they are vital lifelines connecting isolated communities to essential services. Unfortunately transport networks are not 100 per cent reliable. There are many possible sources of degraded network performance, including freak events (such as earthquakes, blizzards and floods); incidents (such as severe traffic crashes, special events, construction works and civil emergencies); day-to-day congestion; and interruptions to passenger and freight transport services due to breakdowns, commercial failure or other reasons. These disruptions cause delays and detours with significant social, economic and environmental consequences. A common response to reduced performance or perceived risk is to upgrade key links by adding more capacity, but this simply makes the network more vulnerable to failure of those key links. An alternative approach could be to add additional links to the network, which may normally be redundant but provide alternative routes when a key link fails. At the urban network level there may be many such latent alternative links – in the local street system. At the regional or national strategic network level this is unlikely to be the case – extra links may then have to be provided – but this may add unnecessary cost to the provision of transport infrastructure. The question is then which additional links should be provided and where should they be located. This paper seeks to develop analytical and visualisation tools that may assist in this planning process.

Much of the previous research on reliability has focused on congested urban road networks and the probability that the network will deliver a required standard of performance. At the national strategic level, accessibility, regional coverage and inter-urban connectivity are the primary considerations. In these sparse networks, vulnerability of the network is more important than 'reliability' because of potentially severe adverse consequences of network degradation. The transport planner might thus prefer to seek opportunities to reduce vulnerability.

This chapter develops the concept of network vulnerability. It begins by reviewing the current state of research into network reliability, and then proposes extensions and adaptations to the reliability concepts that are more appropriate for strategic-level multi-modal transport systems. Several alternative definitions for vulnerability are proposed. The chapter also discusses the development of algorithmic and visualisation tools that may be used to identify specific 'weak spots' in a network, where failure of some part of the transport infrastructure would have the most serious effects on access to specific locations and on overall system performance. Finally, it describes potential applications of network vulnerability concepts, and proposes directions for further research.

NETWORK RELIABILITY

Road network reliability became an important research topic in transport planning during the 1990s, although some elements had been the subject of research interest for some time before that. The Kobe earthquake of 1995 and its aftermath stimulated an interest in connectivity reliability. This is the probability that a pair of nodes in a network remains connected (i.e. there continues to exist a connected path between them) when one or more links in the network have been cut. Bell and Iida (1997, 179-185) provided an analytical procedure for assessing connectivity reliability, and a summary of the procedure is given in Iida (1999).

Subsequent research was directed at degraded networks, usually urban road networks subject to traffic congestion, in which the network remained physically intact but the performance of one or more links could be so severely affected by congestion that their use by traffic is curtailed. This has led to the definition of two additional forms of reliability: travel time reliability and capacity reliability, as described below.

Travel time reliability considers the probability that a trip between an origin-destination pair can be completed successfully within a specified time interval (Bell and Iida, 1997, 191-192). This can be affected by fluctuating link flows and imperfect knowledge of drivers when making route choice decisions (Lam and Xu, 2000). One measure of link travel time variability is the coefficient of variation of the distribution of individual travel times (Asakura and Kashiwadani, 1991). The general assumptions are that this distribution is normal and that the travel times on adjacent links are statistically independent (Iida, 1999). Earlier empirical research (Richardson and Taylor, 1978, Taylor, 1982) supported these assumptions, although with the qualification that in uncongested networks the distribution of individual travel times may be positively skewed and is better represented by a log-normal distribution. Measures of travel time variability are useful in assessing network performance in terms of service quality provided to travellers on a day-to-day basis (Yang *et al*, 2000). Thus travel time variability can be seen as a measure of demand satisfaction under congested conditions (Asakura, 1999).

A supply-side measure of network performance in congested networks is capacity reliability, introduced by Chen *et al* (1999) and applied by Yang *et al* (2000). Capacity reliability is defined as the probability that a network can successfully accommodate a given level of travel demand. The network may be in its normal state or in a degraded stated (say due to incidents). Chen *et al* (1999) defined this probability as equal to the probability that the reserve capacity of the network is greater than or equal to the required demand for a given capacity loss due to degradation. Yang *et al* (2000) indicated that capacity reliability and travel time reliability together could provide a valuable network design tool.

Further research on network reliability is being pursued to develop this design tool. In addition, there is a need for further research to properly specify travellers' responses to uncertainty (e.g. Bonsall, 2000, van Zuylen and Kikuchi, 2002, Kikuchi and van Zuylen, 2002, Chen and Zhaowang, 2001) so that reliability research can be used to properly inform developments of new driver information systems and to influence the design of new traffic control systems. Taylor (2000) suggested that network reliability concepts could be applied in the design and evaluation of traffic calming schemes.

NETWORK VULNERABILITY

Standard approaches to network reliability focus on network connectivity and travel time and capacity reliability. This provides valuable insights into certain aspects of network performance but reliability arguments based on probabilities and absolute connectivity may obscure potential network problems. For example, D'Este and Taylor (2001) used the example of the Australian national highway network to illustrate the potential consequences of the severance of certain road transport connections in this network. In this example, the overall network remained connected and the probability that the route in question is cut by flood or other natural cause is extremely small, so the travel time and capacity reliability is high. Therefore standard measures of network reliability may not indicate any major problem with the network. However the consequences of network failure may be substantial. Nicholson and Dalziell (2001) pointed to similar circumstances in their study of the regional highway network in the centre of the North Island of New Zealand.

These examples illustrate the concept of network vulnerability and the difference between network reliability and vulnerability. The concept of vulnerability is related to the consequences of link failure, irrespective of the probability of failure. In some cases the probability of link failure may be low and the effect on overall network performance may be small, but the adverse social and economic impact on particular sections of the community may be sufficiently large to indicate a major problem warranting remedial action. For example, consider the impact on a rural community of loss of access to markets for its produce and to vital human services (such as a hospital). Low probability of occurrence and network performance elsewhere does not offset the consequences of a network failure. Thus network reliability and vulnerability are related concepts but while reliability focuses on connectivity and probability, vulnerability is about network weakness and consequences of failure.

The first step is to formulate a working definition of vulnerability. The concept of vulnerability can be applied to the connection between a particular origin and destination; or to access from a particular location to other parts of the network; or to the network as a whole. The following definition provides a starting point for applying the concept to network analysis and diagnosis.

> *DEFINITION*
> *A node is vulnerable if loss (or substantial degradation) of a small number of links significantly diminishes the accessibility of the node, as measured by a standard index of accessibility.*

Therefore vulnerability can be defined in terms of the overall quality of access from a given node to other parts of the network. Accessibility to facilities, goods, services and opportunities has long been recognised as a significant outcome of transport, and the concepts of accessibility have been explored over many years. Several alternative definitions of accessibility have been proposed, and while there are some differences between these, their common characteristic is to identify accessibility as a measure of the spatial distribution of activities about a point, adjusted for the ability, need and desire of individuals and groups to overcome separation. Some of the possible definitions of accessibility include:

- observed or expected travel cost between two points
- the opportunity presented to an individual to take part in a particular activity or set of activities
- the number of opportunities reached within a given travel time or distance
- the ease by which activities can be reached from a given location using a particular transport system
- the possibility to reach a given location within an acceptable amount of time, money and effort with respect to a specific policy

The formal definition of specific accessibility indicators may best be seen through a binary classification of the indices as either *relative accessibility indices* or *integral accessibility indices* (Morris *et al*, 1978). Relative accessibility describes the degree of connection between two given points (e.g. a given location and a specific activity), which may be assessed in terms of distance, travel time or travel cost. It is commonly used, for instance, in determining the locations of emergency services (e.g. fire, ambulance or police) where the location of the nearest facility is of prime importance. The relative accessibility A_{ij} between two points i and j may be given as

$$A_{ij} = C_{ij} \quad (1)$$

where C_{ij} is the separation (distance, time or cost) between the points. In terms of network vulnerability, this definition of accessibility measures the loss of utility of travel between two nodes as a consequence of network degradation. This type of vulnerability may be classified as *connective vulnerability*.

On the other hand, integral accessibility describes the interconnections between a given point and all other points or activities within a region. In essence, integral accessibility (AI_i) is given by the summation of the relative accessibility of the location i over all points j

$$AI_i = \sum_j A_{ij} \tag{2}$$

Integral accessibility may then be assessed by a simple measure of travel separation, as implied by equations (1) and (2), which is suitable for tasks such as the location of emergency services (where the closest service location is often of most importance). However, the measure does not include any allowance of the level of provision of the facility at any one site, or allow for user choice in the selection of a given facility (Parolin *et al*, 1994). Composite measures of integral accessibility that include supply demand measures and spatial separation have been devised, with two such measures being firstly, the Hansen index (e.g. Davidson, 1977), given by

$$AI_i = \sum_j B_j f(C_{ij}) / \sum_j B_j \tag{3}$$

in its normalised form, where $f(C_{ij})$ is an impedance function, and B_j is the amount or intensity of an activity at j, and secondly, the Black-Conroy cumulative distribution index, given by

$$AI_i(T) = \int_0^T N_i(t)dt \tag{4}$$

where $AI_i(T)$ is the cumulative number of opportunities within T travel cost units from location i and $N_i(t)$ is the absolute number of opportunities at t travel cost units from i (Black and Conroy, 1977). This index has been rarely used, perhaps because of the possibly arbitrary choice of the limiting travel cost units (T) (Neimeier, 1997), but it has the advantage of providing a continuous measure of the accessibility surface from a given location. This surface could be used to assess the relative accessibility of a given location over a region, and choice of different T values would provide contours of the surface. Variations of these contours under different conditions of network degradation could form the basis for assessing overall changes in network vulnerability and the relative effects for different locations within the network. Vulnerability based on estimated changes in accessibility indices may be termed *access vulnerability*.

Vickerman, Spiekermann and Wegener (1995) provided a practical example of the application of specific accessibility indices for testing the effects of network changes in their study on the proposed development of connected high-speed railway networks in the countries of the European Union.

IDENTIFYING NETWORK WEAKNESSES

The purpose of analysing network vulnerability is firstly, to be able to anticipate points of weakness where the network is vulnerable and network failures will have substantial adverse effects; and secondly, to be able to suggest remedial measures such as adding links to the network to make it more robust. It is tempting to believe that a transport network is most vulnerable simply where link flows are greatest but

- alternative routes may be available providing a new equilibrium pattern of flows at little reduction in overall network performance; and
- considering aggregate flows may obscure significant vulnerabilities in connections between particular origins and destinations.

Therefore the key to diagnosing network vulnerability is the development of analytical tools for identifying network weaknesses. Network vulnerabilities may exist that are not obvious from a scan of network activity patterns under normal operating conditions.

The definition of network vulnerability emphasises the consequences of degradation of the network. In other words, if the 'best' path through the network is no longer available, how much worse is the second-best option, or the third best, and so on. This suggests an approach based on *n*th best paths though the network or on constrained shortest path algorithms, but in general, algorithms for these problems are inefficient and are not included in standard transport network modelling software packages. For a review of algorithms for *n*th-best and constrained path problems, see D'Este (1997).

An alternative starting point is probabilistic route choice algorithms, such as those based on the logit model. According to this model, a traveller will choose a particular path from the set of available paths from the required origin to destination on the basis of the utility of that path compared to the alternatives. The measure of utility is travel time or other appropriate generalised cost. The probability of using a particular path will then depend on its relative utility.

This argument can be extended to individual links. The probability of using a particular link is a measure of the utility[1] of paths through that link compared to paths through alternative links. Note that for a network without loops, the probabilities for links that comprise a network cut will sum to unity. Therefore if the probability of using a particular link is low then there exist other links with similar or 'better' paths. However if the probability is high then paths through alternative links are inferior. The higher the probability of using a particular link, the greater is the difference between its utility and the utility of paths through alternative links. If that link is cut then the network performance will degrade significantly. In other words, the link probabilities provide a measure of the relative performance of alternative paths and hence of the consequences of network failure. It follows that logit-based assignment algorithms, such as Dial (1971), can form the basis of a heuristic method for identifying vulnerable links in a transport network. A number of alternative approaches exist for implementing this approach, including Bell's method and the conditional probability method.

Bell's Algorithm

An alternative to Dial's formulation of logit-based assignment was proposed by Bell (1995). Bell showed that the Floyd-Warshall shortest path algorithm (Floyd 1962, Warshall 1962) could be extended to logit-based multi-path assignment. Bell proposed two algorithms, the second and more general can be summarised as follows. Consider a matrix of weights $\mathbf{W}$, initialised as follows:

if a link joins node m to n then $w_{mn} = \exp(-\alpha\, c_{mn})$, otherwise $w_{mn} = 0$.

where c_{mn} is the cost of travel from m to n and α is a choice parameter that reflects the traveller's sensitivity to path cost. With a value of α close to zero, trips are assigned widely across available paths with little regard to cost, while as α increases, trips are increasingly assigned to the shortest path. Having constructed the matrix of initial weights, the next step is to calculate

$$\mathbf{U} \leftarrow (\mathbf{W} - \mathbf{I})^{-1} - \mathbf{I} \tag{5}$$

then $\mathbf{U}$ is the final matrix of link weights for the choice process. These weights can then be converted into link probabilities, as described by Van Vliet (1981). The probability that a trip from i to j uses the link joining node or centroid r to node or centroid s

[1] Utility in this chapter is merely a general measure of the usefulness (or indeed impedance) of use of a specific link in the network for a given trip from an origin to a destination in the network. We could equally use generalised cost, travel time or travel distance, but utility may reflect a more general concept and is therefore a preferred term.

$$P_{ijrs} = \frac{u_{ir} w_{rs} u_{sj}}{u_{ij}} \tag{6}$$

Using this basic relationship, similar formulas can be constructed for turning proportions or probabilities of using other combinations of links. The advantages of Bell's formulation are that it:

- calculates link weights for trips between all origins and all destinations simultaneously, and
- is valid whether or not there are loops in the network.

The matrix of link probabilities $\mathbf{P}_{ij}$ derived using relationship (6) provides an indicator of the overall vulnerability of trips between the given origin and destination and also of the location of key links. If $\mathbf{P}_{ij}$ has elements close to unity then this is a good indicator of potential network vulnerability because it suggests that there is a significant difference in amenity between the best and next best routes.

Bell's method is elegant but may suffer from two disadvantages. Firstly, it requires the path dispersion factor (α) to be non-zero, i.e. it needs $\alpha > 0$. This presents a small drawback in comprehensive network studies as the case of $\alpha = 0$ for which all available paths are equally probable may be of interest. Secondly, the matrix inversion method may be less efficient in large networks with significant numbers of links. An alternative approach that is computationally efficient even in large networks and that can accommodate zero values of α is the conditional probability method.

Conditional Probability Method

This method is closely related to the original logit assignment algorithms proposed by Dial (1971) and subsequently modified by Taylor (1979).

Assuming that the probability of use of path *R(i, j)* connecting origin *i* to destination *j* is directly proportional to the likelihood of use of all links *e* in the path, we may write

$$\Pr\{R(i,j)\} = K \prod_{e \in R(i,j)} g(e) = KG(R(i,j)) \tag{7}$$

where *K* is a constant, *g(e)* is the *link likelihood function* and *G(R)* is the *path likelihood function* and $0 \leq g(e) \leq 1$. A suitable functional form for *g(e)* is

$$\begin{aligned} g(e) &= \exp(-\alpha z(e)) && \text{if } e \text{ is on an acceptable path} \\ g(e) &= 0 && \text{otherwise} \end{aligned} \tag{8}$$

where $z(e) \geq 0$ is the difference between the travel cost incurred in using link e to travel from *r* to *s* and the minimum cost path between *r* and *s* (Dial, 1971; Taylor, 1979). If $z(e) = 0$ then link *e* is on a minimum path. Taylor (1979) defined an acceptable path to be one on which each succeeding node is closer to the destination than its predecessor[2]. If $V(r, j)$ is the minimum path potential of node *r* with respect to destination *j*, i.e. the minimum travel cost from *r* to *j*, then the link *e* will be on an acceptable path if $V(r, j) > V(s, j)$. The cost difference $z(e)$ is given by

$$z(e) = V(s, j) + c(e) - V(r, j) \qquad (9)$$

where $c(e)$ is the travel cost on link *e*. If *e* is on a minimum path from *i* to *j* then $z(e) = 0$ as $V(r, j) = V(s, j) + c(e)$ by definition.

Equation (7) provides a conceptual definition of the probability of use of a path $R(i, j)$ between *i* and *j*. The probability that link *e* will be used for travel between *i* and *j* is given by summing the path probabilities of all paths from *i* to *j* that use *e*, i.e.

$$\Pr\{e, (i, j)\} = \sum_{R(i,j):e \in R(i,j)} \Pr\{R(i, j)\} \qquad (10)$$

This result is conceptually simple but difficult to use in practice, because it requires definition of all of the acceptable paths between *i* and *j*. However, the conditional probability of a trip from *i* to *j* using *e* given that it passes through node *r* ($\Pr\{e, (i, j)|r\}$ may be found using an efficient recursive algorithm without the need for explicit determination of all acceptable paths. It may be shown (e.g. see Taylor, 1979) that this conditional probability may be computed recursively using the link weight function $w(e)$:

$$w(e) = g(e) \qquad \text{if } s = j \text{, the destination node} \qquad (11)$$

$$w(e) = g(e) \sum_{l \in \beta(s)} w(l) \qquad \text{for all other } s$$

where the summation $\sum_{l \in \beta(s)} w(l)$ is the sum of the weight functions of all links *l* which can be used to leave node *s*, i.e. the set of links $\beta(s)$. It then follows that

$$\Pr\{e, (i, j) \mid r\} = \frac{w(e)}{\sum_{l \in \beta(r)} w(l)} \qquad (12)$$

given that $w(e)$ may be computed recursively by considering each node *s* in forward topological order (i.e. increasing value of path potential $V(s, j)$) from the destination *j*.

---

[2] This definition of a reasonable path is logically consistent with the scale of the regional and national strategic level networks under discussion here. In dense urban road networks with turn penalties and turn bans an alternative model form is needed with a revised definition of a 'reasonable path', as discussed in Taylor (1979, 1999a). In the equivalent dense network model a reasonable path is one in which every (directional) link in the path is closer to the destination than the preceding (directional) link.

Use of Link Probabilities in Network Scans

The individual link probabilities P_{ijrs} determined by one of the above methods may be used to provide an indicator of where to look for the key links where the connection is most vulnerable. In general, the higher the link probability, the greater the adverse impact if that link is broken. This suggests that candidates for the source of network vulnerability will be links with probability higher than a prescribed threshold λ. In general, the loss of any link with $P_{ijrs} > 0.5$ will adversely affect network performance but the effect may be minimal. In practice, a higher value of λ may be more appropriate. Recommendations for efficient values of the threshold λ and the choice parameter α are topics for further research.

Having assessed the vulnerability of connections between a particular origin and destination (relative accessibility), the vulnerability of overall access of a particular node (integral accessibility) can then be evaluated by repeating the process for all destinations reachable from that node. Therefore calculating link probabilities using a logit-based multi-path assignment algorithm provides a heuristic technique for identifying both types of nodal vulnerabilities.

This discussion has considered the analysis of network vulnerability in terms of the spatial or topological configuration of the network, by defining a measure of vulnerability based on access of a given node to all other nodes. This measure may be seen as similar to the simple measure of integral accessibility defined by equation (2). Further research is needed to consider the use of accessibility-based measures of overall network vulnerability, which introduce some measure of access to opportunities located in different amounts at different locations in the network, perhaps using measures similar to the integral accessibility measures defined by equations (3) and (4).

WORKED EXAMPLES

Two examples using different networks are provided as illustrations of the use of vulnerability in the assessment of network weaknesses. The first is a simple test network analysed using Bell's algorithm. The second is a network model of the UK national railway system, which is analysed using the conditional probability method.

Test Network Example

This example illustrates the identification of links where a network is potentially vulnerable. Consider the network shown in Figure 1. The number shown alongside each link is a measure of the link cost.

Figure 2 shows the link probabilities after undertaking a connective vulnerability scan of the network, using Bell's algorithm with $\alpha = 0.1$ (costunits^{-1}). Link probabilities less than 0.3 are shown with a thin dashed line, between 0.3 and 0.6 with a solid line, and link probabilities greater than 0.6 are shown with a thick line. As shown in this figure, there is only one link with a high probability. The connection between the indicated origin and destination is likely to be vulnerable to degradation of this link, but relatively insensitive to loss of other links. If this key link is cut then the minimum and expected cost of travel increases by around 25 per cent, whereas if any other link is cut, there is little (if any) impact on network performance.

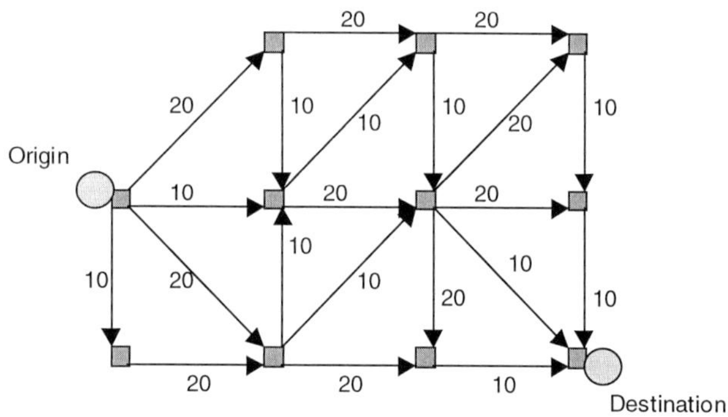

Figure 1 Test Network Example

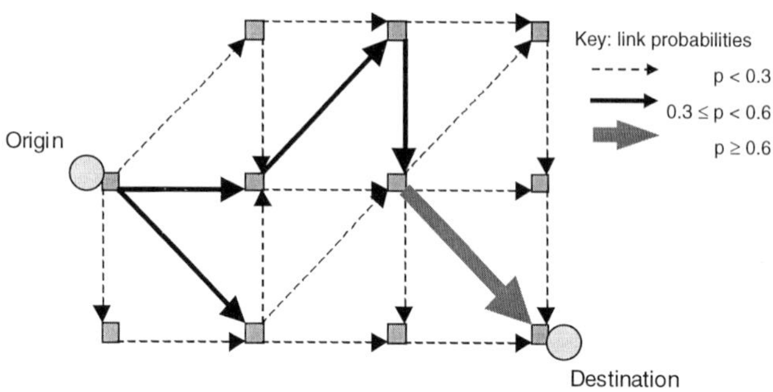

Figure 2 Link Probabilities (arrow style indicates probability of use)

This example illustrates the application of the vulnerability concepts using the technique described above. It also indicates that all links on the shortest path are not necessarily points of vulnerability. If there is sufficient redundancy in the network then diversion to alternate paths can take place with little degradation of network performance.

UK Rail Network Example

The UK national railway network is the subject of considerable discussion in that country at the time of writing. Throughout 2001 it has been a network experiencing ongoing difficulties with service reliability, and poses a number of questions about possible future infrastructure investments to improve national connectivity by rail (e.g. see Heath, 2001). A model of this network was developed for vulnerability studies, and a trial analysis undertaken. This analysis considered the connective vulnerability of the network for travel from Edinburgh (origin node 57) to Folkestone (Harbour) (destination node 117). Figure 3 provides a picture of the network model. The chosen destination is the gateway to Europe through the Channel Tunnel. The analysis is based on the link conditional probability method described above. The analysis was run for two definitions of travel cost: (1) travel distance and (2) estimated travel time. The threshold link probability value (λ) for the vulnerability scan was at 0.5. For travel distance, the path dispersion factor was set at $\alpha = 0.1\ km^{-1}$ ($\alpha = 0.0001\ m^{-1}$). There are two alternative paths available in this case, as indicated in Figure 3. For travel time, three values of α were used: 10, 0.001 and 0 (min^{-1}). These three values give two different path sets: α = 10 restricts travel to the minimum path only, for the other two values there are two alternative paths, see Figure 4. Setting α equal to zero means that all acceptable paths will have equal probabilities of use. Note the differences between the acceptable path sets in Figure 3 (distance based) and Figure 4 (time based, $\alpha = 0.001$ and $\alpha = 0$).

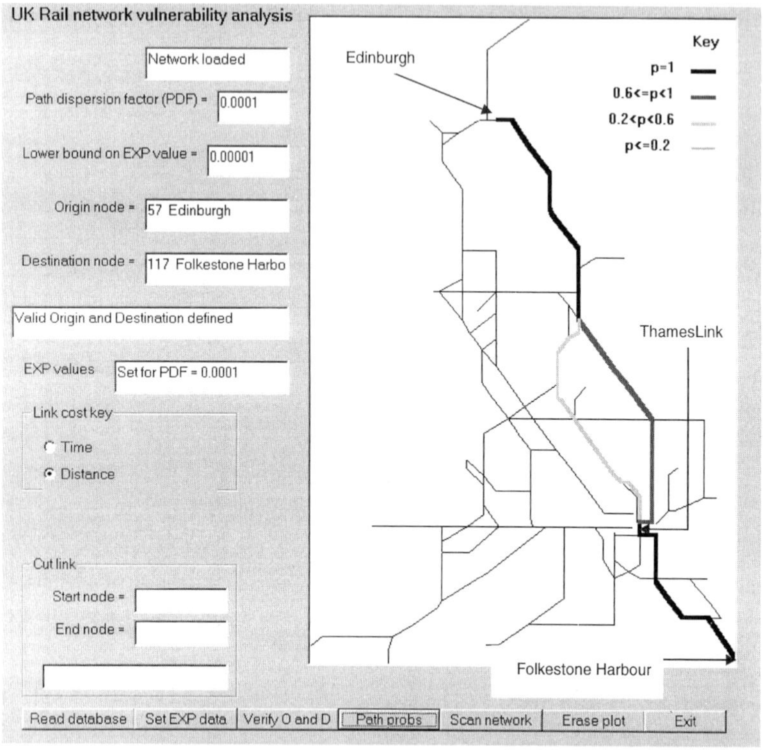

Figure 3 Alternative paths between Edinburgh and Folkestone in the UK national rail network, for connective vulnerability based on travel distance and path diversion factor $\alpha = 0.1$ km^{-1}

Table 1 summarises the results of the connective vulnerability scan based on travel distance. This table shows all links with link probabilities equal to or exceeding the threshold value of 0.5. It also shows the impacts on minimum path travel cost and expected travel cost in the degraded network when each of these links are cut. This is done by listing the minimum and expected path travel costs for the degraded network in which the specified link has been cut. In terms of the impacts of degradation, the path may be seen as consisting of seven segments. The most critical path segment – as expected – is the single link (116, 117), which is the only link accessing the destination node (117, i.e. Folkestone Harbour). Without this link the destination is unconnected.

Table 1 Connective vulnerability scan for UK rail network, with travel cost given by distance

Origin: Edinburgh (57)	Destination: Folkestone (117)		$\alpha = 0.1$ (km^{-1})		$\lambda = 0.5$
Travel distance (km)	Minimum path cost (MP) = 630.712		Expected path cost (EP) = 633.137		
	Vulnerability scan: degrade network by cutting link				
Links (node1-node2- … -nodeN)	Link prob	Minimum path cost (MPD)	Expected path cost (EPD)	MPD/ MP	EPD/ EP
57-56-55-54-53-52-51-50-49-48-47-46-45	1.000	697.342	699.809	1.106	1.105
45-44	1.000	658.016	660.189	1.043	1.043
44-43-42-41-40-39-38-3-163	0.682	638.333	638.333	1.012	1.008
163-158	1.000	723.119	725.586	1.147	1.146
158-6-110	1.000	644.383	646.808	1.022	1.022
110-120-119-118-116	1.000	653.152	655.577	1.036	1.035
116-117	1.000	∞	∞	∞	∞

For the other six path segments alternative paths can be found, and the minimum path and expected travel costs (distances) in the degraded networks increase by between one and 15 per cent when compared to those in the full network. The most vulnerable path segment in the scan, with increases of 15 per cent in both minimum path and expected travel costs, is the single link (163, 158) which represents the 'ThamesLink' north-south connection in the London area.

The vulnerability scan analysis was repeated using travel time as the path cost. Figure 4 shows the acceptable path set for estimated travel time in the network. Tables 2 and 3 show the results of the analysis for the cases where α = 10 min^{-1} and $\alpha = 0$ respectively. These values provide upper and lower bounds on the connective vulnerability analysis for travel times.

Table 2 shows the link probabilities and the effects on cutting each of these links when α = 10 min^{-1}. There is a single minimum travel time path and the table shows all of these links, which have probability of use of unity. There are seven separate path segments, as found in Table 2 for the path distance analysis, and although there are similarities with the path segments found in that case the segments for the travel times are not identical. Once again, link (116, 117) is the most critical –it is the only connecting link from the rest of the network to the destination node. Minimum path and expected travel costs are identical (there is only one path) and the increases in travel time on five of the path segments in the degraded networks are between 14 and 17 per cent for all links. For the path segment comprising the sole link (45, 44), the increase in travel time is five per cent, indicating that this is the least vulnerable path segment.

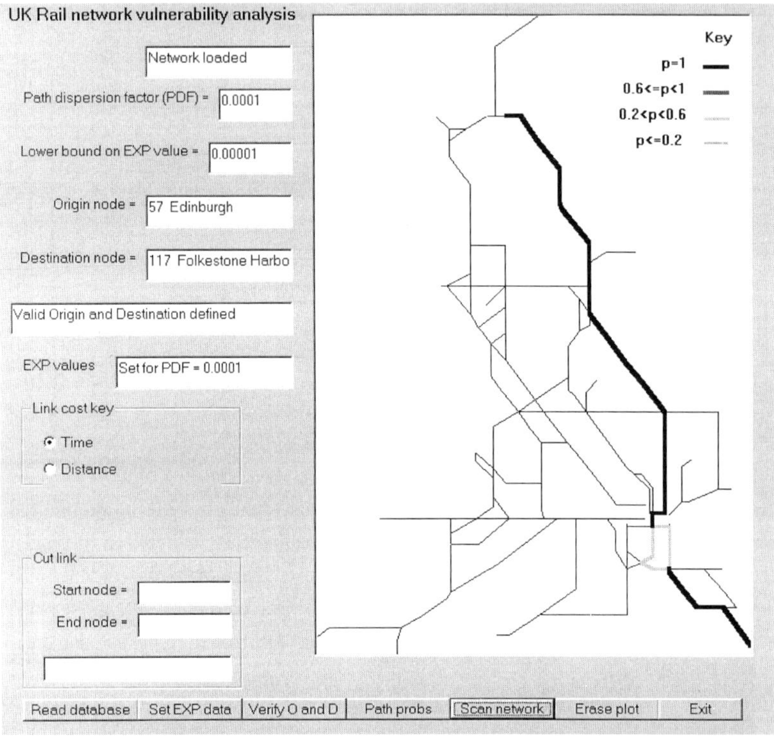

Figure 4 Alternative paths between Edinburgh and Folkestone in the UK national rail network, for connective vulnerability based on travel time

Table 3 shows the scan results for the case where $\alpha = 0$, i.e. when all alternative paths are equiprobable. In this case there is the maximum potential for diversion away from the minimum cost path. Ten path segments can be identified in this analysis. The results for changes in minimum path costs in the degraded networks are the same as those for the minimum path choice only (see Table 2), as should be expected. However, the expected path cost changes are much higher in this case, ranging between five and 29 per cent (Table 3). Path segment (sole link) (45, 44) is the least vulnerable part of the network, as in Table 2. Path segment (110-120-119-118-116), to the south of London, yields an increase in travel time when degraded of 14 per cent. All other segments yield increases in expected path travel costs of at least 23 per cent. This suggests that vulnerability based on travel time is significant in this network.

Table 2 Connective vulnerability scan for UK rail network, with travel cost given by estimated travel time and $\alpha = 10\ \text{min}^{-1}$

Origin: Edinburgh (57)	Destination: Folkestone (117)		$\alpha = 10\ (\text{min}^{-1})$		$\lambda = 0.5$
Travel time (min)	Minimum path cost (MP) = 327.1		Expected path cost (EP) = 327.1		
	Vulnerability scan: degrade network by cutting link				
Links (node1-node2- ... -nodeN)	Link prob	Minimum path cost (MPD)	Expected path cost (EPD)	MPD/ MP	EPD/ EP
57-56-55-54-53-52-51-50-49-48-47-46-45	1.000	381.7	381.7	1.167	1.167
45-44	1.000	342.8	342.8	1.048	1.048
44-43-42-41-40	1.000	378.9	378.9	1.158	1.158
40-39-38-3-163	1.000	380.9	380.9	1.164	1.164
163-158-6-110	1.000	381.7	381.7	1.167	1.167
110-120-119-118-116	1.000	373.9	373.9	1.143	1.143
116-117	1.000	∞	∞	∞	∞

Table 3 Connective vulnerability scan for UK rail network, with travel cost given by estimated travel time and $\alpha = 0\ \text{min}^{-1}$

Origin: Edinburgh (57)	Destination: Folkestone (117)		$\alpha = 0\ (\text{min}^{-1})$		$\lambda = 0.5$
Travel time (min)	Minimum path cost (MP) = 327.1		Expected path cost (EP) = 333.5		
	Vulnerability scan: degrade network by cutting link				
Links (node1-node2- ... -nodeN)	Link prob	Minimum path cost (MPD)	Expected path cost (EPD)	MPD /MP	EPD /EP
57-56-55-54-53-52-51-50-49-48-47-46-45	1.000	381.7	381.7	1.167	1.286
45-44	1.000	342.8	349.3	1.048	1.047
44-43-42-41-40	1.000	378.9	411.9	1.158	1.235
40-39-38-3-163	1.000	380.9	413.4	1.164	1.239
163-158	1.000	381.7	429.4	1.167	1.286
158-6-110	0.500	381.7	429.4	1.167	1.286
158-157-106-156-110	0.500	381.7	429.4	1.167	1.286
110-120-119-118-116	1.000	373.9	380.3	1.143	1.140
116-117	1.000	∞	∞	∞	∞

APPLICATIONS

National Strategic Transport Network

As mentioned earlier in this paper, accessibility is the dominant consideration in the planning and development of transport networks at the national level and outside the major urban centres. For example in the sparsely settled regions typical of ex-urban Australia, transport infrastructure such as roads or railways has been provided at a minimum level to provide regional coverage and inter-urban connectivity. In these sparse networks, vulnerability of the network is perhaps more important than reliability because of potentially severe consequences to transport services of network failure. This problem may be exacerbated in remote areas where the sole highway and railway links run in close proximity. A disaster affecting one mode's network is also likely to affect the other, so that cross-modal substitution is not necessarily available either.

The techniques developed in this paper can provide insights into the robustness of the national network, or conversely to its vulnerability to failure of key links. In the case of an extensive network, it is generally impractical to conduct a detailed risk assessment of all network links. Therefore the ability to identify potential 'weak spots' in the network provides an opportunity to better target risk assessment and planning of possible remedial measures. If potential weaknesses in the network can be anticipated and adaptations made to the network, then the consequences of failure of key links can be substantially reduced. The result would be a transport network that provides a better and more robust level of performance.

Incident Risk Assessment

Although the development of the concept of network vulnerability has focused mainly on sparse non-urban networks, the techniques can be applied in an urban context, specifically in relation to vulnerability to degradation of network performance as a result of incidents, such as

- accidents, breakdowns and disabled vehicles, road blockages by emergency services operations, severe weather, natural disasters or distractions caused by events taking place along the roadside ('rubber-necking'); and
- planned events such as road works and special events.

In North America, it has been estimated that incidents cause about half of all urban traffic congestion. Pfefer and Raub (1998) and Rust-PPK (1996) indicated that congestion resulting from incidents comprises some 40 - 60 per cent of all congestion, with incident-based congestion having a multiplying effect - a one minute blockage may cause a congestion episode of six to ten minutes duration. Further, incidents on multi-lane roads are believed to cause significant reductions in traffic capacity. For instance, incidents involving the blockage of one lane may reduce carriageway capacity by perhaps 50 per cent (see Pfefer and Raub, 1998). Even incidents that merely block the road shoulder may reduce capacity by 15 - 20 per cent (Rust-PPK, 1996). The number, type and duration of incidents have a major effect on the performance of the road system. This effect is highly accentuated during peak traffic periods. The effects of incidents are compounded by secondary accidents and by exponentially growing traffic delays.

Central to the objective of managing the road network to ameliorate the effect of incidents is the ability to identify locations with high incident vulnerability. The value of anticipating and hence avoiding potential problems has been recognised in many disciplines. In particular, safety auditing has become a standard feature of road design and assessment. Safety auditing involves examining all aspects of the road environment to identify safety risks and recommend measures to rectify the problems. This principle can be extended to incident risk assessment.

The aim of incident risk assessment is to identify locations or road links that are especially vulnerable to congestion caused by incidents. The degree of vulnerability will be influenced by a combination of factors, notably

- traffic volume per lane and volume to capacity ratio; and
- availability of alternative routes and permeability of the network (Taylor 1999b, 2000).

Therefore there are two inter-connected sources of incident vulnerability. The first is the sheer volume of traffic. Locations with this type of vulnerability are easy to identify. The second source of incident vulnerability is associated with the inherent structure of the transport network and/or the way that it is managed. Locations with this sort of vulnerability can be more difficult to identify and the effects may be harder to predict. The techniques for analysis of connective vulnerability as described in this chapter provide a framework for diagnosing this second type of vulnerability to incidents.

Incident risk assessment can also be used to evaluate the expected vulnerability of proposed road links to traffic disruption caused by incidents. Incident risk assessment during the planning stages of major road projects should become a routine component of the planning process. It will allow incident management plans to be developed in advance where required, and may lead to design changes to reduce vulnerability to incidents. For example, extra exit ramps may be constructed or an automated incident detection system installed in strategic road links assessed to have a high incident risk.

CONCLUSIONS

The purpose of this chapter is to introduce the concept of network vulnerability and describe some preliminary research on definitions, techniques and applications. A network is considered to be vulnerable if degradation of a small number of network links, perhaps a single link, causes significant reduction in network performance. The aftermath of earthquakes, floods, and traffic incidents illustrate the vulnerability of transport network.

There are three key questions to be addressed in the context of network vulnerability:

- is the link between two locations or the overall accessibility of a location vulnerable to network degradation?

- if so, where is the network most vulnerable?
- what can be done to reduce the degree of vulnerability?

Answers to these questions provide the basis for making transport networks more robust in terms of delivering more consistent performance. This paper has developed a framework for addressing these questions and proposed an heuristic technique based on logit-based multi-path assignment for identifying potential vulnerability. The question of techniques for developing efficient response, perhaps by adding new links to reduce the vulnerability of a network, remains for future research. Other key topics for further research include:

- the development of improved indicators of network vulnerability
- the development of more precise techniques for identifying 'weak' links that make the network most vulnerable
- tools for network diagnosis and auditing of transport networks, including visualisation tools
- tools for identifying the best response to identified vulnerabilities
- the inherent robustness/vulnerability of different network topologies
- appraisal of the practical application of the vulnerability indictors to real world networks.

ACKNOWLEDGEMENTS

The authors acknowledge the support of Australian Research Council. Dr D'Este expresses his gratitude for the ongoing support and encouragement from the Transport Systems Centre of the University of South Australia.

REFERENCES

Asakura, Y. (1999). Evaluation of network reliability using stochastic user equilibrium. *J. Adv. Transpn.*, **33** (2), 147-158.

Asakura, Y. and Kashiwadani, M. (1991). Road network reliability caused by daily fluctuation of traffic flows. *Proc 19th PTRC Summer Annual Meeting*, University of Sussex, Brighton, July, 73-84.

Bell, M. G. H. (1995). Alternatives to Dial's logit assignment algorithm. *Transpn. Res.,* **29B**, 287-295.

Bell, M. G. H. and Iida, Y. (1997). *Transportation Network Analysis*. John Wiley and Sons, Chichester.

Black, J. A. and Conroy, M. (1977). Accessibility measures and the social evaluation of urban structure. *Environment and Planning A,* **9**, 1013-1031.

Bonsall, P. W. (2000). Travellers' response to uncertainty. In: *Reliability of Transport Networks* (M. G. H. Bell and C. Cassir, eds.), pp.1-10. Research Studies Press, Baldock.

Chen, A. and Zhaowang, J. (2001). Travel time reliability with risk-sensitive travellers. *Extended Abstracts of the 1st Int. Symp. on Transportation Network Reliability*. Kyoto University, July, 43-46.

Chen, A., Yang, H., Lo, H. K. and Tang, W. H. (1999). A capacity related reliability for transportation networks. *J. Adv. Transpn.,* **33** (2), 183-200.

Davidson, K. B. (1977). Accessibility in transport/land use modelling and assessment. *Environment and Planning A,* **9**, 1401-1416.

D'Este, G. M. (1997). Implementing a hybrid route choice procedure in a network modelling context. *J. East Asian Soc. for Transport Studies,* **2** (3), 737-720.

D'Este, G. M. and Taylor, M. A. P. (2001). Modelling network vulnerability at the level of the national strategic transport network. *J. East Asian Soc. for Transport Studies,* **4** (2), 1-14.

Dial, R. B. (1971). A probabilistic multipath traffic assignment model which obviates path enumeration. *Transpn. Res.,* **5**, 83-111.

Floyd, R. W. (1962). Algorithm 97: shortest path. *Comm. of the ACM,* **5**, p.345.

Goodwin, P. B. (1992). A quality margin in transport. *Traff. Eng. & Control,* **33** (12), 661-665.

Heath, D. L. (2001). CrossRail: introduction. *Proc Inst. Civil Engnrs. – Transport,* **147** (2), 59-69.

Iida, Y (1999). Basic concepts and future directions of road network reliability analysis. *J. Adv. Transpn.,* **33** (2), 125-134.

Kikuchi, S. and van Zuylen, H. (2002). Modeling of traveler's uncertainty and anxiety. *The Network Reliability of Transport,* 133-154.

Lam, W. H. K. and Xu, G. (2000). Calibration of traffic flow simulator for network reliability assessment. In: *Reliability of Transport Networks* (M. G. H. Bell and C. Cassir, eds.), 139-157. Research Studies Press, Baldock.

Morris, J. M., Dumble, P. L. and Wigan, M. R. (1978). Accessibility indicators for transport planning. *Papers of the Australian Transport Research Forum,* **4** (1), 41-73.

Neimeier, D. A. (1997). Accessibility: an evaluation using consumer welfare. *Transpn.,* **24** (4), 377-396.

Nicholson, A. and Dalziell, E. (2001). Risk evaluation and management: a road network reliability study. *Extended Abstracts of the 1st Int. Symp. on Transportation Network Reliability*. Kyoto University, July, 127-130.

Parolin, B., McDougall, W. and Nichols, M. (1994). Modelling rural accessibility. *Papers of the Australasian Transport Research Forum*, **19**, 45-62.

Pfefer, R. C. and Raub, R. A. (1998). Effects of urban roadway lane blockage on capacity. *ITE Journal* **68** (2), 46-51.

Richardson, A. J. and Taylor, M. A. P. (1978). A study of travel time variability on commuter journeys. *High Speed Ground Transpn. J.*, **12** (1), 77-99.

Rust-PPK (1996). *Incident management and driver information strategy*. Report to Roads and Traffic Authority of New South Wales, Sydney.

Taylor, M. A. P. (1979). Evaluating the performance of a simulation model. *Transpn. Res.* **13A** (3), 159-173.

Taylor, M. A. P. (1982). Travel time variability – the case of two public modes. *Transpn. Sc.*, **16** (2), 517-521.

Taylor, M. A. P. (1999a). Dense network traffic models, travel time reliability and traffic management. I: General introduction. *J. of Adv. Transpn.* **33** (2), 218-233.

Taylor, M. A. P. (1999b). Dense network traffic models, travel time reliability and traffic management II: application to reliability. *J. of Adv. Transpn.*, **33** (2), 235-251.

Taylor, M. A. P. (2000). Using network reliability concepts for traffic calming - permeability, approachability and tortuosity - in network design. In *Reliability of Transport Networks* (M. G. H. Bell and C. Cassir, eds.), pp.217-242. Research Studies Press, Baldock.

Van Vliet, D. (1981) Selected node-pair analysis in Dial's assignment algorithm. *Transpn. Res.*, **15B**, 65-68.

van Zuylen, H. and Kikuchi, S. (2002). Traveller's behaviour under uncertain conditions. *The Network Reliability of Transport*, 155-170.

Vickerman, R., Spiekermann, K. and Wegener, M. (1995). Accessibility and economic development in Europe. *Proc EF/EC Euroconference on 'European Transport and Communication Policies on European Networks'*. Espinho, Portugal, 17-23 April.

Warshall, S. (1962). A theorem on Boolean matrices. *J. of the ACM*, **9**, 11-12.

Yang, H., Lo, H. K. and Tang, W. H. (2000). Travel time versus capacity reliability of a road network. In *Reliability of Transport Networks* (M. G. H. Bell and C. Cassir, eds.), pp.191-138. Research Studies Press, Baldock.

3

RISK EVALUATION AND MANAGEMENT: A ROAD NETWORK RELIABILITY STUDY

Alan Nicholson, University of Canterbury, Christchurch, New Zealand
Erica Dalziell, vivas ltd, Strategic Risk Management Consultants, London, UK

ABSTRACT

Risk evaluation and management methods are used to assess the reliability of a New Zealand inter-urban road network, which is subject to closures due to snow and ice, earthquakes, volcanic activity and road accidents. Using the probabilities and consequences of closures of various durations, the expected annual costs of closures are determined for each hazard. The benefit-cost ratios for various risk mitigation options are also identified. The importance of considering both the probabilities and consequences of closures is discussed.

It is found that a 'high-frequency, low-consequence' hazard (snow and ice) has the highest expected annual cost, and higher than for the 'low-frequency, high-consequence' hazards (earthquakes and volcanic activity). It is also found that not allowing for elasticity in the demand for travel (e.g. trips being cancelled or postponed because of road closures) leads to inaccurate estimates of the costs of closure.

It is concluded that it is important to allow for the behavioural responses of network users when estimating the costs of closures, and that more attention should be given to 'high-frequency, low-consequence' hazards, in order to maximise the economic benefits of expenditure on risk mitigation.

INTRODUCTION

The transportation system is one of several lifeline systems, whose vulnerability to damage and disruption during major disasters (e.g. earthquakes and storms) is of considerable concern in some countries, such as New Zealand (NZ) and Japan. Hopkins et al. (1991) found that the transportation system is the most important lifeline system, largely because damage to it inhibits repairs to the other lifeline systems.

The cost of partial or complete closure of high flow links in a road network can be massive. For instance, the cost of motorist delay associated with the closure of just one facility (Interstate Highway 10) in Los Angeles after the 1994 Northridge earthquake was almost US$1 million per day (Yee et al. 1996). This was after detours, carpool lanes, and rail and bus service enhancements had been established. The socio-economic costs due to commercial traffic disruption and loss of business were not included.

The economies in developed countries depend heavily on their transportation systems, and that dependence is likely to increase, given the trend towards "just-in-time" production methods, which rely on fast and reliable transportation systems. This suggests that studies of the impacts of road closures should not be confined to rare, high impact events (e.g. earthquakes), for which the closure duration may be measured in weeks or months; frequent short closures should also be considered.

The study described here involved applying risk evaluation and management methods to the Central North Island (New Zealand) road network. Figure 1 shows the portion of the network where closures generally occur, that is, the Desert Road section of State Highway 1 (SH1), and the nearby alternative routes (SH4, SH47 and SH49).

SH1 is the main route between Wellington (the capital) and Auckland (the largest urban area), and is thus strategically important. Figure 2 shows the full network, which extends about 270 km and 240 km in the north/south and east/west directions, respectively.

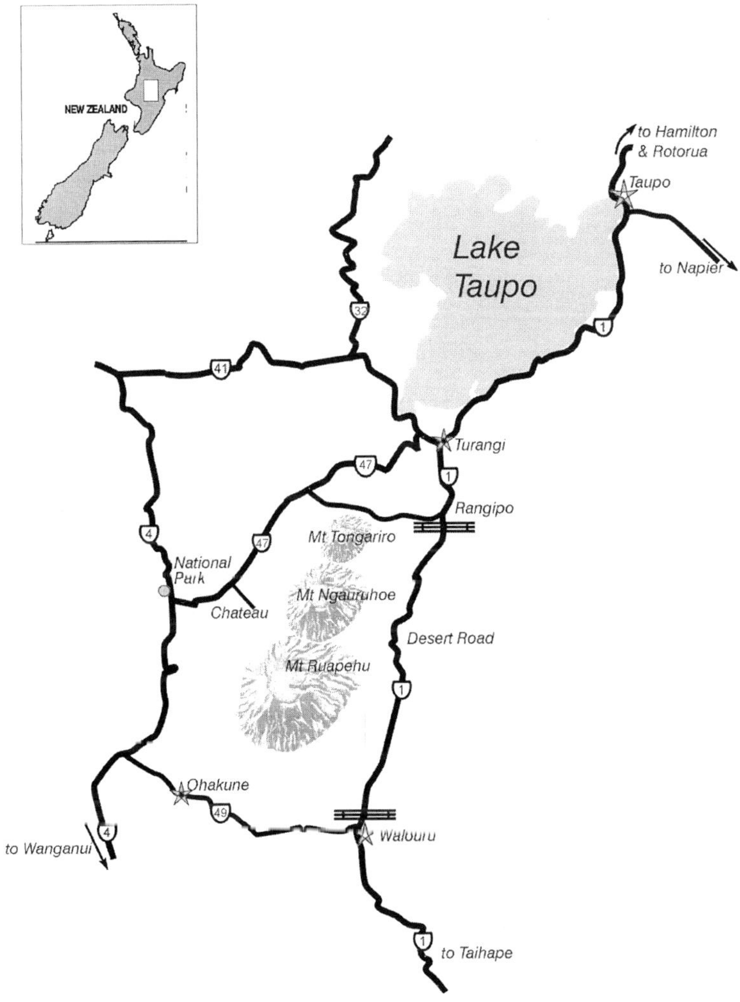

Figure 1 Desert Road Locality Map

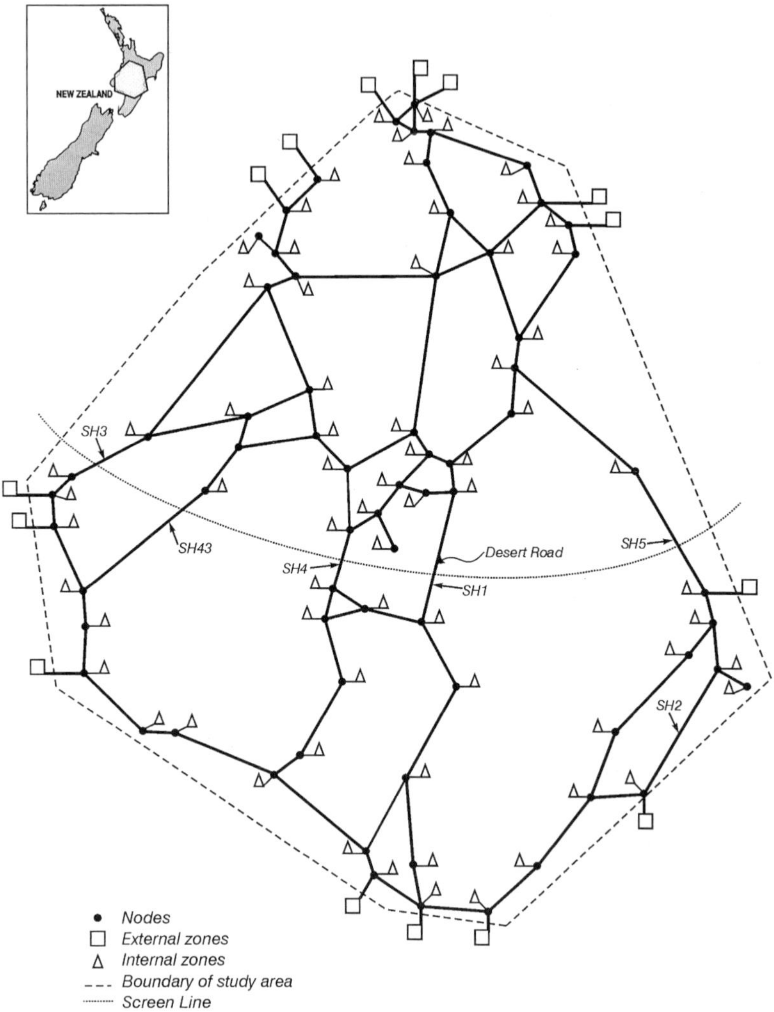

Figure 2 Network Model for Study Area

All trips between Wellington and Auckland pass through the network, which includes all towns in the region, and all the State Highways and other roads used for travel through the region and/or between the towns in the region. All links are two-lane two-way roads, and nearly all carry between 500 and 5000 vehicles/day (two-way flow), so the network is not congested.

The study focused on evaluating the risk of closure of the Desert Road section of SH1 and the nearby alternative routes, and then investigating how the risk should be managed.

RISK EVALUATION AND MANAGEMENT

There are two main dimensions of risk, namely the probabilities and consequences. A frequently mentioned third dimension is the 'context', and this is important when considering the acceptability of risk. Here the measure of risk is simply the sum of the products of the closure probabilities and the economic costs of closures (i.e. the expected annual economic cost of closure). Note that unless otherwise stated, all costs are in July 1997 NZ$, which were equivalent to approximately US$0.66 in July 1997.

Risks can be classified in various ways. For instance, risk can be:
(1) observed (i.e. based on empirical data);
(2) estimated (i.e. based on study of the mechanism of occurrence);
(3) perceived (i.e. based on intuition or belief);
(4) real (i.e. the actual or objective risk).
Experts generally focus on the observed risks for frequent events and the estimated risks for rare events. The general public, however, base their decisions on the perceived risks, which are subjective, for both rare and frequent events. The real risks are unknown unless one is omniscient.

Alternatively, risks can be categorised according to the probability and consequence:
(1) low probability and minor consequence;
(2) high probability and minor consequence;
(3) low probability and major consequence;
(4) high probability and major consequence.
As expected, the first and last types of risk generally receive the least and most attention respectively. Greater attention is generally given to the third type of risk than is given to the second risk type, even when the expectations are much the same, because of the greater level of hysteria, scrutiny and potential embarrassment when the consequences are major.

When considering the risk associated with events causing fatalities, it is common practice to use 'FN-curves', to plot the frequency F per unit time (usually one year) of events causing N or more fatalities (Evans, 1994). The lower the FN-curve is (i.e. the lower the value of F for a given value of N), the better the situation. The UK Health and Safety Commission (1991) has proposed that if

$$F > (0.1/N)$$

the frequency (per year) of events causing N or more fatalities is intolerable. If

$$F < (0.0001/N)$$

the frequency is negligible, while for

$$(0.0001/N) < F < (0.1/N)$$

the risk should be reduced if the benefit exceeds the cost.

FN-curves are traditionally plotted on double logarithmic scales, with the above threshold being straight lines. It is clear from the expressions above that when the FN-curves are plotted on ordinary scales, then the thresholds are non-linear (they are hyperbolic). That is, the thresholds imply that the threshold values of frequency F decline rapidly as N increases above 1, and approach zero asymptotically as N gets large. These thresholds are thus consistent with the greater attention given to events with a low probability and major consequence than events with a high probability and a minor consequence, even when the expectations are equal.

The appropriateness of placing a greater weighting on major consequence events is very doubtful. As noted by Evans (1994) in relation to transport safety, it leads to too much effort and resource being allocated to reducing the risk of rare accidents where large numbers of people are killed, and too little being done to reduce the risk of common accidents where small numbers are killed. The practice of concentrating upon rare disasters may well lead to not minimising the number of fatalities for the given resources.

The risk evaluation process generally involves:

(1) establishing the context (i.e. the technical, financial, legal, social and other criteria for assessing the acceptability of risk);
(2) identifying the hazards (i.e. the potential causes of closure);
(3) analysing the risks (i.e. identifying the probabilities, consequences and expectations);
(4) assessing the risks (i.e. deciding which risks are acceptable and which are unacceptable).

If any risk is unacceptable, the risk needs to be managed. This generally involves:

(1) treating the unacceptable risks, using the most cost-effective treatment options;
(2) monitoring and reviewing the risks (ie. evaluating and revising treatments).

In New Zealand, decisions on road network improvement proposals are based upon cost-benefit analysis. Hence, a risk is considered unacceptable if there is a mitigation option that reduces the risk enough to give a benefit (i.e. a reduction in the expected cost of closures over a 25 year period) sufficiently larger than the cost of implementing the option. The minimum benefit-cost ratio was 4.0 at the time of the study (i.e. 1998).

In general, risk management options are: avoid the risk; reduce or mitigate the risk (i.e. reduce the probabilities and/or consequences); transfer the risk (including risk sharing). In this study only risk mitigation was pursued, as the other options were not considered feasible.

HAZARD IDENTIFICATION

The Desert Road and nearby roads are subject to a wide range of hazards. Firstly, they are well above sea level (1130 metres for the Desert Road) and are subject to closure in extreme weather conditions due to the presence of snow and ice. Secondly, they are located close to three active volcanoes (about 15 km for the Desert Road), the most active of which has an average return period of 45-135 years for an eruption that would deposit sufficient ash on the roads to cause closure. Some roads may also be closed because of bridge damage caused by lahars, as one volcano has a substantial crater lake.

New Zealand is a seismically active region, and a large intensity earthquake has the potential to damage the bridges along the Desert Road and nearby roads, while seismic shaking may also cause liquefaction of soils and damage to the road surface. Finally, the Desert Road carries about 4000 vehicles/day, about 15% of which are heavy commercial vehicles. An accident involving a heavy vehicle (especially one carrying hazardous materials) or several cars may block the road and effectively close the road link.

RISK ANALYSIS

The risk analysis involved estimating, for each hazard and for the range of possible event magnitudes, both the probability and the consequence of occurrence. The consequence included the cost of remedial work (e.g. bridge repairs) and the extra cost for users whose travel is affected. The user costs depend on the closure duration, the availability of alternative routes, and the extra cost associated with using the alternative routes (or the loss of utility if users cancel or postpone their travel).

For some hazards, events of sufficient magnitude to cause road closure are rare, and historical data to assist estimation of the probability of closure duration are limited. It was necessary to adopt a unique approach for assessing the risk of each hazard. This involved identifying the generating mechanisms for each hazard, to establish ways to estimate the frequency of occurrence and the potential consequences of the events. Monte Carlo simulation was used to estimate the frequency distribution for the closure duration for each hazard. Details of the methods are given in Dalziell (1998) and Dalziell and Nicholson (2001).

The economic costs of closure depend on the reason for the closure, because of correlation between the conditions of the routes through the network (i.e. whether they are open or closed), particularly for an event (e.g. major volcanic eruption) with effects spread over a large geographic area. It was therefore necessary to assess the probability of the Desert Road and nearby alternative routes being closed simultaneously, and 22 closure scenarios (i.e.

combinations of the Desert Road with zero or more of the nearby alternative roads being closed) were considered. It was also necessary to consider the interactions of the different hazards (e.g. volcanic eruptions and earthquakes are related and are not independent). Some interactions are one-way (e.g. snow and ice increase the probability of a traffic accident, but accidents have no effect on the occurrence of snow and ice).

The costs imposed on the NZ economy by closures will depend on how users respond. A SATURN computer model, which assumes users seek to minimise their perceived costs, was used to estimate traffic flows and user costs. When links in the network are closed, trips are redistributed throughout the network. The costs on the new routes will exceed the costs on the original routes, if the numbers of trips remains constant. As the cost of travel increases however, some users may cancel their trip. To account for this, an elastic assignment of trips was done, using the power law relationship (Van Vliet and Hall 1995):

$$T_{Cij} = T_{Oij}(C_{Oij} / C_{Cij})^P$$

where T_{Cij} and C_{Cij} are the number of trips and the travel cost (respectively) between an origin zone i and a destination zone j when one or more road links are closed, T_{Oij} and C_{Oij} are the corresponding number of trips and travel cost when all links are open, and P is the elasticity parameter.

The total cost of road closures was assumed to be the sum of the changes in the vehicle operating, occupant time and accident costs, plus the lost user benefit if trips are cancelled, with the unit costs being the standard values for economic appraisal in New Zealand (Transfund NZ 1997).

RISK ASSESSMENT

Traffic flow data for a nine-day closure of the Desert Road were available, and it was found that setting the elasticity parameter P to 3.0 gave a good agreement between the predicted flows (with the Desert Road closed) and the observed flows. This implies an elasticity of demand of about -2.9, which is a relatively high value and probably reflects more than just the increase in the price of travel. Drivers may perceive that if the Desert Road is closed due to snow and ice, then the weather will be unsuitable for the many recreational pursuits in the area and the driving conditions on the nearby alternative routes will also be dangerous, and may be deterred from making trips.

Table 1 shows the predicted effects of the Desert Road being closed, for both inelastic demand ($P = 0$) and elastic demand ($P = 3$), for the whole network (Figure 2). The user cost is simply the sum of the vehicle operation and user time costs, the unit costs being NZ$0.341 per veh-km and NZ$0.355 per minute (Transfund NZ, 1997).

Table 1 The Effects of Closure on Travel Within the Study Network

Measure of Effect	Desert Rd Open	Desert Rd Closed (P=0)	Desert Rd Closed (P=3)
Trips (veh-trips/h)	1,956	1,956	1,891
Travel (veh-km/h)	299,890	306,610	286,250
User Cost (veh-NZ$/h)	180,580	185,210	172,610
Mean Speed (km/h)	81.6	81.0	81.3

If demand elasticity was ignored, the veh-km of travel and user cost were predicted to increase by about 2.2% and 2.6% respectively, with no change in the number of trips. For an elastic demand, however, the number of trips, the veh-km of travel and the user cost were predicted to decrease by 3.3%, 4.5% and 4.4% respectively.

Table 2 shows the predicted effects of the Desert Road being closed, for both inelastic and elastic demand, for flows across the screen line (Figure 2). Allowing for demand elasticity, the number of trips across the screen line was predicted to decrease by 7.4% (a larger reduction than for trips in the whole network). This is not surprising given that many trips within the network do not involve using the Desert Road.

For an inelastic demand, about 81% of the 155 veh/h using the Desert Road when it is open was predicted to divert to SH4, with the more distant alternative routes getting the remaining 19%. For an elastic demand, about 39% of the 155 veh/h was predicted to not cross the screen line, with about 55% and 6% using SH4 and the more distant alternatives, respectively.

Table 2 The Effect of Closure on Flows (veh/h) Across the Screen Line

Crossing Point	Desert Rd Open	Desert Rd Closed (P=0)	Desert Rd Closed (P=3)
SH1	155	0	0
SH4	100	225	185
Other	555	585	565
Total	810	810	750

Table 3 shows the effects on user and accident costs of four closure scenarios (22 were considered), along with the lost user benefit when trips are cancelled (this is simply the number of cancelled trips times the average of the costs with and without closures). The predicted user cost reductions arise primarily from the expected reduction in trips when there are closures, while for those users not cancelling their trips, there will be less traffic around and any congestion effect will be less (this is likely to be negligible in an non-congested network such as in this study, but may be quite important in a congested network). The predicted reductions in accident costs arise primarily from traffic flow reductions (accidents were predicted using accident versus traffic flow relationships), while some of the reductions might arise from traffic shifting from roads with a high accident rate (accidents per veh-km) to roads with a lower accident rate.

Table 3 The Costs (NZ$/hr) of Various Road Closure Scenarios (Elastic Demand)

Closures	User Cost	Accident Cost	Lost User Benefit	Total	Closure Cost
None	180,590	43,017	N/A	223,607	N/A
SH1	172,610	40,858	18,129	231,597	7,990
SH1&4	164,412	38,570	43,498	246,480	22,870
SH1&47	165,220	38,846	33,887	237,953	14,350
SH1&49	169,944	39,783	26,058	235,785	12,180

The total cost for each closure scenario is the sum of the user and accident costs, plus the lost user benefit, while the closure cost for each closure scenario is the total cost for the scenario minus the total cost for no closures. Hence the hourly cost to the NZ economy is about $8000 if only the Desert Road is closed and about $23,000 when SH4 also closed. When the Desert Road and all nearby roads are closed, the closure cost is about $31,000.

The reduction in user benefits (due to trips being cancelled) is the inverse of the case of an increase in user benefits when new trips are made after a road is improved. In the latter case, the average benefit for the new traffic is the average of the costs without and with the improvement (Neuburger, 1971). Similarly, the average dis-benefit associated with cancelled trips is the average of the costs without and with the closures.

It is worth noting that it was only after the lost user benefits (due to trips being cancelled) had been taken into account that an economic loss due to road closure was predicted.

Monte Carlo simulation gave the cumulative frequencies of the expected annual cost of closures for each hazard (Figure 3). Snow and ice was found to be the most important hazard, being expected to cost the NZ economy about \$1.9 million per year, compared to about \$1.5, \$0.3 and \$0.2 million for earthquakes, traffic accidents and volcanic eruptions, respectively. Given these costs, the risks (especially those due to snow and ice, and earthquakes) were considered unlikely to be acceptable, so risk mitigation options were pursued.

RISK MANAGEMENT

Mitigation options may affect the probability and/or cost of road closure. For this study, a number of mitigation options were considered, including:

(1) applying salt (NaCl) or calcium magnesium acetate (CMA);

(2) installing a road weather information system (RWIS) in conjunction with applying chemicals;

(3) using Bailey bridges or fords past earthquake-damaged bridges;

(4) increasing the seismic resistance of vulnerable bridges (retrofitting).

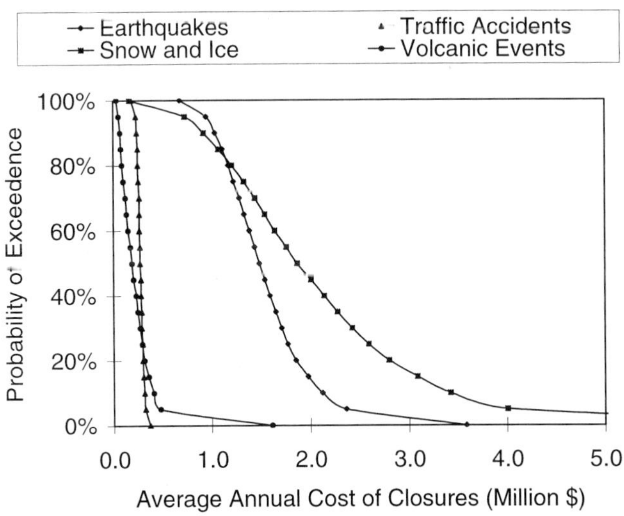

Figure 3 Closure Cost Exceedance Probabilities for Each Hazard

The benefit of a mitigation option is the risk (or expected cost) of road closure without the mitigation in place, minus the risk (or expected cost) of road closure with the mitigation in place. This was compared to the sum of the capital and maintenance costs of the mitigation, to assess the worth of the mitigation. To reflect the uncertainty that exists regarding the probabilities and consequences of the hazards, and the effects of the mitigation options, Monte Carlo simulation was used to obtain a probability distribution of the benefit-cost ratio for each mitigation option (Figure 4).

It can be seen that for all the above mitigation options the expected benefit-cost ratio is high or very high. The probability of the benefit-cost ratio not exceeding the threshold for implementation (4.0 in 1998) is only about 20% for applying CMA and virtually zero for the other three options. The road controlling authority for State Highways has commenced applying CMA and is investigating installing a RWIS, but is unlikely to apply salt because the Desert Road is within a National Park and the environmental effects of salt from the road getting into streams are currently considered unacceptable.

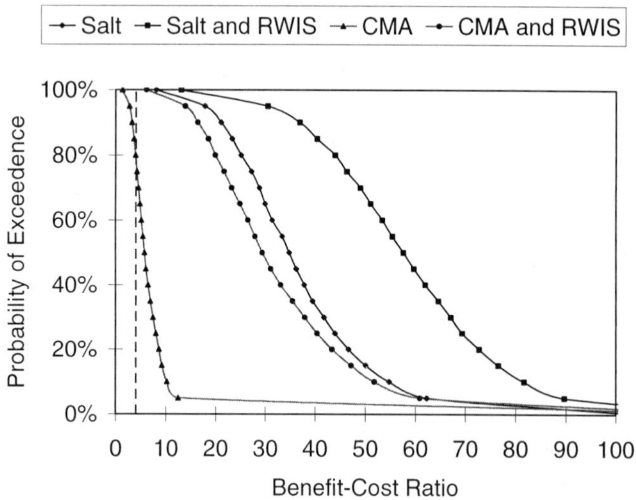

Figure 4 Benefit-Cost Ratio Exceedance Probabilities for Snow/Ice Mitigation

DISCUSSION

When roads are closed, the behavioural responses open to users include:

(1) cancelling their trip (i.e. a change in trip generation);
(2) postponing their trip (i.e. a change in the temporal distribution of trips);
(3) choosing another destination (i.e. a change in the spatial distribution of trips);
(4) travelling by train or plane (i.e. a change mode split);
(5) choosing a different route (i.e. a change in the traffic assignment).

Not all the possible behavioural responses have been considered in this study. The elastic demand procedure has allowed only for trips being cancelled, choosing a different destination (the proportion of trips from origin zone i to destination zone j can change, as the costs of travel to the possible destination zones change by different amounts when closures occur), and choosing a different route. In addition, the equilibrium traffic assignment method assumes that drivers know what routes will be open when they commence their trips and that their routes will not change once their trips commence. For this study, where most of the closures were expected to be for substantial periods, this was considered reasonable. For networks subject to short- term closures, however, an equilibrium assignment procedure is unlikely to be appropriate, as it is unlikely that equilibrium will be achieved.

The funds available for risk mitigation are generally limited, and the goal of risk evaluation and management (applied to network reliability) should be to get the biggest possible economic return (i.e. reduction in the socio-economic cost of network disruption) for that expenditure on mitigation. As noted by Evans (1994) for transport safety expenditure, the traditional focus on rare big disasters rather than frequent little disasters militates against achieving this goal.

Estimating the occurrence probabilities for a range of possible event magnitudes for each hazard has involved considerable effort and expertise (a detailed analysis of the mechanisms of hazard occurrence is required). It can therefore be argued that a method for identifying those circumstances where a full risk evaluation is required would be useful.

One might seek to identify those links or nodes that appear 'critical' because the consequences of their failure are very great. This will tend to perpetuate the tradition of placing greater emphasis upon rare big disasters rather than frequent small disasters, even when the expected costs are very similar. One might also end up doing risk evaluations for networks where there are such 'critical' links or nodes, only to find that the probabilities of failure are such that mitigation works are not warranted.

Doing risk evaluations for networks with 'critical' links and nodes means that resources may not be available for doing risk evaluations for networks which do not include such links or nodes, but for which mitigation works would be warranted. That is, there are two types of error (i.e. doing a risk evaluation when mitigation works are not warranted, and not doing a risk evaluation when mitigation works are warranted).

In that respect, the situation is analogous to a problem in road safety, namely identifying hazardous locations for which there is a good chance of identifying remedial treatments that will be economically attractive. In such situations, one can often end up reducing the probability of one type of error at the expense of an increase in the probability of the other error type (Nicholson, 1988).

It would be desirable to have some network-checking procedure that identifies when a detailed risk evaluation and management study is appropriate, but does not simply re-arrange the probabilities of the two types of error. Perhaps the best option will be to take account of both the probabilities and consequences from the beginning, even if they are only preliminary estimates. These can be refined later if analysis reveals the network reliability measure is sensitive to them.

CONCLUSIONS

The research shows that for the Central North Island road network, snow and ice is the most important hazard, costing the NZ economy about $1.9 million/year compared to $1.5 million/year for earthquakes (the next most important hazard). These are the tangible user costs of road closure; there will be other, less readily quantified effects (e.g. the social impacts associated with community isolation and shifting of business revenue). The risks should be mitigated (i.e. not accepted), as the ratio of the expected tangible benefits of the mitigation options to the expected costs of the mitigation options are very large.

It is particularly interesting that for the network in this study, a 'high-frequency, low-consequence' hazard dominates a 'low-frequency, high-consequence' hazard, given the much greater attention traditionally given to the latter. This suggests that in risk evaluation and management, we should be more even-handed with respect to the two types of hazard, especially if we truly seek to maximise the economic returns from expenditure on risk mitigation.

The study has shown the importance of allowing for the behavioural responses of users, particularly the elasticity of travel demand, in order to get an accurate estimate of the socio-economic impact of road closures. Finally, the equilibrium method used in this study is appropriate only when closures occur for a substantial period. For short-term closures, it will probably be necessary to use a method such as microsimulation, to take account of closures and route changes after a trip has begun.

ACKNOWLEDGEMENTS

The financial support of Transfund NZ and Transit NZ for the research is gratefully acknowledged.

REFERENCES

Dalziell, E.P. (1998). *Risk Assessment Methods in Road Network Evaluation.* PhD Thesis, Department of Civil Engineering, University of Canterbury, Christchurch, New Zealand.

Dalziell, E.P. and Nicholson, A.J. (2001). Risk and Impact of Natural Hazards on a Road Network. *ASCE Journal of Transportation Engineering*, **127(2)**, 159-166.

Evans, A.W. (1994). Evaluating public transport and road safety measures. *Accident Analysis and Prevention*, **26**, 411-428.

Health and Safety Commission (1991). *Major Hazard Aspects of the Transport of Dangerous Substances.* HMSO, London, United Kingdom.

Hopkins, D.C., Lumsden, J.L. and Norton, J.A. (1991). Wellington's Lifeline in Earthquakes Project: An Outline of a Major Review. American Society of Civil Engineers, *Proceedings of Third US Conference on Lifeline Earthquake Engineering*, Los Angeles, 774-804.

Neuburger, H. (1991). User Benefit in the Evaluation of Transport and Land Use Plans. *Journal of Transport Economics and Policy*, **5(1)**, 52-75.

Nicholson, A.J. (1988). Towards a Comprehensive Strategy for Accident Reduction and Prevention. NZ Road Traffic Safety Research Council, *Proceedings of Road Traffic Safety Seminar, Vol.2*, Wellington, New Zealand, 12-25.

Transfund NZ (1997). *Project Evaluation Manual.* Transfund NZ, Wellington, New Zealand.

Van Vliet, D. and Hall, M.D. (1995). *SATURN 9.2 Users Manual.* Institute for Transport Studies, University of Leeds, Leeds, United Kingdom.

Yee, A., Leung, S.K. and Wesemann, L. (1996). The 1994 Northridge Earthquake: A Transportation Impact Overview. Transportation Research Board, *Research Circular 462*, 7-19.

4

RELIABILITY ASSESSMENT ON SEARCHING TIME FOR PARKING IN URBAN AREAS

William H. K. Lam, Department of Civil and Structural Engineering, The Hong Kong Polytechnic University, Hong Kong, P.R China
Mei-Lam Tam Department of Civil and Structural Engineering, The Hong Kong Polytechnic University, Hong Kong, P.R China

ABSTRACT

This paper attempts to assess the parking reliability with taking into account drivers' travel choice behaviors under various traffic conditions. Apart from the well-known travel time reliability, a parking reliability is newly defined as the probability that the drivers' average searching time for parking is less than a given threshold. This is particularly important under conditions of shortage of parking spaces in urban areas. A Monte Carlo simulation approach, which incorporates a combined trip distribution and assignment model with explicit elastic demand function, is proposed to estimate the two reliability measures of road network (i.e. travel-time and parking reliabilities). A numerical example is used to illustrate the applications of the reliability measures and the proposed approach.

INTRODUCTION

Parking provision should not be just the "solution" to parking problems but should also be instrumental in influencing parking demands at strategic areas. It is well recognized that shortage in parking supply and increase in parking charges would have impacts on searching times for parking as well as choices of travel to and from central areas. To respond to the growing parking problems in central areas with limited parking spaces, there is a need to introduce some sort of performance measures for assessing the adequacy of parking facilities particularly in the urban areas. It becomes important as increasing populations, coupled with economic growth, produce travel demands that exceed the existing parking facilities in Hong Kong.

Hong Kong, a city of more than 6.9 million people and a land area of only 1097 square kilometers, is one of the most densely populated cities in the world, with residential densities of approximately 39,000 persons per square kilometer of developed land. The rapid development of Hong Kong and great concentrations of people focused in the urban areas, have put tremendous pressure on the existing road network and the parking facilities. The Government of Hong Kong Special Administration Region (SAR) has recently completed the Second Parking Demand Study (PDS-2) to refine the existing Parking Demand Model and to recommend new parking measures for alleviating the parking problems in urban areas.

Parking is a common problem for most car owners and road users in urban areas. A shortage of parking space increases the searching time for a parking space and induces traffic congestion. It also causes illegal roadside parking and traffic accidents due to a reduction of road space. On the other hand, excess parking space in some areas represents a waste of resources and can induce traffic, causing further congestion. Such parking supply problems are caused by inaccurate prediction of car ownership/usage and parking demand. Parking controls are seen importantly to regulate and restrain car use, and provide parking facilities according to the estimated level of car ownership and parking demand (Coombe et al., 1997). Therefore, Lam et al. (1999) has investigated the effects of balancing the demand and supply of parking spaces in urban areas. It was found that a balance of parking demand and supply leads to the most effective result on reduction of the total network travel time. Recently, Shoup (1999) claimed that the minimum parking requirements increase the supply and reduce the price but not the cost of parking. It was concluded that pricing curb parking rather than requiring off-street parking would improve urban design and reduce urban traffic congestion etc.

In order to respond to the increasing demands on parking facilities, a decision on either expansion of the road system (including parking facilities) or restraint of road traffic by some sort of fiscal measures is required. The former is related to the network design problems (NDPs) that become important as increasing populations coupled with economic growth produce travel demands which exceed the existing capacities of road networks and of parking facilities, while resources available for expanding these capacities remain limited. In the past decade, a number of studies have investigated the NDP formulations and solution procedures in two different forms: continuous and discrete (e.g. references, Friesz et al., 1992; Yang and Bell, 1998; Tam and Lam, 1999). Solutions to the NDP then require selecting a set of links or link capacity increases that minimize the objective function such as the total network costs and construction costs if the travel demand is fixed. Alternatively, electronic road pricing (ERP) is considered to be one of the most efficient approaches to reducing road traffic congestion, and will be implemented in many road network systems (Small, 1992). Some developments in model formulation and solution procedures for the congestion ERP problem under queuing network equilibrium conditions have been presented by Yang and Lam (1996). In their approach, the bi-level model considers a subset of links subject to toll charges and the toll charges are subject to given lower and upper bounds. However, the travel demand is assumed constant and fixed.

Although ERP is attractive in terms of economic benefits, it has not yet been implemented in practice (except Singapore). It is because of the political reasons; ERP is unlikely to be implemented in Hong Kong in the near future. On the other hand, there is currently interest in implementation of the Parking Information System (PIS) and Time-varying Parking Charges (TPC) for improving the efficiency of the parking facilities (Lam et al., 1998). In general, introduction of PIS or TPC would induce changes in traffic flow and journey time over the network. More importantly, implementation of PIS or TPC to a congested network without considering the response of network users may actually increase network-wide congestion. Therefore, prediction of traffic flow via a sound behavioral model is essential to the network design process.

In order to estimate the degree of congestion on road network, a traffic assignment technique is generally used to predict link flows in a road network. Traditionally, it was assumed in the NDP models that the origin-destination (OD) demand is given and fixed, and the drivers' route choices are characterised by a user equilibrium (UE) principle or the stochastic user equilibrium (SUE) assumption (Sheffi, 1985). In general, the NDP involves long term investment made in a road network, which in turn influences the future travel demand. Thus, it is more realistic to consider the elasticity of travel demand in the NDP. The UE model with elastic demand can be adopted for determining the OD travel demand endogenously given road capacity expansions. Boyce and Janson (1980) presented the NDP with a combined trip distribution/assignment (CDA) model (Evans, 1976; Lam and Huang, 1992) to explicitly consider the impacts of network improvement on trip distribution. In their model, trip productions and attractions are insensitive to zonal accessibility, and the influence of social factors is not considered. However, trip productions and attractions may actually vary due to the effect of accessibility measures (Ortúzar and Willumsen, 1994). The advantage of incorporating elastic trip productions and attractions is that the model can reflect the response of road users to traffic congestion and the availability of transport facilities (e.g. road capacities and parking spaces), which includes changing the time of day at which a journey is made and switching to an alternative mode or not making the journey at all.

On the other hand, there are few analyses of the effect of parking information on road network, and only series of researches by Kurauchi et al. (1998a, 1998b) are available at present. In their work, by using dynamic simulation model, the effect of parking information is varied when the locations of variable message signs are different. Due to the impacts of PIS market penetration and TPC on the benefits to both the equipped and unequipped drivers in the network, total vehicular demand might vary in response to the change in the PIS benefits resulting from improved driver decisions and to the change in parking charges. This overall demand elasticity will be modeled by using an elastic modal split trip end model, thereby allowing for mode shift of travelers among public transit, equipped and unequipped private vehicles. In addition, the destination, mode and route choice behaviors will also be considered in the proposed model (Lam et al., 1999; Tam and Lam, 1999).

In this paper, a new definition on parking reliability is introduced for assessing the network performance. **Parking reliability** is defined as the probability of the ability to reach an available parking space at a given destination from an origin within a given searching time for parking. This performance index is introduced to assess the benefits of drivers unequipped and equipped with PIS under TPC environment. This is particularly important under conditions of shortage of parking spaces in urban areas. Reliability has long been a key topic in management and operation studies as well as system engineering (Bell and Iida, 1997). Most of the previous related studies aimed to investigate the reliability assessment on road network such as connectivity and travel-time reliability. However, little attention had been given to the searching-time reliability for parking particularly in urban areas. Unexpected exogenous delays may be caused by shortage of parking spaces in urban areas, which can in turn cause further knock-on delays in viscous cycles as more vehicles are circulating in the urban roads for searching parking spaces.

Contrary to these previous studies, this paper attempts to develop an approach to assess the parking reliability with taking into account drivers' travel choice behaviors under various traffic conditions. In addition to the travel-time reliability, a new measure, referred to as parking searching-time reliability, is further defined in order to investigate comprehensively of the road network performance.

MEASURES OF RELIABILITY

Smooth traffic flow in a road network can be interrupted by events such as earthquakes, floods, traffic accidents, and adverse weather, for periods varying from a few hours to a few years. Among natural disaster, slope failures are particular prevalent in Hong Kong, because many major highways connecting suburban residential towns and the urban area are located in mountainous areas. In addition, urban road networks could exhibit high levels of congestion during peak-hours even without extraordinary events. It is thus important to study the reliability of road-based transport facilities including road capacities and parking spaces.

Existing reliability studies of road networks are mainly limited to connectivity and travel-time reliability and may not be sufficient for a comprehensive network performance measure. Connectivity reliability is concerned with the probability that the network nodes remain connected. A special case of the connectivity reliability is the terminal reliability, which concerns the existence of a path between a specific OD pair (Iida and Wakabayashi, 1989). For each node pair, the network is considered successful if at least one path is operational. A path consists of a set of components (e.g., roadways or links) which are characterized by zero-one variable denoting the state of each link (operating or failed). Capacity constraints on the links are not accounted for when finding the connectivity reliability. This type of connectivity reliability analysis might be suitable for abnormal situations, such as earthquakes, but there is an inherent deficiency in the sense that it only allows for two operating states: operating at full capacity or complete failure with zero capacity. This binary state approach limits the application to everyday situations where links are operating in-between these two extremes. The approach may thus result in misleading results of reliability and risk assessment of a road network.

Another measure of network reliability is the travel-time reliability or the coefficient of variation of travel time (Asakura and Kashiwadani, 1991). This is concerned with the probability that a trip between a given OD pair can be made successfully within a specified interval of time. This measure is useful to evaluate network performance under normal daily flow variations. Bell et al. (1999) proposed a sensitivity analysis based procedure to estimate the variance of travel time arising from daily demand fluctuations. Asakura (1996) extended the travel-time reliability to consider capacity degradation due to deteriorated roads. He defined travel-time reliability as a function of the ratio of travel times under the degraded and non-degraded states. This type of reliability can be used as a criterion to define the level of service that should be maintained despite the deterioration of certain links in the network. When the ratio is close to unity, it is essentially operating at ideal capacity. On the other hand, when it approaches infinity, the destination is not reachable because certain links are so severely degraded. This extreme case is consistent with network connectivity reliability.

Although most of the previous studies regarding road network reliability have focused on the above travel-time reliability, it is noted that a single measure will not be capable of estimating reliability of road-based transport facilities (parking spaces and road capacities) comprehensively. For example, in a road network with perfect travel-time reliability, some drivers may not be able to find a parking space at their destinations due to shortage of parking spaces and have to circulate on streets for searching the spare parking space, particularly during the commuting rush hours in urban areas. For these drivers, their actual journey times are obviously larger than that they expected. Hence they may not perceive that such road system is reliable. Therefore, we propose a new definition of parking reliability that is defined as the probability of the ability to reach an available parking space at a given destination from an origin within a given searching time for parking. Both of the travel-time reliability (excluding parking times) by OD pair and the parking reliability at destination should be used to assess the reliability of the road-based transport facilities.

In the paper, the mean and standard deviation of travel times and parking times for all OD pairs are calculated using the simulation results. It is assumed that the travel times or parking times T_n is a normally distributed random variable as $T_n \sim N(\mu_n, \sigma_n^2)$, where μ_n is the mean and σ_n is the standard deviation of the travel times or parking times calculated. Subsequently, the probability that the travel times or parking times within the threshold t can be expressed as

$$\Pr\{T_n \leq t\} = \Phi\left(\frac{t-\mu_n}{\sigma_n}\right). \tag{1}$$

APPROACH OF ESTIMATING RELIABILITY

In this section, we will propose a simulation procedure to estimate the above two reliability measures (i.e. travel-time and parking reliabilities). Be aware that since our simulation procedure incorporates an elastic demand CDA model, which reduces the need of detailed simulations, it may achieve a relatively fast assessment of reliability with less computation efforts.

In this paper, we will employ Lam et al. (1999)'s model with some modification in the simulation procedure, in which the variations of link travel times and searching times for parking at destination are explicitly considered. The mathematical formulation of the model is presented below.

$$\underset{\mathbf{v},\mathbf{t}}{Minimise} \sum_{a \in A} \int_0^{v_a} c_a(x)dx + \frac{1}{\alpha}\sum_{r \in R}\sum_{s \in S} t_{rs}(\ln t_{rs} - 1) + \sum_{s \in S} \int_0^{\sum_{r \in R} t_{rs}} d_s(z)dz \tag{2}$$

subject to

$$\sum_{k \in K_{rs}} f_{rs}^k = t_{rs}, r \in R, s \in S \tag{3}$$

$$\sum_{s \in S} t_{rs} = G_r, \quad r \in R \tag{4}$$

$$\sum_{r \in R} t_{rs} = A_s, \quad s \in S \tag{5}$$

$$v_a = \sum_{k \in K_{rs}} f_{rs}^k \delta_{rs}^{ak}, a \in A \tag{6}$$

$$f_{rs}^{k} \geq 0, k \in K_{rs} \tag{7}$$

$$t_{rs} \geq 0, r \in R, s \in S \tag{8}$$

The functions of trip generation and attraction are expressed as $G_r = \gamma_0 z_r + \gamma_1 a_r + \gamma_2 e_r$ and $A_s = \lambda_0 z_s' + \lambda_1 e_s + \lambda_2 p_s$, where G_r is the trip generation at origin r, A_s is the trip attraction at destination s, a_r is population at origin r, e_r/e_s is employment at origin r or destination s, p_s is the available parking space at destination s, z_r is the accessibility measures for producing car trips in origin r that defined as $z_r = \max\{0, \ln \sum_s e_s \exp(-\theta(g_{rs} + d_s))\}$, z_s' is the accessibility measures for attracting car trips in destination s that defined as $z_s' = \max\{0, \ln \sum_r a_r \exp(-\theta(g_{rs} + d_s))\}$, g_{rs} is the travel time for OD pair rs, d_s is the average parking searching-time at destination s and $\gamma_0, \gamma_1, \gamma_2, \lambda_0, \lambda_1, \lambda_2, \theta$ are the pre-determined parameters.

Given the road network configuration and OD demands, the above analytical model may find essentially deterministic equilibrium traffic flows, travel times, parking delays and/or searching times for parking. However, the potential sources of disruption to road network are numerous such as incident, adverse weather and road works. It is such disruption that degrades the network's capacity and makes it unreliable. As a result of change of the journey times by OD pair (including both travel times by OD pair and searching times for parking at destination), the OD demands will be changed accordingly. Therefore, it is reasonable to assume that travel times on road links and searching times for parking are random and follow different distributions respectively. To investigate their effects on reliability measures, we can perform a Monte Carlo simulation to estimate the variations of link travel times and searching times for parking, and then evaluate the aforementioned reliability measures. The simulation procedure can be described as follows:

Step 1: Set sample number $n := 1$.

Step 2: Generate the variations of link travel times and searching times for parking.

Step 3: Perform the elastic demand CDA model with the generated input vectors.

Step 4: Collect statistics from the CDA model to construct reliability distribution functions.

Step 5: If sample number n is less than the required sample size $n_{\max}$, then increment sample number $n := n + 1$ and go to Step 2. Otherwise, stop.

Variations of Link Travel Time and Searching Time for Parking

The UE approach is based on the assumption that all the drivers have perfect knowledge of network traffic conditions. However, in the real world, this is not the case. There is a difference between the stochastic and actual average travel times by OD pair. Thus, the stochastic travel times on link *a* can be expressed as

$$C_a = c_a + \varepsilon_a \tag{9}$$

in which C_a is the stochastic travel time on link *a*, c_a is the actual average travel time on link *a*, and ε_a is the variation of travel time on link *a*. Different assumptions can be made for the distribution of ε_a. In this paper, the assumption of a normal distribution of $\varepsilon_a \sim N(0, \omega c_a)$ has been made in which ω is the coefficient of variance for stochastic variation of travel time.

Similarly, fluctuations do exist in the searching times for parking at the destination. The searching times for parking (including parking fee for at least one hour) and its variation can be expressed as

$$D_s = d_s + \xi_s \tag{10}$$

where D_s is the stochastic searching time (as well as parking fee) for parking at destination *s*, d_s is the actual average searching time (as well as parking fee) for parking at destination *s*, and $\xi_s \sim N(0, \beta d_s)$ is the variation of searching time for parking at destination *s*, in which β is the coefficient of variance for the stochastic variation of searching time for parking.

Numerical Example

To demonstrate the Monte Carlo simulation procedure, we will apply it to a simple road network that is similar to the one adopted in the previous related studies with taking account the effect of parking problem. The network shown in Figure 1 consists of 4 zones which represented by dotted line circles and connected to road nodes by two-way links, 6 road nodes, 16 internal nodes in which nodes 13, 14, 17, 18, 21, 22, 25 and 26 are parking lots, 14 one-way external links, and 20 two-way internal links of which 8 are parking links. The average link travel times are calculated from the BPR (Bureau of Public Roads) type formula

$$c_a = c_a(0) + k(\frac{v_a}{s_a})^p \tag{11}$$

where s_a is the capacity of link a, v_a is the link flow, $c_a(0)$ is the free flow travel time, and p and k are parameters of link a. The characteristics for the 14 external links in the study network are given in Table 1.

The generalized time for parking delay can be considered as the free-flow parking access time, plus the searching time for an available parking space plus the parking charge (at least to pay 1-hour parking fee). The free-flow access time is the minimum time taken to reach the parking facilities in a given zone under free-flow conditions. A similar parking searching-time function adopted by Lam et al. (1999) is used in this paper. The parking duration of cars has not been considered in the study.

$$d_s = d_s(0) + h(\frac{q_s}{p_s})^n + \frac{f_s}{t} \qquad (12)$$

Table 1 Link Data for the Example Network

Link number	Free-flow travel time	Capacity	Parameters	
A	$c_a(0)$ (min.)	s_a (pcu/hr)	p	k
1, 3	7.38	2000	4.0	0.15
2, 8	6.48	1600	4.0	0.15
4, 6	5.64	2000	4.0	0.15
5, 10	2.10	2000	4.0	0.15
7, 13	5.28	1600	4.0	0.15
9, 11	5.10	2000	4.0	0.15
12, 14	5.52	2000	4.0	0.15

where p_s is the available parking space at destination s, q_s is the trip attractions by car at destination s, $d_s(0)$ is the free-flow access time for parking, f_s is the parking charge for at least 1-hour parking, t is the pre-determined value of time perceived by travelers, and n and h are parameters of destination s. The parking data for the 4 destination zones is given in Table 2. The pre-determined value of time, t, is assumed to be \$5/min.

The coefficients of variance for the stochastic variations of link travel times and searching times for parking are set arbitrarily as $\omega = 0.1$ and $\beta = 1.0$. Larger values of ω and β represent greater variations between the stochastic and actual average travel times and searching times for parking.

With these pre-defined coefficients of variance for the stochastic variation of link travel times and searching times for parking, the reliability distribution can be constructed on the basis of the statistics obtained from the Monte Carlo simulation. As a result, the travel-time and parking-time reliabilities can be determined from the distributions with specified time intervals.

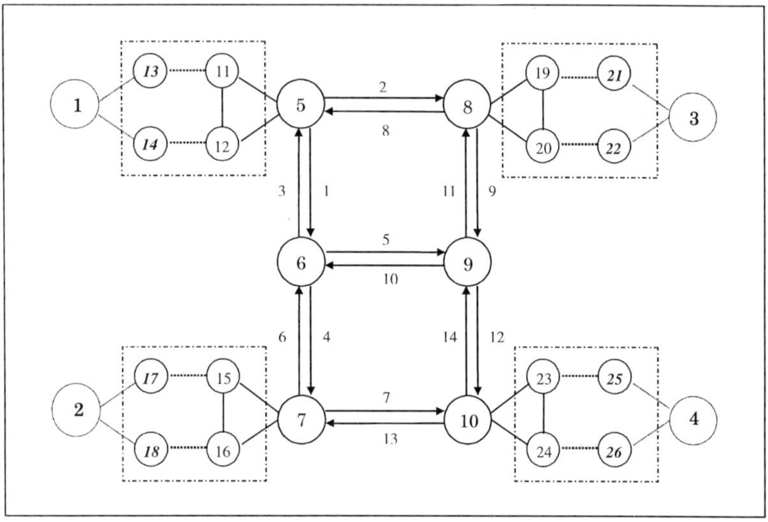

Figure 1 Example Network

Table 2 Parking Data for the Example Network

Destination	Free-flow access time	Parking space	Parking charge	Parameters	
s	$d_s(0)$ (min.)	p_s	f_s ($)	n	h
1	6.00	2800	25.0	4.0	0.15
2	5.10	3000	20.0	4.0	0.15
3	4.80	3150	22.0	4.0	0.15
4	6.50	2650	28.0	4.0	0.15

Preliminaries

Before evaluating the reliabilities of the example network, convergence tests on the sample sizes of the Monte Carlo simulation were performed. Three simulations of sample sizes of 1,000, 2,000 and 5,000 are examined and the resultant total travel-time reliability (excluding parking times) and total parking-time reliability (not including parking fee) are shown in Figures 2 and 3 respectively. The total travel-time and total parking-time reliabilities are referred to the total time of all trips in the entire road network. It is found that, in Figures 2 and 3, the reliability distributions using 1,000 samples show no significant difference from the ones using 2,000 and 5,000 samples. Thus, a sample size of 1,000 is used for all subsequent simulations.

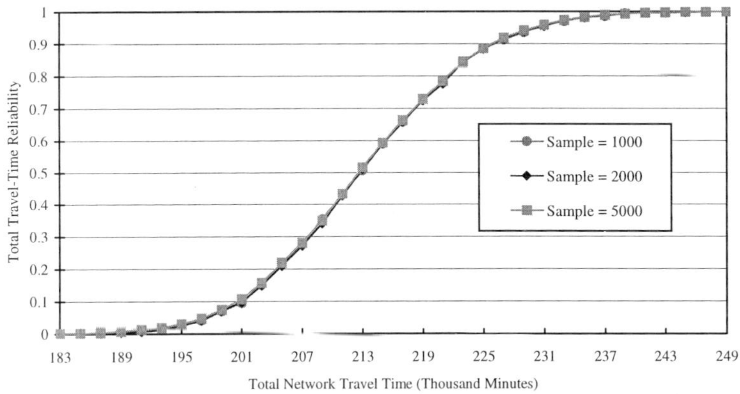

Figure 2 Effect of Sample Sizes on Total Travel-Time Reliability

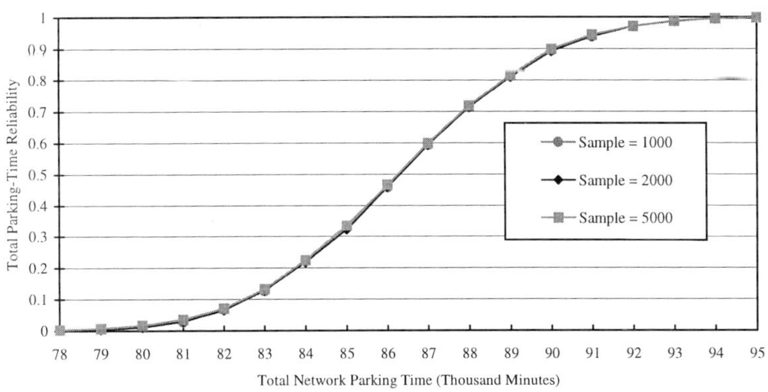

Figure 3 Effect of Sample Sizes on Total Parking-Time Reliability

Reliability Assessment

In order to compare the reliability of searching times for parking at destinations under different circumstances, two sensitivity tests have been undertaken. In the first test, different values of coefficients of variance for parking times have been set. As aforementioned, the coefficients of variance for the stochastic variations of link travel times and searching times for parking are $\omega = 0.1$ and $\beta = 1.0$. This case is denoted as the base case. In the first sensitivity test, ω keeps being 0.1 but β is varied from 1.5 to 4.0. Figures 4 and 5 show the total parking-time reliability (not including parking fee) and the total travel-time reliability (excluding parking times) for the study network using different coefficients of variance, respectively. It can be seen in Figure 4 that for the total parking searching-times less than 71,000 minutes, the reliability is 0; while for those larger than 99,000 minutes, the reliability is 1. In other words, it is highly unlikely that the total searching times for parking are less than 71,000 minutes but should be less than 99,000 minutes in the example network. On the other hand, it can be observed from Figure 5 that the total travel time for this road network should be in a range of 165,000 minutes to 255,000 minutes. Figures 4 and 5 also show that the total parking-time and total travel-time reliabilities are increased in higher minutes region when the coefficient of parking variation (β) is decreased. However, the reliabilities are reduced with decreasing β in lower minutes region.

In the second sensitivity test, the values of coefficients of variance for link travel time and searching time for parking are kept as 0.1 and 1.0 (denoted as base case), respectively, and only the parking charges are changed. This test is to examine the effects of parking charges on the reliability of total parking times together with the total travel times. Figures 6 and 7 show the total parking-time reliability (not including parking fee) and the total travel-time reliability (excluding parking times) for the study network using different parking fees, respectively. It was found that parking-time reliability is increased with increasing parking charges. This is because increases in parking charges would reduce the desire for trip making and so improves the total travel-time reliability. In addition, travelers can be easier to find parking spaces so to improve the parking reliability.

Furthermore, the average travel-time and parking-time estimates at specific probability for the above two sensitivity tests have been obtained and the results are presented in Table 3. Table 3 gives the time estimates at the probability of 10%, 50% and 90% that a driver can reach an available parking space at a given destination for the base case and various sensitivity tests. In addition, a sensitivity test was carried out to test the effect of OD travel demand on travel-time and parking-time reliabilities. It was found that both travel-time and parking-time reliabilities are decreased when OD demand increases, however, the rates of changes in parking-time and travel-time reliabilities are not significant.

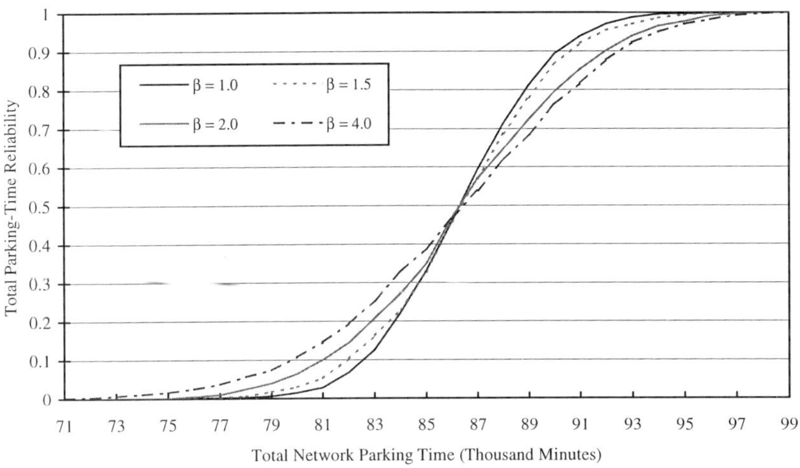

Figure 4 Total Parking-Time Reliability with Different Coefficients of Parking Variation

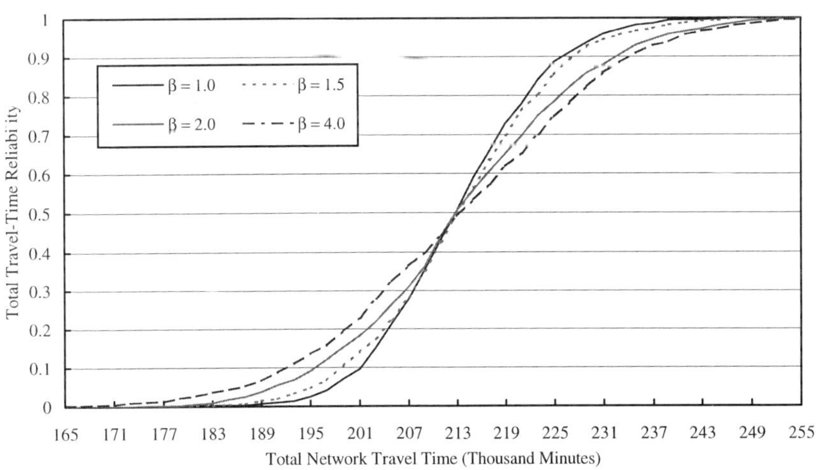

Figure 5 Total Travel-Time Reliability with Different Coefficients of Parking Variation

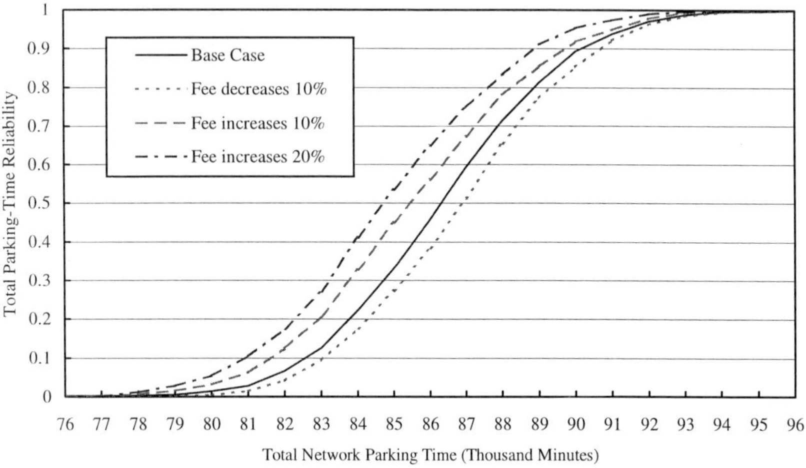

Figure 6 Total Parking-Time Reliability with Different Parking Fees

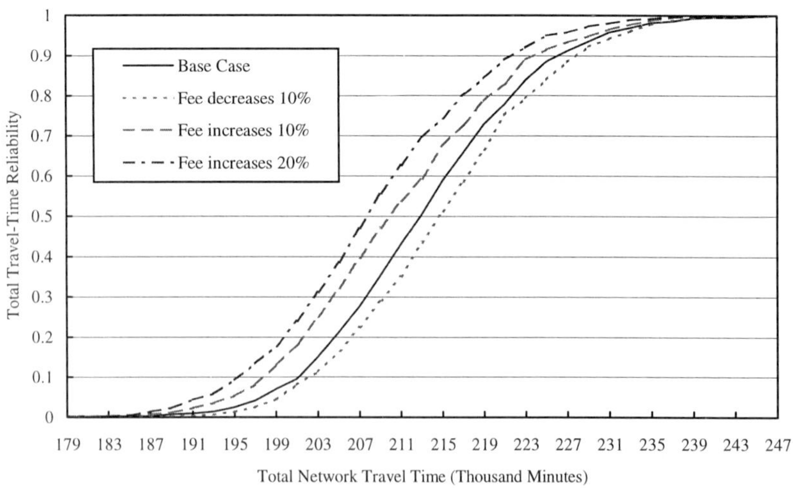

Figure 7 Total Travel-Time Reliability with Different Parking Fees

Table 3. Average Travel-Time and Parking-Time Estimates at Specific Probabilities

	(1) Average Travel-Time (min)			(2) Average Parking-Time (min)			(1)+(2) Average Network-wide Travel Time (min)		
	10%	50%	90%	10%	50%	90%	10%	50%	90%
Base Case	13.83	14.11	14.43	5.70	5.72	5.75	19.53	19.83	20.18
$\beta = 1.5$	13.81	14.12	14.45	5.69	5.72	5.75	19.50	19.84	20.20
$\beta = 2.0$	13.73	14.12	14.57	5.69	5.72	5.76	19.42	19.84	20.33
$\beta = 4.0$	13.69	14.13	14.61	5.68	5.72	5.77	19.37	19.85	20.38
–10% of Fee	13.86	14.15	14.46	5.70	5.72	5.75	19.56	19.87	20.21
+10% of Fee	13.77	14.05	14.38	5.69	5.72	5.74	19.46	19.77	20.12
+20% of Fee	13.72	14.00	14.32	5.69	5.71	5.74	19.41	19.71	20.06

CONCLUSIONS

A Monte Carlo simulation procedure is proposed in this paper to assess the parking reliability. Apart from the well-known travel-time reliability, a parking reliability has been defined and evaluated with taking into account the drivers' travel choice behaviors under various traffic conditions. These reliability measures are particularly useful to assess the impacts of parking information system on the network performance. The research will continue to investigate the applicability of the proposed procedure to real road network together with the model calibration and validation. Parking duration will also be considered in order to reflect the real performance of road network and parking facilities.

ACKNOWLEDGEMENT

The work described in this paper was supported by a research grant from the Institute of Systems Science of Japan (A/C H-ZF21) and a research grant from the Hong Kong Research Grant Council (PolyU 5040/01E) to the Hong Kong Polytechnic University.

REFERENCES

Asakura, Y. (1996). Reliability measures of an origin and destination pair in a deteriorated road network with variable flows. *Proceeding of the 4th Meeting of the EURO Working Group in Transportation.*

Asakura, Y. and M. Kashiwadani (1991). Road network reliability caused by daily fluctuation of traffic flow. *Proceedings of the 19th PTRC Summer Annual Meeting*, Brighton, 73-84.

Bell, M.G.H. and Y. Iida (1997). In: *Transportation Network Analysis*, Chapter 8: Network reliability, pp.179-192. John Wiley & Sons, Chichester.

Bell, M.G.H., C. Cassir, Y. Iida and W.H.K. Lam (1999). A sensitivity based approach to reliability assessment. *Proceedings of the 14th International Symposium on Transportation and Traffic Theory*, Jerusalem, Israel, 20-23 July, 1999, 283-300, Pergamon, Oxford.

Boyce, D.E. and B.N. Janson (1980). A discrete transportation network design problem with combined trip distribution and assignment. *Transportation Research*, **14B**, 147-154.

Coombe, D., P. Guest, G. Schofield and A. Skinner (1997). The effects of parking control strategies in the city of Bristol. *Traffic Engineering and Control*, **38**, 204-208.

Evans, S.P. (1976). Derivation and analysis of some model for combining trip distribution and assignment. *Transportation Research*, **10**, 37-57.

Friesz, T.L., H.J. Cho, N.J. Mehta, R.L. Tobin and G. Anandalingam (1992). A simulated annealing approach to the network design problem with variational inequality constraints. *Transportation Science*, **26**, 18-26.

Iida, Y. and H. Wakabayashi (1989). An approximation method of terminal reliability of a road network using partial minimal path and cut set. *Proceedings of the 5th World Conference on Transport Research*, Yokohama, 367-380.

Kurauchi, F., Y. Iida and Y. Yoshiya (1998a). Experimental analysis on drivers parking behavior under information provided. Paper presented in the 8th World Conference on Transport Research, Antwerp, Belgium.

Kurauchi, F., Y. Iida and Y. Yoshiya (1998b). Dynamic traffic simulation model for evaluating parking guidance and information system. *Proceedings of the 3rd Conference of the Hong Kong Society for Transportation Society*, Hong Kong, 227-235.

Lam W.C.H., R.Y.C. Fung, S.C. Wong and C.O. Tong (1998). The Hong Kong parking demand study. *Proceedings of the Institution of Civil Engineers -Transport*, **129**, 218-227.

Lam, W.H.K. and H.J. Huang (1992). A combined trip distribution and assignment model for multiple user classes. *Transportation Research*, **26B**, 275-287.

Lam, W.H.K., M.L. Tam, H. Yang and S.C. Wong (1999). Balance of demand and supply of parking spaces. *Proceedings of the 14th International Symposium on Transportation and Traffic Theory*, Jerusalem, Israel, 20-23 July, 1999, 707-731, Pergamon, Oxford.

Ortúzar, J. de D. and L.G. Willumsen (1994). *Modelling Transport*. John Wiley & Sons, Chichester.

Sheffi, Y. (1985). *Urban Transportation Networks: Equilibrium Analysis with Mathematical Programming Methods*. Prentice-Hall, Englewood Cliffs, NJ.

Shoup, D. (1999). The trouble with minimum parking requirements. *Transportation Research,* **33A**, 549-574.

Small, K.A. (1992). Special Issue: Congestion pricing. *Transportation*, **19**, 287-291.

Tam, M.L. and W.H.K. Lam (1999). Analysis of demand for road-based transport facilities. *Transportation Research Record*, **1685**, 73-80.

Yang, H. and M.G.H. Bell (1998). Models and algorithms for road network design: a review and some new developments. *Transport Reviews*, **18**, 257-278.

Yang, H. and W. H. K. Lam (1996). Optimal road tolls under conditions of queueing and congestion. *Transportation Research*, **30A**, 319-332.

5

An Evaluation of the Reliability of Travel Time in Road Networks Based on Stochastic User Equilibrium

Hiroshi Inouye, Himeji Institute of Technology, Himeji, Japan

Abstract

In this paper, a model to estimate travel time reliability is proposed assuming a stochastic user equilibrium. Travel time reliability is defined as the probability that travel time between the origin and destination does not exceed a standard travel time corresponding to each service level. Also it is shown how to calculate travel time reliability in a large-scale network. This model is applied to the road network in the Hanshin area to evaluate new lines in the national road plan from the viewpoint of travel time reliability. Some important links and lines in the future road network are evaluated from the viewpoint of travel time reliability.

Introduction

In planning and evaluating a future road network, traffic flow on the network, the economic condition in the region and the effect on the surrounding areas have been analysed so far. Recently, it has become important to evaluate road networks by simple indices for use in network planning.

As the road network has expanded, users' desires have extended to various kinds of road traffic information. On the other hand, it has become possible for the administrator of the road to operate and manage roads as a network. They are requested to provide reliable and stable traffic services for users, whence, the need to introduce the idea of reliability for the evaluation of road networks.

This paper proposes a measure of travel time reliability in a road network based on the assumption that traffic flow on a road network conforms to a stochastic user equilibrium. This model was applied to the evaluation of new lines in the Hanshin area, comparing the change of travel time reliability due to the blockage of a link or the removal of a new line.

EXISTING STUDIES

The idea of reliability of a road network is proposed formerly from the perspective that our country is subject to natural disasters. In most of the early researches, measures that represent the connection of a road network in the event of a natural disaster were proposed, to design a road network resistant to natural disasters.

Masuya (1984) analysed the connection of a network following an earthquake. Moritsu (1984) dealt with the reliability of a road network following a disaster from the viewpoint of optimum network design. Wakabayashi and Kameda (1992) analysed traffic service in San Francisco Bay Area before and after Roma-Prieta earthquake.

Through the experience of Hanshin-Awaji earthquake in 1995, we learned that it is not sufficient to assure only the connectivity of the road network. It is important to fulfil its function to some degree also following a natural disaster. Furthermore, the change of traffic demand and travel behaviour following a natural disaster should be considered. Therefore, there remain some problems in estimating reliability of a road network.

It is quite hard to forecast the occurrence of natural disasters, so provision of reliable and stable road services in ordinary times is a more important policy goal of local governments. Iida and Wakabayashi (1988), Iida, Wakabayashi and Yoshigi (1988) and Iida, Wakabayashi and Fukushima (1989) defined the probability that smooth travel is assured between node pairs when a link is blocked due to a traffic jam as *connectivity reliability* between nodes, and showed how to calculate this by the systems engineering method. Wakabayashi and Iida (1992), Wakabayashi, Iida and Inoue (1993) defined the probability that a driver can reach his destination within given time as *travel time reliability*, and showed how to calculate this from probabilistic distributions of link flow. Asakura, Kashiwadani and Kumamoto (1989) showed how to calculate travel time reliability on a road network supposing probabilistic variation of OD traffic flow. Also, Chen and Yang (1999) showed how to calculate travel time reliability as a function of capacity variation.

With respect to this research, some problems remain such as how to model the variation of traffic flow and travel time on a road network reasonably or how to calculate reliability for a large-scale network.

EXTRACTION OF IMPORTANT LINKS

Stochastic User Equilibrium

In this paper, it is assumed that traffic flow on a network conforms to a stochastic user equilibrium. The stochastic user equilibrium concept allows for the variation of perceptual travel time in route choice. However, there is some doubt whether stochastic user equilibrium assignment holds precisely for a road network. Many researchers regard it as superior to deterministic user equilibrium assignment, which assumes perfect information.

Stochastic user equilibrium assignment was proposed by Daganzo and Sheffi (1977). They defined the equilibrium state of traffic flow on a network as a stochastic user equilibrium when every user chooses his route such that perceived travel time between origin and destination is minimum, but perceived travel time on a link varies randomly across users. The problem becomes probit type if perceived travel time follows normal distribution and logit type if it follows Gumbel distribution. When OD traffic demand matrix $\{ q_{rs} \}$ is given, the condition of stochastic user equilibrium is represented as follows.

$$f_k^{rs} = q_{rs} P_k^{rs} \quad \forall r, s, k \tag{1}$$
$$P_k^{rs} = \Pr(C_k^{rs} < C_l^{rs}, \forall l \neq k \in K_{rs}) \tag{2}$$

where

C_k^{rs} : the perceived travel time on the route k between origin r and destination s,
f_k^{rs} : traffic flow on the route k between origin r and destination s,
P_k^{rs} : probability that route k is chosen between origin r and destination s,
q_{rs} : OD traffic demand between origin r and destination s,
K_{rs} : set of route between origin r and destination s.

In probit model, it is assumed that perceived link travel time T_a follows the normal distribution with average equivalent to actual trip time t_a and variance βt_a .

The next equations should hold as a matter of course.

$$\sum_k f_k^{rs} = q_{rs} \qquad \forall r, s \tag{3}$$

$$C_k^{rs} = \sum_a T_a \delta_{ak}^{rs} \qquad \forall r,s,k \tag{4}$$

$$t_a = t_a(x_a) \qquad \forall a \tag{5}$$

where

$$t_a = \mathrm{E}[T_a] \qquad \forall a$$

$$x_a = \sum_{rs}\sum_k f_k^{rs}\delta_{ak}^{rs} \qquad \forall a$$

Sheffi and Powell (1982) and Daganzo (1982) showed that stochastic user equilibrium is equivalent to the following optimisation problem:

$$\min_x z(x) = -\sum_{rs} q_{rs}\, \mathrm{E}[\min_{k\in K_{rs}}\{C_k^{rs}\} \mid c^{rs}(x)] + \sum_a x_a t_a(x_a) - \sum_a \int_0^{x_a} t_a(\omega)d\omega \tag{6}$$

where

c^{rs} : the actual travel time on the route k between origin r and destination s.

As this optimisation problem contains operation of expectation, it is impossible to use it directly for solving the problem. Sheffi and Powell (1981) showed a practical solution by the method of successive averages for probit type and logit type problems.

Extraction of Important Links

Important links are selected to analyse the reliability of the road network. They are extracted based on their function in the road network under practical traffic conditions. When traffic flow on a network follows stochastic user equilibrium, the objective function (6) takes the minimum. Then, the equation (7) holds (Sheffi and Powell, 1982).

$$\frac{\partial}{\partial x_b}\{\sum_{rs} q_{rs}\, \mathrm{E}[\min_k C_k^{rs}]\} = [\sum_{rs}\sum_k q_{rs} P_k^{rs} \delta_{bk}^{rs}]\frac{dt_b}{dx_b} \tag{7}$$

As it can be supposed that the perceived travel time supplies negative utility to users, the total negative utility in the network is defined by the sum of perceived travel time:

$$\overline{U} = \sum_{rs} q_{rs} C_{rs} \tag{8}$$

Then, the contribution of any link flow in the total negative utility can be defined by its marginal utility:

$$M\overline{U_b} = \frac{\partial}{\partial x_b}\{\overline{U}\} = \frac{\partial}{\partial x_b}\{\sum_{rs} q_{rs} C_{rs}\} = \frac{\partial}{\partial x_b}\{\sum_{rs} q_{rs}\, \mathrm{E}[\min_k C_k^{rs}]\} \tag{9}$$

Then, equation (9) results in:

$$M\overline{U_b} = [\sum_{rs}\sum_{k} q_{rs} P_k^{rs} \delta_{bk}^{rs}] \frac{dt_b}{dx_b} = x_b \frac{dt_b}{dx_b} \tag{10}$$

Equation (10) means that the contribution of any link flow in the total negative utility is equivalent to the product of its link flow and derivative of link travel time function. It represents the contribution of any link flow in the traffic service of a road network under practical conditions. Whence, it can be used as one of the indices to extract important links in a network when the reliability of a road network is analysed.

ESTIMATION OF TRAVEL TIME RELIABILITY

Definition of Travel Time Reliability

Travel time reliability is defined on the basis that traffic flow on a network follows a stochastic user equilibrium. The probit type model is used in this study because it can modify the defect of the logit type model such as IID characteristics.

The actual travel time on the route k between origin r and destination s is given by:

$$c_k^{rs} = \sum_a t_a \delta_{ak}^{rs} \qquad \forall r, s, k \tag{11}$$

When a trip is extracted between any origin and destination at random, its actual travel time can be assumed as dispersed probabilistic variable because it varies by route. Then, the probabilistic density function of actual travel time c^{rs} is given by:

$$f(c_k^{rs}) = \Pr(c^{rs} = c_k^{rs}) = P_k^{rs} \tag{12}$$

Travel time reliability is defined as the probability that travel time does not exceed standard travel time between its origin and destination. When standard travel time between origin r and destination s is denoted by c_*^{rs}, the probability that actual travel time c^{rs} does not exceed standard travel time c_*^{rs} is written by:

$$F(c_*^{rs}) = \Pr(c^{rs} \le c_*^{rs}) = \sum_{k, c_k^{rs} \le c_*^{rs}} P_k^{rs} \tag{13}$$

whence, this probability gives travel time reliability between origin r and destination s. Also its weighted average by OD traffic demand is defined as overall travel time reliability in a network.

$$R = \sum_{rs} q_{rs} F(c_*^{rs}) \Big/ \sum_{rs} q_{rs} \qquad (14)$$

Definition of Standard Travel Time

How to set the standard travel time between each origin and destination is an issue. There is no reasonable way to set it in the previous studies. Then, it is defined relatively using service level for the standard network and standard traffic conditions.

When origin and destination traffic demand $\{q_{rs}^*\}$ is loaded to the standard network N^* on the basis of stochastic user equilibrium, the average of travel time μ_*^{rs} and its standard deviation σ_*^{rs} is obtained for each origin and destination. Then μ_*^{rs}, $\mu_*^{rs}+\sigma_*^{rs}$, $\mu_*^{rs}+2\sigma_*^{rs}$ and $\mu_*^{rs}+3\sigma_*^{rs}$ are defined as standard travel time in service level A, B, C and D, respectively.

The change of travel time reliability between origin and destination, also the overall travel time reliability in a whole network, can be computed when any part of the network is changed, for example when a new principal road is added to a network or any part of the network is blocked by disaster.

Computational Method

To estimate travel time reliability for a given network and OD traffic volume, the stochastic user equilibrium assignment should be solved first. There are some methods to solve stochastic user equilibrium assignment problem. The Monte-Carlo simulation approach using random numbers and the method of successive average are practical methods in probit type model. As routes are not enumerated clearly in practical traffic assignment, the proportion of route choice cannot be estimated specifically. Therefore, the stochastic user equilibrium assignment problem is solved first and the average and standard deviation of each link are obtained. Then, samples of travel time between each origin and destination are generated using the Monte-Carlo simulation approach. Based on samples of travel time, travel time reliability between each origin and destination pair is estimated. Also, overall travel time reliability in a network is estimated by a weighted average.

The computational procedure is summed up as follows:

Phase 1: Solve the stochastic user equilibrium assignment problem for given OD traffic demand and network, using the Monte-Carlo simulation approach and the method of successive average.

Phase 2: Based on the solution of stochastic user equilibrium, generate perceived travel times using the Monte-Carlo simulation approach. Search for the minimum perceived travel time route and request real travel time on the route. Thus, make samples of actual travel times.

Phase 3: Compute travel time reliability between origin and destination based on samples of actual travel time. Then, compute overall travel time reliability in the network.

APPLICATION TO THE NETWORK IN THE HANSHIN AREA

Outline of the Application

In the earthquake disaster in the Hanshin area, citizens' living and industrial activities were widely influenced by the blockage of transportation facilities. It can be easily imagined that the influences would be decreased if they could minimise the blockage of principal roads. It is important to make a reliable road network robust to natural disaster. Whence, the evaluation of travel time reliability in a road network is very important in the planning of disaster prevention and future road network. In the Hanshin area, some new highway lines are examined to make the network more robust to disaster. But, constructing these highways is very expensive and effects the region. They should be evaluated in a many-faceted way and given a priority to construct. Then, as an index of the evaluation of new lines, travel time reliability was estimated for some scenarios.

The network consists of principal roads including urban and inter urban expressways that cover the Hanshin area, Kobe, Ashiya, Nishinomiya and other surrounding cities. The standard network includes all of the principal roads authorised by a high-ranking national road plan. The eleventh National Road Plan. It consists of 571 nodes, 843 links and 98 centroids.

Link travel time function is set to be a BPR type:

$$t_a(x_a) = t_{a0}\{1 + 2.62(x_a / c_a)^4\} \tag{15}$$

where

t_{a0} : travel time at zero flow

c_a : traffic capacity

The expressway charge is converted to the equivalent travel time assuming the time value is 52 (yen/min.). OD traffic demand matrix (98×98) is the authorized data set in the Eleventh National Road Plan (Figure 1).

The Change of Reliability by the Blockage of an Important Link

The change of travel time reliability is estimated assuming that important links are blocked by the disaster. The important links in the standard network are selected based on link flow and marginal utility to total negative utility.

Travel time reliabilities are estimated for the standard network and for the networks when one of the important links is blocked by disaster. The procedure to calculate travel time reliability is as follows.

First, stochastic user equilibrium in the standard network is solved through the method of successive average with one time of inner iteration and 100 times of outer iteration. β is assumed to be 0.5. Based on the solution, travel times between each origin and destination pair are generated by the Monte-Carlo simulation approach and their average and standard deviation are calculated.

Next, the stochastic user equilibrium problem is solved for the case when an important link in the network is blocked, in the same way as for the standard network. Then, travel time between each origin and destination pair is sampled by the Monte-Carlo simulation approach.

Based on samples of travel times, travel time reliability is computed for each origin and destination pair.

Travel time reliabilities for the standard network and for the cases when an important link is blocked are shown in Table 1. Overall travel time reliability for the standard network is 0.789 at service level A, 0.926 at service level B and 0.979 at service level C. The variation of travel time stays almost in the region below $\mu+2\sigma$.

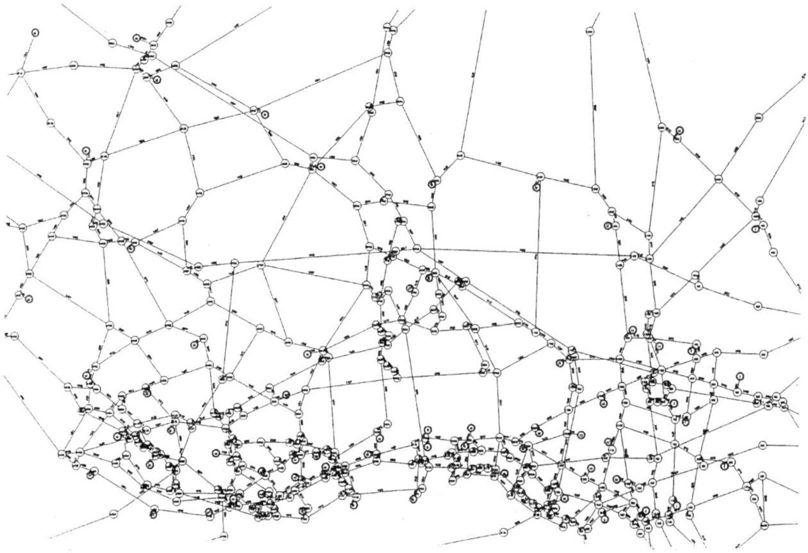

Figure 1 Road Network in the Hanshin Area

From Table 1, it is found that links on the national expressway line such as Meishin Expressway, 2nd Meishin Expressway or Chugokudo Expressway have the largest influence on the travel time reliability in the network. When they are blocked, travel time reliability in the network is decreased by 10 to 16 percent in service level A or B as compared with the standard network. Next to them, links on urban expressway line and national highway line, such as Hanshin Expressway Kobe Line and National Highway Route 2, have a larger influence. When they are blocked, the travel time reliability decreases by 10 to 13 percent as compared with the standard network.

Consequently, they are the most important links, which influence a wide area of the network and decrease travel time reliability if they are blocked. They should be managed prior to any other roads in the network.

Table 1 The Change of Reliability by the Blockage of an Important Link

Link no.	Name of the line	Traffic volume (pcu/h)	Travel time (min.)	Travel time reliability		
				Service level A	Service level B	Service level C
	Standard	-	-	0.789	0.926	0.979
1061	Meishin Expressway	202,900	3.27	0.641	0.772	0.850
1063	Meishin Expressway	150,900	15.93	0.666	0.808	0.884
1023	Chugokudo Expressway	149,100	6.06	0.629	0.783	0.859
1025	Chugokudo Expressway	102,800	7.65	0.658	0.805	0.876
9015	2nd Meishin Expressway	99,800	13.13	0.621	0.770	0.844
9019	2nd Meishin Expressway	50,800	9.13	0.680	0.822	0.887
3087	2nd Shinmei Expressway	101,900	7.90	0.690	0.827	0.884
2001	Kakogawa Bypass	109,400	5.90	0.707	0.833	0.895
3053	Hanshin Exp. Kobe Line	92,100	3.08	0.673	0.810	0.886
3055	Hanshin Exp. Kobe Line	103,200	3.01	0.662	0.818	0.895
3097	Hanshin Exp. Wangan Line	188,600	1.62	0.662	0.789	0.865
9411	Hanshin Exp. Wangan Line	72,800	6.25	0.662	0.822	0.883
3137	Hanshin Exp. Kitakobe Line	94,300	2.55	0.694	0.837	0.894
9473	Hanshin Exp. Nanboku Line	135,700	11.65	0.661	0.784	0.857
9457	Hanshin Exp. Kanzakigawa Line	195,600	12.81	0.658	0.800	0.874
4027	National Highway 2	92,000	6.30	0.687	0.825	0.885
4049	National Highway 2	106,600	18.67	0.681	0.806	0.848
4095	National Highway 171	135,900	15.55	0.655	0.765	0.823
4171	National Highway 173	36,400	13.50	0.684	0.831	0.896
4253	National Highway 175	49,800	39.94	0.708	0.863	0.910

The Change of Reliability by the Removal of a New Line

In The 11th National Road Plan, some new lines are included to the present road network in the Hanshin area. Here, they are evaluated from the viewpoint of travel time reliability. The method of analysis is as follows:

The network with all of the new lines is taken as the standard network. Travel time reliability is calculated for each case when one of the new lines is removed from the standard network. Traffic demand and other parameters in the network are assumed to be the same as the former case study. The computational result is shown in Table 2, in which travel time on the line, average travel time of all trips and travel time reliability at service level A, B, C and D are shown respectively.

Table 2 The Change of Reliability by the Removal of a New Line

Name of the line	Travel time (min.)	Average travel time (min.)	Travel time reliability			
			Service level A	Service level B	Service level C	Service level D
Normal Network	—	55.9	0.615	0.889	0.969	0.989
2nd Meishin Expressway	25.14	59.6	0.474	0.766	0.885	0.923
Hanshin Exp. Kitakobe Line	15.54	57.0	0.491	0.807	0.910	0.949
Hanshin Exp. Kanzakigawa Line	62.82	58.8	0.427	0.708	0.835	0.904
Hanshin Exp. Nanboku Line	16.37	57.5	0.489	0.765	0.895	0.942
Hanshin Exp. Kobeyamate Line	12.05	58.4	0.457	0.756	0.855	0.900
Hanshin Exp. Higashi Kobe Line	11.37	57.1	0.497	0.805	0.903	0.943
Hanshin Exp. Wangan Line	21.79	57.1	0.454	0.796	0.905	0.949

New urban expressway lines such as Hanshin Expressway Kanzakigawa Line, Kobe Yamate Line and Wangan Line have a large influence on travel time reliability in the network. When they are removed from the standard network, travel time reliability decreases by 10 to 18 percent in service level A or B as compared with the standard network. Among them, Kanzakigawa Line has the largest influence. Kobe-Yamate Line, Wangan Line and 2nd Meishin

Expressway Line follow this. This demonstrates that they are the most important lines to improve travel time reliability in the Hanshin area road network.

Concluding Remarks

In this paper, a model to estimate travel time reliability was proposed through the stochastic user equilibrium approach. Travel time reliability is defined as the probability that travel time between the origin and destination does not exceed some standard travel time corresponding to each service level. Its weighted average by OD traffic demand is defined as overall travel time reliability in a network. Also it was shown how to calculate travel time reliability in a large-scale network. Service level is defined relatively by the range of travel time between each OD pair in the standard (undamaged) network.

This model was applied to the road network in the Hanshin area, to evaluate new lines in the national road plan from the viewpoint of travel time reliability. Some important links and important lines in the future road network were contracted on the point of travel time reliability. Also some interesting facts were found about the change of travel time reliability by the blockage of a link or by the removal of a new line. The method can be applied to actual large-scale networks.

In future researches, the model should be extended to include other possible sources of variability, such as daily and accidental variance of OD traffic volume. Also the model should be improved such that it reflects precisely features of actual traffic behaviour on a road network.

Acknowledgement

The author would like to acknowledge Prof. Iida of Kyoto University and Prof. Bell of Imperial College London, for their meaningful suggestions and encouragement in the study.

References

Asakura, Y., Kashiwadani, M. and Kumamoto, N. (1989). Reliability Measures of Regional Road Networks Dueing to Daily Variance of Traffic Flow, *Infrastructure Planning Review*, **7**, 235-242.

Chen, A., Yang, H., Lo, K.K. and Tang, W.H. (1999). A Capacity Related Reliability for Transportation Networks, *Journal of Advanced Transportation*, **v33n2**, 183-200.

Daganzo, C.F. and Sheffi, Y. (1977). On Stochastic Models of Traffic Assignment, *Transportation Science*, **16(3)**, 253-274.

Daganzo, C.F. (1982). Unconstrained Extremal Formulation of Some Transportation Equilibrium Problems, *Transportation Science*, **16(3)**, 332-360.

Iida, Y. and Wakabayashi, H. (1988). An Efficient Calculation Method to Obtain Upper and Lower Bounds of Terminal Reliability of Road Networks Using Boolean Algebra, *Proceedings of the Japan Society of Civil Engineers*, **395/IV-9**, 75-84.

Iida, Y., Wakabayashi, H. and Yoshigi, T. (1988). An Approximate Evaluation of Road Network Reliability with Partial Minimal Path Sets and Cuts Sets, *Traffic Engineering*, **23**, No.4, 3-13.

Iida, Y. Wakabayashi, H. and Fukushima, H. (1989). A Comparative Study of Approximation Methods of Terminal Reliability Analysis for Road Networks, *Proceedings of the Japan Society of Civil Engineers*, **407/IV-11**, 107-116.

Masuya, Y. (1984). Evaluation Methods of System Serviceability of Road Network During Seismic Disaster, *Traffic Engineering*, **19**, No.5, 3-17.

Moritsu H. (1984). A Fundamental Study on the Method of the Composition of an Optimal Transportation Network, Doctor Thesis.

Nakagawa, S., Wakabayashi H. and Iida, Y. (1996). Simple Method for Reliability Analysis by Searching the N-th Shortest Path and its Application to the Evaluation of Traffic Management Schemes, *Proceedings of Infrastructure Planning*, **13**, 861-868

Sheffi, Y., and Powell W.B. (1981). A Comparison of Stochastic and Deterministic Traffic Assignment over Congested Networks, *Transportation Research*, **15B(1)**, 53-64.

Sheffi, Y., and Powell W.B. (1982). An Algorithm for the Equilibrium Assignment Problem with Random Link Times, *Networks*, **12(2)**, 191-207.

Wakabayashi, T. and Kameda, H. (1992). Analysis of Functional Effects of the 1989 Loma Prieta Earthquake on the San Francisco Bay Area Road Network and Evaluaton of Traffic Management, *Infrastructure Planning Review*, **10**, 103-110.

Wakabayashi, T, and Iida, Y. (1992). Travel Time Reliability of Road Networks with Traffic Management, *Proceedings of Infrastructure Planning*, **15(2)**, 99-102.

Wakabayashi, T, Iida, Y. and Inoue, Y. (1993). Analysis of Traffic Variance in Road Network and Link Reliability Estimation Method Attributed to Stochastic Traffic Variation, *Proceedings of the Japan Society of Engineers*, **460/IV-19**, 35-44.

6

EFFECT OF ROUTE CHOICE MODELS ON ESTIMATION OF TRAVEL TIME RELIABILITY UNDER DEMAND AND SUPPLY VARIATIONS

Anthony Chen, Department of Civil and Environmental Engineering, Utah State University, USA
Zhaowang Ji, Department of Civil and Environmental Engineering, Utah State University, USA
Will Recker, Institute of Transportation Studies, University of California at Irvine, USA

ABSTRACT

Travel time variability has generally been recognized as one of the most important attributes in travelers' route choice decisions. In fact, many empirical studies have indicated that both passengers and freight carriers are strongly averse to travel time variability, because it introduces uncertainty to their route choice decisions. In this chapter, we examine the effect of incorporating travel time variability and risk-taking behavior into the route choice models and its impact on the estimation of travel time reliability under demand and supply variations.

INTRODUCTION

Recent empirical studies (Abdel-Aty *et al.*, 1996; Small *et al.*, 1999; Kazimi *et al.*, 2000; Lam and Small, 2001; Ghosh, 2001) suggest that travelers are interested in not only travel time saving but also reduction of travel time variability. Abdel-Aty *et al.*, (1996) found that travel time variability is one of the important factors in route choice decisions. Specifically, about 54% of the respondents in the survey indicated that travel time variability is either the most important or second most important reason for choosing their daily commute routes. Small *et al.* (1999) found that both passengers and freight carriers are strongly averse to the scheduling mismatches because they cannot predict precisely what their travel time will be. For this reason, they will pay a premium to avoid congestion and to achieve greater reliability in travel times. From the two value-pricing projects in Southern California, Kazimi *et al.* (2000), Lam and Small (2001), and Ghosh (2001) consistently found that travelers are willing to pay quite a substantial amount to reduce variability in travel time. Suffice to say, the valuation of travel time and travel time variability are important factors for travelers when making their route choice decisions under risk or circumstances where they do not know with certainty the outcome of their decisions.

The literature on the value of travel time is abundant and well researched. Majority of the studies focus on empirical estimation of values of travel time using mode choice models and evaluation the marginal rates of substitution between the costs and travel time of alternative modes (Small *et al.*, 1999). However, research on value of travel time variability is relatively new compared to studies on value of travel time. Most of the research focuses on the reliability of public transport systems (primarily conducted in the U.K.) and passenger vehicle travel (primarily conducted in the U.S.). In this study, we are primarily concerned with passenger vehicle travel. Variability introduces uncertainty for travelers such that they do not know exactly when they will arrive at the destination. Thus, it is considered as a risk (or an added cost) to a traveler making a trip. Fundamental research into the behavioral issues of travel time variability, particularly on issues related to perceptions of variability and risk preferences, is needed in order to better understand the effects of including travel time variability in the route choice decision process so that appropriate management and information strategies can be exploited to better utilize network capacity and to improve network performance by influencing the travel patterns of road users.

Although travel time variability has generally been found to be one of the most important attributes, few attempts have been made to make these findings operational. This study continues our previous effort on risk-taking route choice models and its effect on travel time reliability (Chen and Recker, 2001). The risk-taking route choice models examined in this chapter focus on risk perceptions and preferences in making route choice decisions based on the tradeoff between expected travel time and travel time variability. The goal is to examine what the aggregate impact of changes in variability might be on network assignment and how travelers with different risk-taking behaviors respond to the changes caused by supply and demand variations.

ROUTE CHOICE MODELS AND IMPLEMENTATION

The objective of a route choice model in the traffic assignment problem is to describe how individual route choice decisions represented by an origin-destination trip table interact with the transportation network represented by a set of link performance functions so that the resulting network loads are accurately predicted (Sheffi, 1985). In the literature, several route choice models have been proposed which differ in:

(1) Characterization of network travel times (i.e., deterministic or stochastic)

(2) Traveler's knowledge of network travel times (i.e., with or without perception error)

Recently, Chen and Recker (2001) provided a classification scheme of the route choice models under the presence of congestion using network uncertainty and perception error as follows:

Table 1 Classification of Route Choice Models

		Perception Error?	
		No	Yes
Network Uncertainty?	No	DN-DUE	DN-SUE
	Yes	SN-DUE	SN-SUE

where DN = Deterministic Network
SN = Stochastic Network
DUE = Deterministic User Equilibrium
SUE = Stochastic User Equilibrium

Under the above classification scheme, four distinct route choice models are presented in Table 1.

In each model, the travel time for every link on the network is assumed to be an increasing function of the flow of vehicles on the link. Each traveler makes a rational route choice decision based on minimizing some criteria related to average travel times or some disutility measure based on average travel times and their variances.

The widely accepted DN-DUE route choice model originally proposed by Wardrop (1952) and formulated as a mathematical program by Beckman *et al.* (1956) is based on strong assumptions that the network travel times are deterministic for a given flow pattern and that all travelers are perfectly aware of the travel times on the network and always capable of identifying the shortest travel time route. To overcome the deficiencies of the deterministic model, some researchers have proposed different *stochastic user equilibrium* models (i.e., DN-SUE) to relax the assumption of perfect knowledge of network travel times, allowing travelers to select routes based on their perceived travel times (Daganzo and Sheffi, 1977; Fisk, 1980;

Sheffi and Powell, 1982). Due to variations in travelers' perceptions of travel times, travelers do not always end up picking the correct shortest travel time route. However, both DN-DUE and DN-SUE models ignore network uncertainty and assume that the network is deterministic. Since travel time variability is not considered in the route choice decision, all travelers in the DN-DUE and DN-SUE models are implicitly assumed to be risk neutral.

Given that many recent empirical studies have indicated the significance of travel time variability in making route choice decisions, it is necessary to incorporate network uncertainty in the route choice models. Link travel times are treated as random variables. For a given set of flows, there is a probability distribution associated with the link travel times, which describes the variations in the travel times experienced by the travelers on the network. Such variations could result from the differences in the mix of vehicle types on the network for the same flow rates, differences in driver reactions under various weather and driving conditions, differences in delays experienced by different vehicles at intersections, etc. Because the travel time variability is included in the model, it allows us to capture how travelers make tradeoff decisions between a route that is longer but has reliable travel time versus another route that is shorter but has unreliable travel time according to their risk taking behaviors. Depending on the behavioral nature of travelers, they are classified as risk averse, risk prone or risk neutral. The risk in this case is the variability associated with network travel times. For instance, a risk averse traveler will trade off a reduction in travel time variability with some increase in expected travel time, whereas a risk prone traveler may choose a route with a greater variability so as to increase the possibility of a smaller travel time. A risk neutral traveler would choose a route based on only expected travel times without consideration of its variability.

For the SN-DUE model, travelers are assumed to have perfect knowledge of the variable nature of network travel times. This model may be suitable for peak-hour traffic where regular commuters have a good idea of the mean and variance of network travel times. However, a truly stochastic model should consider both the traveler perception errors as well as the stochasticity of network travel times. The SN-SUE model that accounts both for variable network travel times as well as a method to model different traveler responses to network travel time variability by assuming different risk taking behaviors was presented in (Mirchandani and Soroush, 1987). This model also accounts for the variations in each individual traveler's perception errors. For detailed information about these four classes of route choice models, please refer to Tatineni *et al.* (1997), Chen and Recker (2001), and Chen, Ji, and Recker (2002).

All four route choice models described above were implemented using the Frank-Wolfe (FW) based algorithm. The main differences are the criterion used to determine the search direction and the one-dimensional search schemes used to determine the step size. Table 2 briefly summarizes the criterion and line search scheme used in the FW based algorithm to implement the four route choice models.

Table 2 Implementation of the Route Choice Models

Route Choice Model	Search Direction	Step Size
DN-DUE	Minimize expected travel time	Bisection
DN-SUE	Minimize expected perceived travel time	1/n
SN-DUE	Minimize expected disutility	1/n
SN-SUE	Minimize expected perceived disutility	1/n

The implementation of the DN-DUE and DN-SUE models is based on the FW algorithm and the method of successive averages (MSA) algorithm, respectively. The details of these two algorithms are well described in the excellent text by Sheffi (1985): pages 119-120 for the FW algorithm and page 327 for the MSA algorithm. What we emphasize here is the implementation for the two route choice models with network uncertainty and risk taking behavior. Both SN-DUE and SN-SUE models can also be solved using the MSA algorithm with a different criterion used to perform the stochastic network loading procedure (see page 301 in Sheffi, 1985) to determine the search direction. The criterion used in the SN-DUE and SN-SUE models basically depends on (1) average travel time and travel time variability, (2) assumption of perception error distribution, and (3) disutility function describing travelers' risk-taking behavior.

1. Average link travel times are assumed to be flow-dependent. The link travel time is estimated using the standard BPR (Bureau of Public Road) function. Because of lack of data, travel time variability is assumed to be flow-independent (i.e., based on link length or free-flow travel time) and is a random variable with a pre-determined distribution (e.g., normal or gamma).

2. For the SN-SUE model, each traveler i is assumed to have a variable perception error $\xi_i \sim N(\mu_i, \theta_i)$ where $\mu_i \sim N(0, \tau)$ and $\theta_i \sim G(\alpha, \beta)$ (Mirchandani and Soroush, 1987). This variable perception error allows each individual traveler to experience a different travel time for a given set of flows. This is different from the probit-based DN-SUE model, which only accounts for the randomness of the travelers' perceived travel times (i.e., $\xi_i \sim N(0, \tau)$) and treats the randomness of link travel times in the form of expected values.

3. As mentioned above, travel time variability introduces uncertainty for travelers such that they do not know exactly when they will arrive at the destination. Thus, it is considered as a risk (or an added cost) to a traveler making a trip. To model the travelers' attitudes towards risk, disutility functions are needed to represent different risk-taking behaviors. Following the study by Tatineni *et al.*, (1997), we also use the exponential functional form

to describe risk averse and risk prone behaviors. For the risk neutral behavior, we simply use the DN-SUE model, which minimizes the perceived expect travel time and ignores travel time variability. Suppose that the travelers' disutility has the following forms:

$$\text{Risk Averse:} \quad DU(T) = a_1\,[\exp(a_2\,T) - 1] \qquad (1)$$

$$\text{Risk Prone:} \quad DU(T) = b_1\,[1 - \exp(-b_2\,T)] \qquad (2)$$

where T is the travel time in minutes for a given route, a_1, a_2, b_1, and b_2 are positive parameters to be estimated. The parameters a_2 (b_2) is a measure of risk aversion (proneness). As a_2 (b_2) gets larger, the traveler becomes more averse (seeking) to risk. The details of the estimation of the parameters can be found in Tatineni (1996). An advantage of using the exponential functional form to represent risk-taking behavior is that the disutility associated with a route can be estimated by summing the link disutilities on a given route (see Tatineni (1996) on the estimation of link disutilities using the moment generating function). This allows the classical Dijkstra-type shortest path algorithm to be used in finding the minimum expected disutility route in the SN-DUE model and the minimum expected perceived disutility route in the SN-SUE model.

TRAVEL TIME RELIABILITY EVALUATION PROCEDURE

Reliability is generally defined as the probability that the system of interest has the ability to perform an intended function or goal (Ang and Tang, 1990). Currently, there are several performance measures for assessing transportation network such as *connectivity reliability* (Iida and Wakayabashi, 1989), *travel time reliability* (Asakura and Kashiwadani, 1991; Asakura, 1996; Bell *et al.*, 1999), *system and origin-destination (OD) sub-system reliability* (Nicholson and Du, 1997; Du and Nicholson, 1997), *capacity reliability* (Chen *et al.*, 1999, 2000, 2002), *travel demand satisfaction reliability* (Lam and Zhang, 2000). For a detailed review of these performance measures, see Lam and Zhang (2000).

In this paper, we are primarily interested in the effect of route choice models on estimating travel time reliability under demand and supply variations. Travel time reliability is concerned with the probability that a trip between a given OD pair can be made successfully within a given time interval and at a specified level-of-service. There are two potential measures - path travel time reliability and OD travel time reliability - that are of interest to the travelers and traffic managers (Bell *et al.*, 1999). Path travel times are computed by summing up the link travel time on a given route. For the OD travel times, they are computed as a weighted average of the path travel times, where the weights are the path flows.

- Path travel time reliability is defined as the probability that the travel time of a given path is within an acceptable threshold.
- OD travel time reliability is defined as the probability that the weighted average travel time of a given OD pair is within an acceptable threshold.

Travelers are more concerned about the path travel time reliability, because it directly affects their route choice decisions. OD travel time reliability measures all relevant paths used by travelers to define an aggregate measure for the level-of-service between a given OD pair. Thus, traffic managers or planners can use the OD travel time reliability as a proxy to evaluate the performance of a given OD pair (i.e., OD sub-system reliability in terms of travel time). Similar to the system and OD sub-system reliability measure defined by Nicholson and Du (1997), the OD travel time reliability can also be extended to the system level to measure the performance of the whole network.

The travel time reliability evaluation procedure presented below is based on a Monte Carlo simulation framework developed by Chen and Recker (2001). Link capacities (C_a) and OD demands (Q_{rs}) are treated as random variables. Assuming that the random variables C_a and Q_{rs} follow a probability distribution, a random variate generator is used to generate the values of C_a for each link and Q_{rs} for each OD pair that preserve the provided distribution properties (e.g., mean, variance, correlation, etc.). For each set of link capacities and OD demands generated, four different route choice models discussed above are used to estimate the OD and path travel time reliabilities.

The travel time reliability procedure is described as follows:

Step 0: Set sample number $k := 1$.

Step 1: Generate a vector of OD demands $\mathbf{Q}_k = \{\ldots, Q_{rs}, \ldots\}$ and/or a vector of link capacities $\mathbf{C}_k = \{\ldots, C_a, \ldots\}$.

Step 2: Perform the DN-DUE, DN-SUE, SN-DUE, and SN-SUE route choice models with same the OD demand vector $\mathbf{Q}_k$ and/or link capacity vector $\mathbf{C}_k$.

Step 3: Collect statistics from Step 2 to compute OD and path travel time reliabilities.

Step 4: If sample number k is less than the required sample size $k_{\max}$, then increment sample number $k := k + 1$ and go to Step1. Otherwise, terminate the procedure.

NUMERICAL EXPERIMENT

The travel time reliability evaluation procedure is demonstrated in this section using a small network depicted in Figure 1. The network consists of 5 nodes, 7 links, and 2 OD pairs. The base demand for OD pairs (1,4) and (1,5) are 15.0 and 18.8, respectively.

The free-flow travel times for each link on the network as well as the ideal capacity values are shown in Table 3. The link travel time function used in the route choice models is the standard Bureau of Public Road (BPR) function given below:

$$t_a = t_a^f \left(1 + 0.15\left(\frac{v_a}{c_a}\right)^4\right) \quad (3)$$

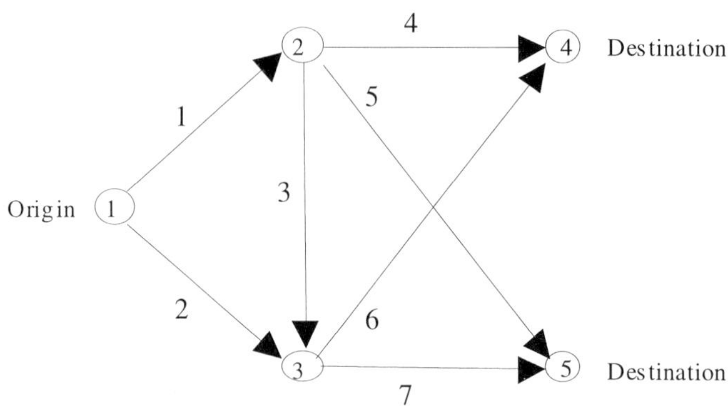

Figure 1 Test network

Here v_a, t_a^f, and c_a are the flow, free-flow travel time, and capacity on link *a*, respectively. In order to model the risk-taking route choice models (i.e., SN-DUE and SN-SUE), we assume that all links in the network have some variability associated with their travel times. For link 1, the variability (e.g., standard deviation[1]) of the travel time is assumed to be 100 percent of the link's free-flow travel time. For all other links, the travel times are assumed to vary up to 10 percent of the link's free-flow travel time.

[1] Other measure such as the difference between the 95th percentile travel time and median travel time (Lam and Small, 2001) can also be used here.

Table 3 Link Free-flow Travel Times and Ideal Link Capacities

Link #	Free-flow travel time (min)	Capacity (veh/min)
1	4.0	25.0
2	5.2	25.0
3	1.0	15.0
4	5.0	15.0
5	5.0	15.0
6	4.0	15.0
7	4.0	15.0

To estimate the parameters of the disutility function given in equations (1) and (2), we need to make assumptions on how travelers value route travel time according to their risk-taking behavior. For illustration, we just explain the estimation of parameters for the risk averse case. Suppose that all travelers have a disutility of 1 unit for a route that is 5 minutes, and they are indifferent between a route that has a travel time of 3.33 minutes for certain and a route that has an equal chance of having a travel time that is either close to zero minute or is equal to 5 minutes (i.e., an average of 2.5 minutes on this route). Travelers who took the longer travel time route (3.33 minutes) would pay a risk premium of 0.83 minute (i.e., 3.33 – 2.5) to avoid uncertainty. This is a typical behavior for risk averse travelers. Given these two assumptions, we can set up two equations to solve for the values of the two parameters a_1 and a_2 as follows:

$$DU(5) = a_1 [\exp(a_2\, 5) - 1] = 1.0$$
$$DU(3.33) = a_1 [\exp(a_2\, 3.33) - 1] = 0.5$$

Solving the above two equations for a_1 and a_2 give the following disutility function for a risk averse traveler:

Risk Averse: $DU(T) = 0.309 [\exp(0.289\, T) - 1]$

Using similar assumptions for the risk prone travelers, the disutility function for a risk prone traveler is:

Risk Prone: $DU(T) = 1.309 [1 - \exp(-0.289\, T)]$

To simulate the uncertainty of the traffic demands, a normal distribution is used. Three levels of mean demand are specified for each OD pair by varying the base OD demand by ± 25%. For each level of mean OD demand, three levels of standard deviation (σ) of the OD demand is specified to reflect the variation of the OD demand as follows: (1) $\sigma = 1.5(q_{rs}/3)$ for high variation, (2) $\sigma = 1.0(q_{rs}/3)$ for medium variation, and (3) $\sigma = 0.5(q_{rs}/3)$ for low variation.

These three variations of traffic demands can be interpreted as the relative accuracy of each level of mean OD demand. Given that both the mean and standard deviation of OD demands are defined for a total of nine combinations (3 means x 3 standard deviations), random samples for each OD pair can be generated according to the standard normal distribution as follows:

$$Q_{rs} = q_{rs} \pm Z\sigma$$

where:

Q_{rs} = random demand between OD pair (r,s),

Z = random variable generated from N(0,1).

σ = standard deviation of the OD demand.

Road capacity is not deterministic in real life situations. It is influenced by a wide variety of physical and operational characteristics (e.g. roadway type, geometric configuration, mixture of vehicle types, weather conditions, etc.). These factors often degrade the capacity of a road network. To simulate the uncertainty of link capacities, let $\mathbf{C} = \mathbf{C}_0 - \varepsilon$, where $\mathbf{C} = \{\ldots, C_a, \ldots\}$ is a vector of random link capacities, $\mathbf{C}_0 = \{\ldots, C_{a0}, \ldots\}$ is a vector of ideal link capacities, and ε is a vector of random variables representing link capacity degradation. The random link capacity degradation vector (ε) ranges from 0 (i.e., operating at ideal capacity) to $\mathbf{C}_0$ (i.e., complete failure with zero capacity). For simplicity, a uniform distribution is used to generate the random link capacity degradation. The mean link capacity degradation is 25% of the ideal link capacity with three levels of standard deviation. To ensure high accuracy, we use 10,000 samples in our simulation experiment to generate the numerical results for this study.

Two scenarios are constructed to test how different route choice models respond to uncertain environment caused by demand and supply variations and its effect on estimating travel time reliability. Scenario I examines the effect of route choice models on estimating travel time reliability caused by daily fluctuations of demand only; and scenario II considers both demand and supply variations.

Scenario I:

Scenario I considers only demand variations. Link capacities are assumed to be deterministic and operating at the ideal capacities.

OD Travel Time Comparison for Different Route Choice Models

Mean and standard deviation (shown in parenthesis) of OD travel times for 9 combinations of demand variations (3 means x 3 standard deviations) and for four route choice models are shown in Table 4. For the SN-DUE and SN-SUE models, risk averse behavior is used for comparison with the DN-DUE and DN-SUE models that do not consider travel time variability in the route choice process. Except the DN-DUE model where all used paths of a given OD pair have the minimum travel time equal to the OD travel time, the OD travel times for the other three models are computed as a weighted average of the travel times on the different paths, where the weights are the path flows.

Table 4 OD Travel Time Comparisons for Different Route Choice Models under Demand Variations

	OD (1,4)					
Mean/Std. Dev.	DN-DUE			DN-SUE		
	LSD	MSD	HSD	LSD	MSD	HSD
LM	9.2 (0.0)	9.2 (0.1)	9.3 (0.2)	9.2 (0.0)	9.2 (0.1)	9.3 (0.2)
MM	9.3 (0.1)	9.4 (0.2)	9.5 (0.5)	9.4 (0.1)	9.4 (0.3)	9.6 (0.5)
HM	9.6 (0.2)	9.8 (0.6)	10.1 (1.2)	9.7 (0.3)	9.9 (0.6)	10.2 (1.2)
Mean/Std. Dev.	SN-DUE			SN-SUE		
	LSD	MSD	HSD	LSD	MSD	HSD
LM	9.5 (0.0)	9.5 (0.1)	9.5 (0.2)	9.2 (0.0)	9.2 (0.1)	9.3 (0.3)
MM	9.5 (0.1)	9.6 (0.4)	9.8 (0.7)	9.4 (0.2)	9.6 (0.5)	9.7 (0.7)
HM	10.2 (0.6)	10.2 (0.8)	10.4 (1.2)	10.1 (0.5)	10.1 (0.7)	10.3 (1.2)
	OD (1,5)					
Mean/Std. Dev.	DN-DUE			DN-SUE		
	LSD	MSD	HSD	LSD	MSD	HSD
LM	9.2 (0.0)	9.3 (0.1)	9.3 (0.2)	9.2 (0.1)	9.2 (0.2)	9.3 (0.8)
MM	9.4 (0.1)	9.5 (0.4)	9.7 (0.8)	9.5 (0.2)	9.6 (0.4)	9.8 (1.0)
HM	9.8 (0.3)	10.0 (0.9)	10.5 (1.9)	9.9 (0.4)	10.1 (0.9)	10.6 (2.0)
Mean/Std. Dev.	SN-DUE			SN-SUE		
	LSD	MSD	HSD	LSD	MSD	HSD
LM	9.4 (0.0)	9.4 (0.1)	9.5 (0.3)	9.2 (0.1)	9.3 (0.2)	9.4 (0.4)
MM	9.5 (0.2)	9.8 (0.6)	10.0 (1.0)	9.5 (0.3)	9.7 (0.6)	9.9 (0.9)
HM	10.3 (0.7)	10.5 (1.0)	10.9 (1.9)	10.3 (0.6)	10.3 (1.0)	10.7 (1.9)

LM=Low Mean, MM=Medium Mean, HM=High Mean, LSD=Low Standard Deviation, MSD=Medium Standard Deviation, HSD=High Standard Deviation

The general trend for all the models is that as mean demand increases, OD travels times for both OD pairs (1,4) and (1,5) also increase. This is also true for the case of a constant demand and only the standard deviation of OD demand is increased. In terms of the route choice models, the DN-DUE model has the lowest expected OD travel times and the SN-DUE has the highest expected OD travel times. This is because the DN-DUE model is based on the assumption that travelers choose routes that minimize travel times, while the SN-DUE model that assumes risk averse behavior without perception error tries to avoid paths that contain the high variability link and takes longer paths that result in higher OD travel times. For the two SUE models that contain perception error, their OD travel times are in-between those given by the DN-DUE and SN-DUE models. This result can be explained as follows. Because of the perception error, travelers in the DN-SUE model cannot always perceive exactly the shortest travel time path and results in taking paths that are slightly longer with higher expected OD travel times. Similarly, travelers in the SN-SUE model with risk averse behavior and perception error cannot always identify paths that have the least travel time variability, and sometimes choose paths that contain some variability but lower travel times. This evidently results in lower OD travel times than those found by the SN-DUE model.

Effect of Route Choice Models on the Estimation of OD Travel Time Reliability

In Figure 2, we show the OD travel time reliability curves for both OD pairs (1,4) and (1,5) for four route choice models using the medium demand with medium standard deviation (i.e., the MM, MSD cell in Table 6). On the X-axis, we plot the travel time threshold value, which can be used as an indicator for level of service (a smaller value indicates a higher level of service). The Y-axis is the OD travel time reliability, which is defined as the probability that the weighted travel time for a given OD pair is within an acceptable level of service. This measures, as mentioned above, is an aggregate measure of all relevant used paths serving an OD pair. In both OD pairs, the OD travel time reliability under the DN-DUE model is always higher compared to the other three models. Since travelers in the DN-DUE model assume to have perfect information and use travel time as the sole criterion for route choice, it is reasonable to have such results. However, for the DN-SUE model, travelers are assumed to make route choices with imperfect information, hence the OD travel time may be slightly sub-optimal resulting in lower OD travel time reliability. For the SN-DUE and SN-SUE models which assume risk averse behavior, route choices are based on minimizing disutilities and perceived disutilities, respectively, which are based on expected link travel time and travel time variability leading to higher OD travel times or lower OD travel time reliabilities. Thus, it would seem reasonable that the OD travel time reliability is lower when route choices are assumed to be based on imperfect information and even more so when they are based on minimizing disutilities that are a function of factors other than link travel times.

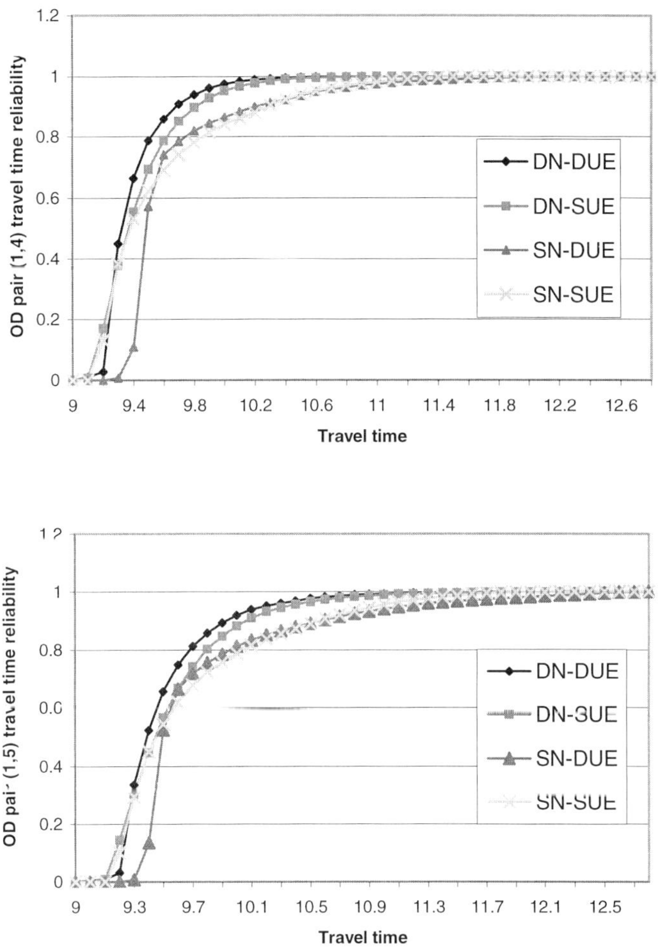

Figure 2 Effect of Route Choice Models on the Estimation of OD Travel Time Reliability under Demand Variations

Effect of Route Choice Models on the Estimation of Path Travel Time Reliability

For brevity, we show only the results of path 1 (link sequence 1->4) and path 2 (link sequence 2->6), which are two major alternate paths for OD pair (1,4) using the medium mean demand and medium standard deviation of demand (i.e., cell (MM,MSD) in Table 4) as an example. Figure 3 shows the effect of different route choice behaviors on estimating path travel time reliabilities. Similar to the OD travel time reliability curves, the X-axis is the travel time threshold value, and the Y-axis is the path travel time reliability which is defined as the probability that the travel time of a given path is within an acceptable threshold value. For the DN-DUE and DN-SUE models, travelers are assumed to make their route choices solely based on average travel times. No consideration is given to travel time variability. The travel time reliability on path 1 and path 2 are approximately equal. For the SN-DUE and SN-SUE models that consider both average travel time and travel time variability, less travelers are allocated to path 1 (1->4) because it contains link 1 which assumes to have a high variability in travel time, while more travelers are allocated to path 2 (2->6) that has less variability in travel time. Because more travelers use path 2, travel time of path 2 is higher than that of path 1. This evidently decreases the travel time reliability of path 2 for the risk averse travelers in the two SN models. The differences in travel time reliability between the route choice models implemented without and with perception error are minor between the DN-DUE and DN-SUE models, but are more visible between the SN-DUE and SN-SUE models. This result seems to suggest that perception error may have an effect on the risk taking behavior. Similar results (not shown here) are also observed for path 4 (1->5) and path 5 (2->6), which are the two major paths for OD pair (1,5).

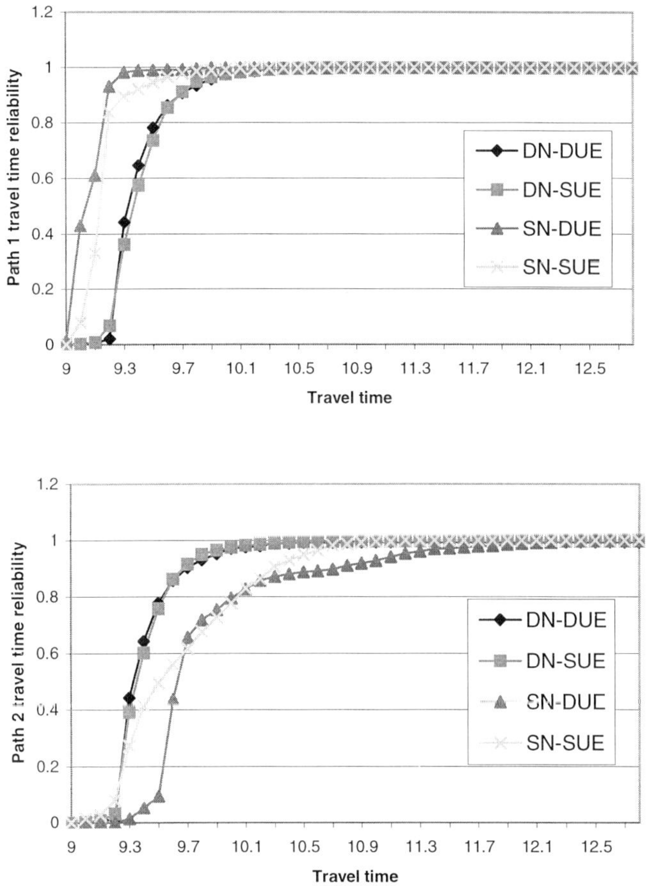

Figure 3 Effect of Route Choice Models on the Estimation of Path Travel Time Reliability under Demand Variations

Scenario II:

Scenario II considers both demand supply variations. Link capacities are assumed to degrade at a mean of 25% of the ideal capacities with three levels of standard deviation. Using the medium demand case with three levels of standard deviation (MM-LSD, MM-MSD, and MM-HSD) and mixing them with the capacity variations (LSD, MSD, and HSD), there are 9 combinations (3 demand variations x 3 supply variations).

OD Travel Time Comparison for Different Route Choice Models

Similar to scenario I, mean and standard deviation (shown in parenthesis) of OD travel times are shown in Table 5. Again, we use the risk averse behavior for the two route choice models that consider network uncertainty. The results are similar to those presented in scenario I. In general, the average OD travel times and standard deviation of OD travel times for both OD pairs and for all route choice models increase as the variations caused by demand and supply increase. In terms of the route choice models, the DN-DUE model has the lowest expected OD travel times, followed by the DN-SUE model and the SN-SUE model, and the SN-DUE model has the highest expected OD travel times. The same reasons explained in Scenario I also apply here.

Effect of Route Choice Models on the Estimation of OD Travel Time Reliability

Similar to scenario I, we show the OD travel time reliability curves in Figure 4 using the medium mean and medium standard deviation of demand and medium standard deviation of capacity case (i.e., cell (MM-MSD,MSD)) in Table 5. Again, the OD travel time reliability under the DN-DUE is the highest, followed DN-SUE, and the two SN models have the lowest OD travel time reliability. This is to be expected since the two DN models directly minimize travel times, albeit that the DN-SUE model is allocating flows based on imperfect information, and the two SN models that assume risk averse behavior are minimizing disutilities which are not solely based on travel times. In general, the OD travel time reliabilities for both OD pairs are significantly lower than those in scenario I. Using OD (1,4) as an example, the OD travel time reliability at a travel time threshold value of 10.4 for the DN-DUE model is 1.0 under scenario I, but it drops to 0.79 under scenario II when supply variation is also considered. Hence, it seems that the uncertainty due to supply variation can significantly contribute to the decrease of travel time reliability.

Table 5 OD Travel Time Comparisons for Different Route Choice Models under Both Demand and Supply Variations

Demand/Supply	OD (1,4)					
	DN-DUE			DN-SUE		
	LSD	MSD	HSD	LSD	MSD	HSD
MM-LSD	9.8 (0.3)	9.8 (0.4)	9.8 (0.4)	9.9 (0.3)	9.9 (0.4)	10.0 (0.5)
MM-MSD	10.0 (0.8)	10.0 (0.8)	10.0 (0.9)	10.1 (0.8)	10.1 (0.9)	10.2 (1.0)
MM-HSD	10.4 (1.6)	10.4 (1.7)	10.5 (1.9)	10.5 (1.6)	10.5 (1.7)	10.6 (1.9)
Demand/Supply	SN-DUE			SN-SUE		
	LSD	MSD	HSD	LSD	MSD	HSD
MM-LSD	10.3 (0.6)	10.3 (0.6)	10.4 (0.7)	10.3 (0.5)	10.3 (0.5)	10.3 (0.5)
MM-MSD	10.4 (0.9)	10.4 (0.9)	10.5 (1.0)	10.3 (0.8)	10.3 (0.9)	10.4 (1.0)
MM-HSD	10.7 (1.5)	10.8 (1.6)	10.8 (1.8)	10.6 (1.5)	10.7 (1.7)	10.7 (1.9)
Demand/Supply	OD (1,5)					
	DN-DUE			DN-SUE		
	LSD	MSD	HSD	LSD	MSD	HSD
MM-LSD	10.0 (0.5)	10.0 (0.5)	10.1 (0.6)	10.1 (0.4)	10.2 (0.5)	10.2 (0.5)
MM-MSD	10.3 (1.2)	10.3 (1.3)	10.4 (1.4)	10.4 (1.2)	10.5 (1.3)	10.5 (1.4)
MM-HSD	10.9 (2.5)	11.0 (2.7)	11.1 (3.0)	11.0 (2.6)	11.1 (2.7)	11.3 (3.0)
Demand/Supply	SN-DUE			SN-SUE		
	LSD	MSD	HSD	LSD	MSD	HSD
MM-LSD	10.7 (0.8)	10.7 (0.8)	10.7 (0.9)	10.5 (0.6)	10.5 (0.6)	10.5 (0.6)
MM-MSD	10.8 (1.2)	10.8 (1.3)	10.9 (1.4)	10.6 (1.2)	10.7 (1.2)	10.7 (1.4)
MM-HSD	11.3 (2.5)	11.3 (2.6)	11.4 (2.9)	11.1 (2.5)	11.2 (2.6)	11.3 (2.9)

MM-LSD=Medium Mean-Low Standard Deviation, MM-MSD=Medium Mean-Medium Standard Deviation, MM-HSD=Medium Mean-High Standard Deviation

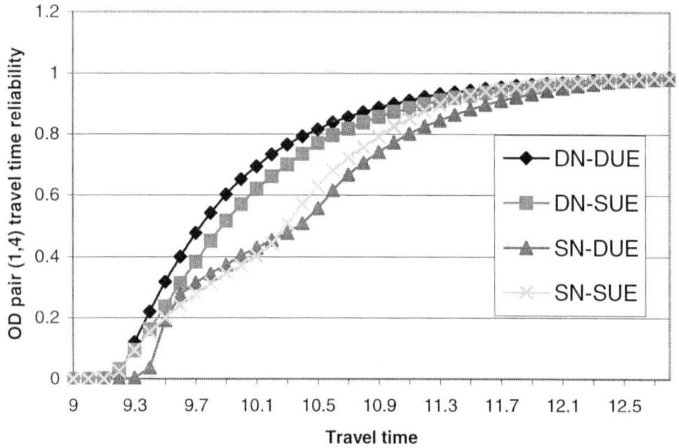

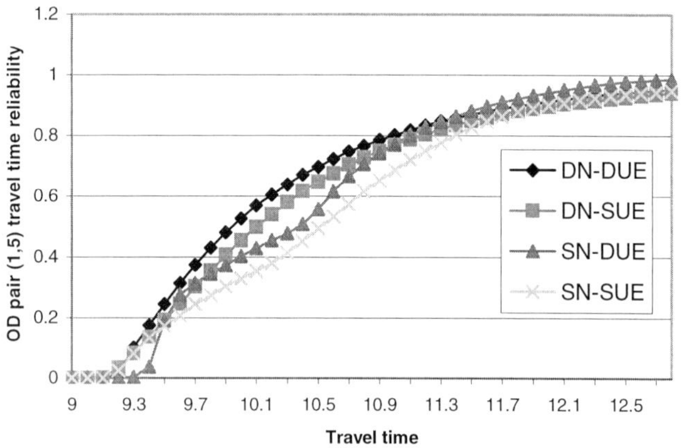

Figure 4 Effect of Route Choice Models on the Estimation of OD Travel Time Reliability under Both Demand and Supply Variations

Effect of Route Choice Models on the Estimation of Path Travel Time Reliability

Again for brevity, we show only the results of path 4 (1->5) and path 5 (2->7), which are two major alternate paths for OD pair (1,5) using cell (MM-MSD,MSD) in Table 6. Figure 5 plots the path travel time reliability curves for all four route choice models on path 4 and path 5.

Despite the paths shown in Figure 5 are for a different OD pair, the results are generally agreed with those presented in Figure 3 of scenario I. That is, the two SN models that assume risk averse behavior allocate less flow to path 4 to avoid the high travel time variability link 1, which is part of path 4. Since path 5 is the only alternative path for OD pair (1,5) that does not contain link 1, the two SN models allocate more flow to path 5, which results in higher path travel time and lower path travel time reliability. When both demand and supply variations are considered, the path travel time reliability curves are in general lower than those in scenario I that considers only demand variations. Using the DN-DUE model with a travel time threshold value of 10.5 minutes as an example, the travel time reliability of path 5 is 0.963 for scenario I (not shown here), but drops to 0.702 for scenario II.

Table 6 Link Flow Comparisons Between Scenario I and Scenario II

	Scenario I				Scenario II			
Link #	DN-DUE	DN-SUE	SN-DUE	SN-SUE	DN-DUE	DN-SUE	SN-DUE	SN-SUE
1	20.89 (2.37)	20.59 (4.29)	12.12 (4.68)	14.64 (2.94)	18.80 (3.51)	19.70 (4.04)	13.70 (5.08)	14.85 (4.84)
2	12.96 (5.78)	13.26 (3.74)	21.72 (3.97)	19.20 (5.58)	15.04 (4.95)	14.14 (4.14)	20.14 (3.92)	19.00 (4.30)
3	4.17 (1.38)	7.51 (1.01)	0.03 (0.02)	3.17 (0.79)	2.38 (1.49)	5.98 (0.85)	0.18 (0.62)	2.21 (1.08)
4	7.41 (2.26)	5.60 (2.08)	5.02 (3.01)	4.75 (2.08)	7.25 (2.48)	5.82 (2.38)	5.82 (2.83)	5.26 (2.71)
5	9.30 (2.83)	7.47 (2.91)	7.07 (3.51)	6.73 (2.85)	9.18 (3.07)	7.90 (3.19)	7.71 (3.49)	7.38 (3.48)
6	7.60 (2.75)	9.41 (2.93)	10.00 (2.35)	10.27 (3.07)	7.77 (2.63)	9.19 (2.70)	9.19 (2.60)	9.76 (2.54)
7	9.53 (3.51)	11.36 (3.36)	11.72 (3.15)	12.10 (3.51)	9.65 (3.30)	10.93 (3.14)	11.12 (3.18)	11.45 (3.01)

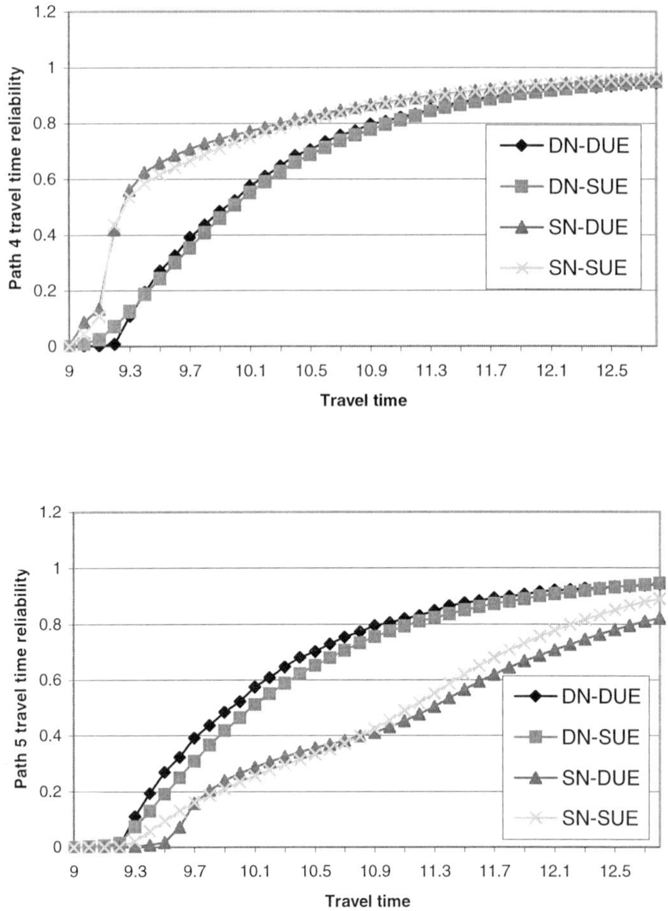

Figure 5 Effect of Route Choice Models on the Estimation of Path Travel Time Reliability under Both Demand and Supply Variations

Comparison between Scenario I and Scenario II

Comparing the results of scenario II with those in scenario I, it is found that the both the mean and standard deviation of OD travel times are higher. The higher OD travel times are due to the 25% capacity reduction on all links in the network, while the larger standard deviation of OD travel times is due the additional variation coming from the supply side. Thus, it seems necessary to account for both demand and supply variations when assessing the performance of a road network.

Link Flow Allocation Comparisons

In Table 6, we compare the link flow allocations between the two scenarios, using the results from the medium demand and medium standard deviation of demand case (i.e., cell (MM,MSD)) in Table 4 and from the medium mean and medium standard deviation of demand and medium standard deviation of capacity (i.e., cell (MM-MSD,MSD)) in Table 5. For brevity, we focus the comparison on link 1 and link 2 which are the only two links emanating from origin 1 (see Figure 1). Link 1 has a lower free-flow travel time but higher variability associated with the travel time, while link 2 has a higher free-flow travel time but lower variability. Hence, one would expect that the DN-DUE model would allocate more flow to link 1, the SN-DUE model assuming risk averse behavior would allocate less flow, and the DN-SUE and SN-SUE models that contain perception errors in their route choice models would allocate flow in-between these two limits. This expectation appears to hold true for scenario I. Because of the 25% capacity degradation in scenario II, the flow allocation between link 1 and link 2 in the DN-DUE model is slightly more evenly distributed compared to those in scenario I. This flow allocation is optimal in terms of travel time minimization, as reflected in Table 5 where the expected OD travel times are lowest among the four route choice models. Because of the perception error in the DN-SUE model, the flow allocation over the network is slightly sub-optimal in terms of higher expected OD travel times (also see Table 7 for the sub-optimal flow allocation to path 3 and path 6) despite that a larger amount of flow is allocated to link 1. For the SN-DUE and SN-SUE models, flow allocations over the network are similar for both scenarios.

Path Flow Allocation Comparisons

For completeness, we also compare the path flow patterns of both scenarios in Table 7. Similar to the link flow allocation comparison, we use the results from the medium mean and medium standard deviation of demand in scenario I (i.e., cell (MM,MSD)) in Table 4 and from the medium mean and medium standard deviation of demand and medium standard deviation of capacity in scenario II (i.e., cell (MM-MSD,MSD)) in Table 5 as an example for the path flow allocation comparison. There are 6 paths in total serving the 2 OD pairs. Paths 1 (1->4), 2 (2->6), and 3 (1->3->6) are for OD pair (1,4), and paths 4 (1->5), 5 (2->6), and 6 (1->3->7) are for OD pair (1,5). Based on the assumptions of the route choice model, it is apparent that each route choice model allocates a different amount of flow to each path. For example, the DN-DUE model seems to allocate more flow to path 1 and path 4, which contain link 1 (lower free-flow travel time but higher travel time variability), while the two SN models that assume risk averse behavior allocate more flow to path 2 and path 5 to avoid link 1. It is also interesting to observe that the SN-DUE model, which assumes to have perfect knowledge of the variable nature of network travel times, allocate significantly less flow to path 3 and path 6 compared to those allocated by the other three route choice models. This is because both path 3 and path 6 contain link 1 (the high variability link) and have higher average travel times. The differences in flow allocation to the paths in the network between the route choice models implemented without and with perception error are also evident from the results presented in Table 7. Thus, it would appear that the different assumptions in modeling route choice behavior could have an effect on the path flow patterns as well as the link flow patterns.

Table 7 Path Flow Comparisons between Scenario I and Scenario II

Path #	Scenario I				Scenario II			
	DN-DUE	DN-SUE	SN-DUE	SN-SUE	DN-DUE	DN-SUE	SN-DUE	SN-SUE
1	7.41 (2.24)	5.60 (2.90)	5.02 (3.01)	4.75 (2.08)	7.25 (2.49)	5.82 (2.38)	5.82 (2.83)	5.26 (2.71)
2	6.01 (3.06)	5.97 (2.75)	10.01 (2.36)	8.69 (3.03)	7.05 (2.66)	6.39 (2.27)	9.17 (2.55)	8.67 (2.59)
3	1.61 (1.39)	3.48 (2.44)	0.01 (0.22)	1.58 (0.42)	0.72 (1.12)	2.81 (0.71)	0.02 (0.63)	1.10 (0.50)
4	9.30 (2.76)	7.48 (3.49)	7.07 (3.51)	6.73 (2.85)	9.18 (3.10)	7.88 (3.17)	7.71 (3.49)	7.38 (3.49)
5	6.96 (3.05)	7.33 (3.74)	11.70 (3.15)	10.51 (3.72)	7.99 (3.15)	7.74 (2.76)	10.97 (3.08)	10.33 (3.05)
6	2.57 (0.94)	4.06 (2.63)	0.02 (0.20)	1.59 (0.56)	1.66 (0.97)	3.19 (0.72)	0.15 (0.88)	1.13 (0.69)

Concluding Remarks and Future Research

In this chapter, we have presented our continuing efforts on modeling risk-taking route choice models and its effect on travel time reliability. A Monte Carlo simulation procedure was used to simulate risk perceptions and preferences in making route choice decisions under an uncertain environment. Numerical results were also presented to examine what the aggregate impact of changes in variability caused by demand and supply variations might have on network assignment and how travelers with different risk-taking behaviors respond to these changes.

In general, the results from the four route choice models presented in Table 1 may have a significant impact on the estimation of travel time reliability measures, and it is important to examine how the different assumptions made in the route choice models affect the calculation of the travel time reliability measures under an uncertain environment. Based on the results of the two scenarios constructed for the numerical experiment, the following conclusions can be drawn:

- As the mean and standard deviation of OD demand increase, the expected and standard deviation of OD travel time also increase.
- The weighted OD travel time is lowest for the DN-DUE model, followed by the DN-SUE and SN-SUE models, and highest for the SN-DUE model.
- OD travel time reliability is lower when route choices are assumed to be based on imperfect information and even more so when they are based on minimizing disutilities that are a function of factors other than travel time.
- Additional variation from capacity degradation can significantly increase not only the OD travel time but also the standard deviation of OD travel time.
- Perception error may contribute to sub-optimal path and link flow allocations since the network assignment is based on imperfect information.

The numerical results, albeit based on a small network, reveal the significance of the route choice model on the estimation of travel time reliability measures. The different assumptions used in modeling route choice behavior could result in different estimates of travel time reliability. However, we note that the conclusions presented above are only relevant to the particular network and parameters used in this study. For future study, we plan to conduct more experiments using larger networks with realistic setting. Another future work is to relax the homogeneous behavior assumption of the travelers in the risk-taking route choice models to model heterogeneous behavior by allowing individual travelers to respond differently according to various degrees of risk aversion to travel time variability. This is important in studying route choice behavior at the individuals level, and how these individual responses to network uncertainty collectively affect the transportation systems performance. One way to account for different responses to travel time variability across individuals is to use the random-coefficient logit model (Ben-Akiva et al., 1993; Bhat, 1998; Train, 1997) to account for the degrees of risk aversion by capturing the taste variation across individuals resulting in differences in their responses to route-specific attributes (e.g., expected travel time and travel time variability) or in differences in their intrinsic preferences for route alternatives. For the risk-taking route choice model, the random-coefficients logit model can capture the hetereogeneity of individuals' risk perceptions and preferences with randomly distributed coefficients on the value of time and value of travel time variability, resulting in a continuous multi-class network equilibrium model to describe the risk-taking route choices of heterogeneous travelers. The goal is to enhance the predictive capability of the route choice model and to expand the behavioral framework of route choice modeling.

ACKNOWLEDGMENT

This research was supported in part by the National Science Foundation (NSF) and the California Partners for Advanced Transit and Highways (PATH) Program, USA.

REFERENCES

Abdel-Aty, M., Kitamura, R., and Jovanis, P. (1996). Investigating effect of travel time variability on route choice using repeated measurement stated preference data. *Transportation Research Record*, **1493**, pp. 39-45.

Ang, A.H.S. and Tang, W.H. (1990). *Probability concepts in engineering planning and design, vol.1 basic principle; vol. II – decision, risk and reliability*. John Wiley & Sons.

Asakura, Y. (1996). Reliability measures of an origin and destination pair in a deteriorated road network with variable flows. In. *Proceeding of the 4th Meeting of the EURO Working Group on Transportation.*

Asakura, Y. and Kashiwadani, M. (1991). Road network reliability caused by daily fluctuation of traffic flow. In. *Proceedings of the 19th PTRC Summer Annual Meeting*, Brighton, pp. 73-84.

Beckmann, M. J., McGuire, C.B., and Winsten, C.B. (1956). Studies in the Economics of Transportation. Yale University Press, New Haven, Connecticut, USA.

Bell, M.G.H., Cassir, C. Iida, Y., and Lam, W.H.K. (1999). A sensitivity based approach to reliability assessment. In. *Proceedings of the 14th International Symposium on Transportation and Traffic Theory*, pp. 283-300.

Ben-Akiva, M., Bolduc, D., and Bradley, M. (1993). Estimation of travel choice models with randomly distributed values of time. *Transportation Research Record*, **1413,** 88-97.

Bhat, C.R. (1998). Accommodating variations in responsiveness to level-of-service measures in travel mode choice modeling. *Transportation Research*, **32A,** 495-507.

Chen, A., Ji, Z.-W., and Recker, W. (2002). Route choice model with variable perception errors and probabilistic travel times. Paper submitted to the 82nd Annual Meeting of the Transportation Research Board, Washington, D.C., January 2003.

Chen, A. and Recker, W. (2001). Considering risk taking behavior in travel time reliability. Paper presented the 80th Annual Meeting of the Transportation Research Board, Washington, D.C., January 2001.

Chen, A., Tatineni, M., Lee, D.-H., and Yang, H. (2000). The effect of route choice models on estimating network capacity reliability. *Transportation Research Record* **1733**, 493-513.

Chen, A., Yang, H., Lo, H.K., and Tang, W. (1999). A capacity related reliability for transportation networks. *Journal of Advanced Transportation*, **33**, No. 2, 183-200.

Chen, A., Yang, H., Lo, H.K., and Tang, W. (2002). Capacity reliability of a road network: An assessment methodology and numerical results. *Transportation Research,* **36B**, 225-252.

Daganzo, C. and Sheffi, Y. (1977). On stochastic models of traffic assignment. *Transportation Research* **14B**, 243-255.

Du, Z.P. and Nicholson, A. (1997). Degradable transportation systems: sensitivity and reliability analysis. *Transportation Research*, **31B**, 225-237.

Fisk (1980). Some Developments in Equilibrium Traffic Assignment. *Transportation Research*, **14B**, pp. 243-255.

Ghosh, A. (2001). Heterogeneity in value of time: Revealed and stated preference estimates from the I-15 congestion pricing project. Paper presented at the 80th Annual Meeting of the Transportation Research Board.

Iida, Y. and Wakabayashi, H. (1989). An Approximation Method of Terminal Reliability of a Road Network Using Partial Minimal Path and Cut Set. In. *Proceedings of the 5th World Conference on Transport Research*, Yokohama, Japan, pp. 367-380.

Kazimi, C., Brownstone, D., Ghosh, A., Golob, T.F., and van Amelsfort, D. (2000). Willingness-to-pay to reduce commute time and its variance: Evidence from the San Diego I-15 congestion pricing project. Paper presented at the 79th Annual Meeting of the Transportation Research Board.

Lam, T. and Small, K. (2001). The value of time and reliability: measurement from a value pricing experiment. *Transportation Research*, **37E**, 231-251.

Lam, W.H.K. and Zhang, N. (2000). A new concept of travel demand satisfaction reliability for assessing road network performance. Paper presented at the Matsuyama Workshop on Transport Network Analysis, August 2000, Matsuyama, Japan.

Mirchandani, P. and Soroush, H. (1987). Generalized traffic equilibrium with probabilistic travel times and perceptions. *Transportation Science*, **21**, pp. 133-152.

Nicholson, A. and Du, Z.P. (1997). Degradable transportation systems: An integrated equilibrium model. *Transportation Research*, **31B**, pp. 209-224.

Sheffi, Y. (1985). *Urban Transportation Networks: Equilibrium Analysis with Mathematical Programming Methods.* Prentice-Hall, Englewood Cliffs, New Jersey.

Sheffi, Y. and Powell, W.B. (1982). An algorithm for the equilibrium assignment problem with random link times. *Networks*, **12**, pp. 191-207.

Small, K.A., Noland, R., Chu, X., and Lewis, D. (1999). Valuation of Travel-Time Savings and Predictability in Congested Conditions for Highway User-Cost Estimation. NCHRP Report 431, Transportation Research Board, National Research Council.

Tatineni, M. (1996). Solution Properties of Stochastic Route Choice Models. Ph.D. Dissertation, University of Illinois at Chicago.

Tatineni, M., Boyce, D., and Mirchandani, P. (1997). Experiments to compare deterministic and stochastic network traffic loading models. *Transportation Research Record*, **1607**, pp. 16-23.

Train, K. (1997). Mixed logit models for recreation demand. In *Valuing the Environment Using Recreation Demand Models* (Kling, C. and Herriges, J.) Elgar Press.

Wardrop, J.G. (1952). Some theoretical aspects of road traffic research. In. *Proceedings of the Institute of Civil Engineers* 2 (1), 325-378.

7

Reliability Assessment on Transit Network Services

Yafeng Yin, Institute of Transportation Studies, University of California, Berkeley, USA
William H. K. Lam, The Hong Kong Polytechnic University, Hong Kong, P.R. China
Hitoshi Ieda, Department of Civil Engineering, University of Tokyo, Tokyo, Japan

Abstract

This paper attempts to assess the transit service reliability with taking into account the interaction between network performance and passengers' travel choice behaviors. Besides the well-known schedule reliability, a waiting-time reliability is newly defined as the probability that the passengers' average waiting time is less than a given threshold. A Monte Carlo simulation approach, which incorporates a stochastic user equilibrium transit assignment model with explicit capacity constraints and elastic frequencies, is proposed to estimate the above two reliability measures of transit service. A numerical example is used to illustrate the applicability of the reliability measures and the proposed approach.

Introduction

As the expectation of society for improvement of quality of life is more demanding, reliability becomes an increasingly important attribute of transport networks. Consequently, more attention has been given to the research on transport network reliability and some results have been procured on the issues relating to the measurement of reliability (e.g., Iida, 1999; Bell, 1999; Chen et al., 1999 and Lam and Xu, 1999), traveller decision-making under uncertainty (e.g., Bonsall, 2000; Fujii and Kitamura, 2000 and Yin and Ieda, 2000), the provision of information to travelers (e.g., Asakura, 1999, Uno and Iida, 2000 and Yin and Ieda, 2001a), the design of transport networks (e.g. Asakura, 2000 and Yin and Ieda, 2001b) and routing or scheduling strategies (e.g., Taniguchi et al., 2000).

Most of the previous related studies aimed to investigate reliability problems of road transport systems. These findings may not be applicable to the scheduled public transport systems, where the systems operate and passengers make their travel decisions based on nominal or posted timetables. Transit systems often fail to provide reliable service due to some unexpected exogenous delays. Such delays may be caused by breakdown or failure of equipment or infrastructure, varying passenger boarding or alighting times and traffic congestion etc. Furthermore, these exogenous delays frequently lead to knock-on delays on other transit vehicles (buses and trains, etc), which can in turn cause further knock-on delays in viscous cycles.

In a recent survey (Frima et al., 1998), reliability was found as one of the most important attributes of transit service quality. Assessment and prediction of transit service reliability has received little attention in the literature. As noted by Carey (1999), although most public transport operators maintain detailed records of punctuality and reliability (in terms of the percentage of services on-time, or the mean arrival or departure deviation), what is often of more interest to transport planners or managers is the ability to predict reliability of the proposed new schedules or services or when considering service changes. Hallowell and Harker (1998) conducted extensive simulations to predict the schedule reliability on a partially double rail line where meet/pass interference is the major cause of delays. Carey (1999) proposed ex ante heuristic measures of schedule reliability that can be used in advance with considering whether an increase in the measure is likely to ensure that the probability of knock-on delays will decrease.

Contrary to these previous studies, this paper attempts to develop an approach to assess the transit service reliability with taking into account the interaction between network performance and passengers' travel choice behaviors. In addition to the traditional schedule reliability, a new measure, termed as passenger waiting-time reliability, is further defined in order to investigate comprehensively transit network performance. Be aware that our model is aimed to assess macroscopically reliability of public transport services in the stage of network design and service planning. The problem of the online operation, like that once an exceptional accident happens, how the schedules are adjusted to recover the transit services as soon as possible, is out of concern here.

This paper has been organized as follows. The next section defines the measures of transit service reliability. In Section 3, a capacity restraint transit stochastic user equilibrium (SUE) assignment model with elastic line frequency is discussed and then a Monte Carlo simulation procedure incorporating the model is suggested to quantify the reliability measures. Section 4 illustrates the approach with a small transit network. Conclusions and future research issues are provided in Section 5.

MEASURES OF TRANSIT SERVICE RELIABILITY

The schedule reliability of public transport services is important in planning, management, operating and marketing of these services (Carey, 1999). It can be defined generally as the ability that the operation of a transit system follows the posted timetable. Various methods have been used to measure the reliability, e.g., arrival error (mean and standard deviation). Measuring such reliability after the event is relatively easy while predicting it at the design stage is more difficult. Existing analytical methods are usually practical for only simple structured systems and simulation methods may become intractable because every line in a transit network must require a detailed simulation to obtain data necessary for estimating the reliability measures.

For the sake of applicability, frequencies of transit lines are employed to estimate schedule reliability in this paper. It is well known that variations of running times at line segments and dwelling times at stations make transit service unpunctual and unreliable. Such delays will also lead to longer cycle times of transit vehicles and correspondingly less frequencies of lines provided that number of vehicles on each transit line is constant, which is likely in practice. Therefore, our schedule reliability for each line is defined as the probability that an actual line frequency is larger than a given threshold, described as follows,

$$SR_l = \Pr\left(f_l \geq \alpha \cdot f_{l0}\right) \quad \forall l \subset L \tag{1}$$

where SR_l = the schedule reliability of line l; f_l = the actual line frequency; f_{l0} = the nominal line frequency; α = a pre-defined threshold value, $0 < \alpha \leq 1$ and L = the set of transit lines. Be aware that SR_l is designed for macroscopic evaluation and it may be less accurate than those traditional schedule reliability measures. For example, in some case that the service is not reliable if we look at the measure of arrival error, but the proposed SR_l may show a perfect reliability.

Although most of the relevant studies have focused on the above schedule reliability, it is noted that such a single measure will not be capable of estimating transit service reliability comprehensively. For example, even in a transit system with perfect schedule reliability, some passengers may not be able to get on the first coming vehicle due to its insufficient capacity and have to wait more time for the next coming one, particularly during the commuting rush hours. For these passengers, their actual travel times are obviously larger than those they expected. Hence they may not perceive such transit service as reliable one. Therefore, we need passengers' waiting-time reliability as another measurement of transit service reliability. It is defined as the probability that passengers' average waiting time is less than a given threshold, written as:

$$WTR_l^n = \Pr\left(w_l^n \leq \beta \cdot w_{l0}^n\right) \quad \forall l \in L, \forall n \in N(l) \tag{2}$$

where WTR_l^n = the waiting time reliability of line l at station n; w_{l0}^n = the average waiting time for passengers boarding on line l at station n according to the nominal schedule under free-flow conditions. And, $w_{l0}^n = \dfrac{\alpha}{f_{l0}}$, $\alpha = 1$ corresponds to an exponential distributions assumed for the vehicle headway and a uniform passenger arrival distribution, and the value $\alpha = 0.5$ is an approximation of a deterministic headway and a uniform passenger arrival distribution (see Lam and Morrall, 1982); β = a pre-defined threshold value, $\beta \geq 1$ and $N(l)$ = the set of stops in line l. w_l^n = the actual waiting time at station n for line l, written as:

$$w_l^n = \frac{\alpha}{f_l} + d_l^n, \ \forall l \in L, \forall n \in N(l) \tag{3}$$

where d_l^n = the overload delay of line l at station n due to the insufficient capacity of line l.

Approach of Estimating Reliability

In this section, we will propose a simulation procedure to estimate the above two transit service reliability measures. Note that since our simulation procedure incorporates a transit assignment model, which reduces the need of detailed simulation, it may achieve a relatively fast assessment of reliability with less computation efforts.

A Transit SUE Assignment Model

Although the transit assignment problem has been studied in the past two decades, most of the models are heuristic and do not consider congestion effects. Recently, De Cea and Fernández (1993) and Wu et al. (1994) have formulated two equilibrium models taking into account the congestion phenomena due to insufficient capacities of transit lines concentrated at transit stops. Lam et al. (1999a) proposed a stochastic user equilibrium model with explicit capacity constraints. They proved that when the constraints are reached, the associated Lagrange multipliers of the mathematical programming model are equivalent to the equilibrium passenger overload delays in the congested transit network. Lam et al. (2000) further extended the model to accommodate elastic line frequency that is related to the passenger flows on transit lines.

In this paper, we will employ the model by Lam et al. (2000) in the simulation procedure. Some concepts employed in the model are introduced as follows.

The common lines problem. A transit network constitutes a set of transit lines and a set of stations (nodes) where passengers can board, alight or transfer. A transit line is a fleet of vehicles that run back and forth between two nodes (terminals) on the urban road or transit network, and can be described by the frequency of the vehicles and the vehicle types (e.g. bus or underground train). A line segment is a portion of a transit line between two consecutive stations.

Different transit lines may run parallel for part of their itineraries with some stations in common. That means there exist different lines available to the passenger between some pair of transfer stations (stops) in a transit network, and thus the passenger must select the lines he will probably use. Chriqui and Robilland (1975) made a reasonable assumption on how the passenger operates a selection: among the set of all lines passing from station 1 to station 2, he will select a subset so as to minimize his total expected travel time (in-vehicle time plus waiting time). Once this subset is selected, the passenger will take the first coming vehicle serving one of the lines in the subset to arrive at 2. Transit lines belonging to the subset are termed as common (attractive) lines and can be picked up by an algorithm suggested by Chriqui and Robilland (1975). Therefore, it is seen that, passengers in the transit network cannot exactly decide beforehand which line to board between any pair of two consecutive transfer stops on their journey. This is why the concept of paths used in road networks cannot be directly applied to transit networks.

For the simplicity of the treatment of the common lines calculation, De Cea and Fernández (1993) defined two different representations for a transit network, i.e., alternatives to the traditional definition in terms of nodes and line segments (**G0**). These are the networks **G1=(N, L)** and **G2=(N, S)**. In **G1**, the set of links **L** contains all the line sections, which are any portion of a transit line between two nodes, not necessarily consecutive, while in **G2**, **S** contains all the route sections which are associated with a set of common (attractive) lines. In both cases, **N** is the set of nodes (transit stops). An example is given below to illustrate these concepts.

Figure 1*a* shows a very simple network with four nodes and four transit lines coded as **G0** in terms of line segment, and Figure 1*b* represents the same network as **G1,** coded in terms of line section and Figure 1*c* further represents it as **G2,** coded in terms of route sections after some common lines calculation. That implies that whether L2 and L3 are included in route section S2, and L3 and L4 included in S5 are dependent on their respective travel times and frequencies. With the concepts of line section and route section, the SUE assignment on transit network **G2** will become similar as that on road network. Therefore, the representation of transit network by De Cea and Fernández (1993) is adopted in this paper.

Capacity Constraint and Elastic Line Frequency. Although the SUE assignment on transit network **G2** is similar as that on road network, it still presents some unique characteristics, such as the congestion effect and line capacity. In the transit network, the *fundamental diagram* is not applicable any more. In order to describe the congestion, De Cea and Fernández (1993) and Wu et al. (1994) consider waiting times at transit stops to be unbounded increasing function of passenger flows. As the same authors acknowledged, their models might lead to the result that some lines are overloaded. The transit line capacity, dependent on the line frequency and vehicle capacity, can be determined more clearly and easily, comparing with road capacity. Since each transit line has an absolute capacity that cannot be exceeded in practice, it is expected that the line capacity constraint should be met more "strictly". The aforementioned limitation is particularly serious when these transit assignment models are used to study transit network with crowed vehicles and congested stations/stops due to insufficient capacity of the transit services. In viewing of this, Lam et al. (1999a) proposed a stochastic user equilibrium model with explicit capacity constraints. They proved that when the constraints are reached, the associated Lagrange multipliers of the mathematical programming model are equivalent to the equilibrium passenger overload delays in the congested transit network.

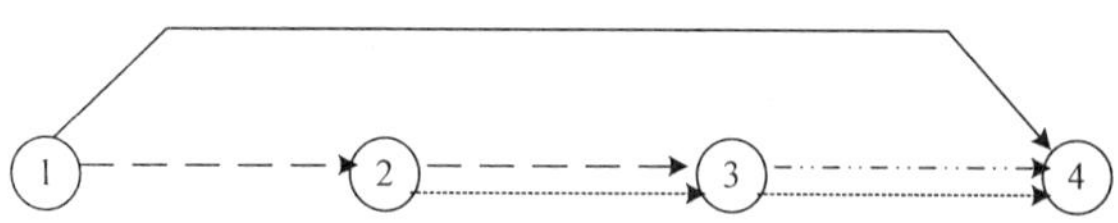

Figure 1a Transit Network Based on Line Segment (G0)

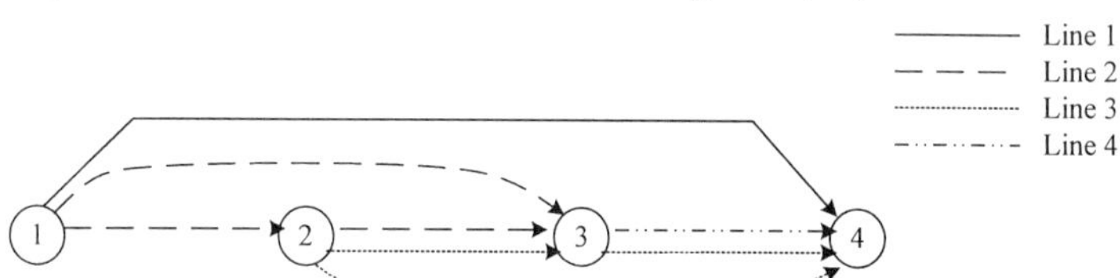

Figure 1b Transit Network Based on Line Sections (G1)

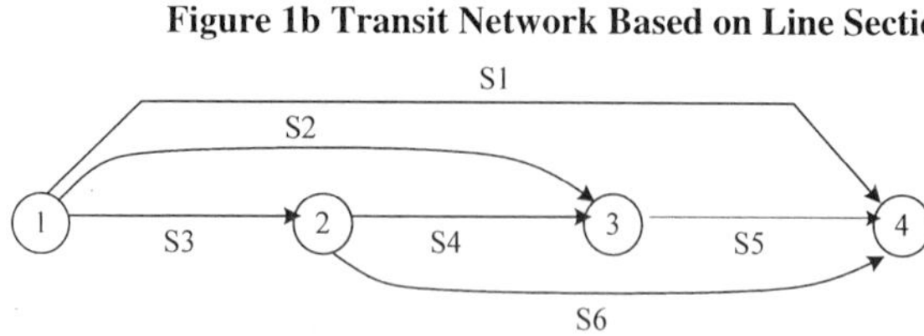

Figure 1c Transit Network Based on Route Sections (G2)

Most of the previous model assumed that the frequency on each transit line is fixed. However, the assumption may not be true in practice, particularly when the transit system is heavily crowded during the peak periods. In a realistic setting, the actual frequency is related to the cycle journey times of transit vehicles, which are dependent on the running times on line segments and dwelling times at stations. Wirasinghe and Szplett (1984) and Lam et al. (1999b) have found that the vehicle dwelling time at a station was governed by the volume of boarding and alighting passengers. Therefore, the actual frequency of each transit line is related to passenger flows. On the other hand, the passenger route choice is obviously influenced by the line frequencies. It can be imagined there exists an equilibration procedure through which equilibrium passenger flows and line frequencies are achieved. Lam et al. (2000) formulated this equilibration procedure as a fixed-point problem, and the above equilibrium flows and frequencies, together with passenger overload delays can be obtained from their model.

In summary, the transit SUE assignment model adopted in this paper is characterized in: 1) The network representation by De Cea and Fernández (1993) is used to deal with the problem of common lines in a transit network; 2) Stochastic user equilibrium route choice behaviors are considered; 3) Each link has a explicit capacity constraint; 4) The link cost consists of three components, namely: in-vehicle time, the waiting time and the equilibrium passenger overload delay; 5) The dwelling time for a transit vehicle is dependent on the boarding and alighting passengers at stations; 6) Consequently, the line frequency and the waiting time are related to the equilibrium passenger flows on transit lines.

A Monte Carlo Simulation Procedure

Given the transit network configuration and OD demands, the above analytical model may find essentially deterministic equilibrium passenger flows, waiting times, overload delays and actual frequency. However, as mentioned above, the potential sources of disruption to transit system routine operations are numerous and it is those disruptions that degrade the transit service quality and make it unreliable. Needless to say, it is reasonable to assume that the OD travel demand and travel times on line segments are random and follow different distributions respectively. To investigate their effects on service reliability, we can perform a Monte Carlo simulation to estimate the impacts of variations of frequencies and waiting times, and then evaluate the aforementioned reliability measures. The simulation procedure can be described as follows:

Step 1:	Set sample number $k := 1$.
Step 2:	Generate vectors of OD travel demands, travel times;
Step 3:	Perform the stochastic user equilibrium transit assignment with the generated input vectors;
Step 4:	Collect statistics to construct reliability distribution functions;

Step 5: If sample number k is less than the required sample size $k_{\max}$, then increment sample number $k := k + 1$ and go to Step 1. Otherwise, stop.

The heuristic algorithm proposed by Lam et al. (2000) is employed to solve the transit assignment in Step 3. The algorithm includes a method of successive average procedure, used to solve the fixed-point problem, and the iterative balancing algorithm by Bell (1995), used to solve a SUE assignment problem with explicit capacity constraint and fixed frequencies in each iteration. In Step 4, we collect the count of successful samples according to the definitions (1) and (2) to construct the distribution of the reliability measures. Depending on the purpose of the study, other statistics could also be collected, such as passenger flows, cycle travel times and others.

NUMERICAL EXAMPLE

Data Input

To demonstrate the Monte Carlo simulation procedure, we apply it to a simple transit network, taken from the previous related studies, as shown in Figure 2. It consists of four transit lines (L_1, L_2, L_3 and L_4) and four nodes where Z and Y are the transfer nodes only. Let L_2^1, L_2^2 be the line segment of $L2$ between A and Z and between Z and Y respectively. Also, let L_3^1, L_3^2 be the line segment of $L3$ between Z and Y and between Y and B respectively. The line data of the transit network in Figure 2 is given in Table 1. Without loss of generality, it is assumed that the in-vehicle travel times between two consecutive nodes are the same regardless of their travel directions. Furthermore, only passengers' travel times are considered in our example as the transit fare is omitted. Since there are various disruptions to the operation of transit network, we assume the running times are stochastic, following truncated normal distributions $0.9t_{l0} \le N(t_{l0}, 0.5t_{l0}) \le 1.5t_{l0}$.

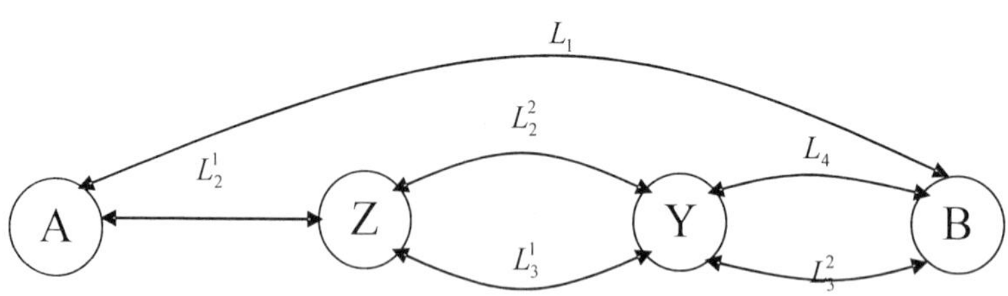

Figure 2 The Example Transit Network

Table 1 Basic Link Data for the Example Transit Network

Basic data	Links					
	L_1	L_2^1	L_2^2	L_3^1	L_3^2	L_4
In-vehicle time t_{l0} (min)	15	7	6	5	5	8
Nominal frequency f_{l0} (vehs/hr)	13	10	10	6	6	12
Vehicle Capacity k_l (pass/veh)	20	20	20	20	20	20

In this example network, there are only two OD pairs ($A \rightarrow B$ and $B \rightarrow A$) and their OD demands are assumed to follow identically a truncated normal distribution $\min(450, N(300,50))$.

Assuming that α is equal to 0.5 when calculating the average waiting time as Eq. (3) and the dispersion parameter of Logit model is 0.5. Meanwhile, the relationship between vehicles' dwelling time and the number of passengers boarding/alighting at stations is considered as follows:

$$dt_l^n = \max(dt_{l0}^n,\ 1.5 + 0.083Bo_l^n + 0.05Al_l^n)$$

where dt_l^n = the actual dwelling time of Line l at stop n; dt_{l0}^n = the minimal (scheduled) dwelling time of Line l at stop n, and we set it as 2 minutes uniformly; Bo_l^n = the number of passengers boarding Line l at stop n; Al_l^n = the number of passengers alighting Line l at stop n.

Preliminaries

Before evaluating the reliability of the example network, we firstly perform the convergence tests on the sample sizes of the Monte Carlo simulation. Three simulations of sample sizes of 1,000, 5,000, and 10,000 are examined and the resultant schedule reliability and waiting-time reliability are shown in Figure 3 and 4 respectively. It is found that in Figure 3 and 4 the reliability distributions using 1,000 samples show no significant difference from the ones using 5,000 and 10,000 samples. Thus, a sample size of 1,000 is used for all subsequent simulations.

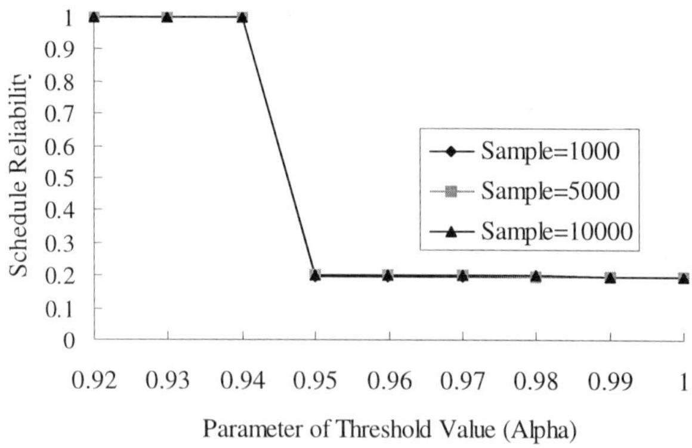

Figure 3 Effect of Sample sizes on the Schedule Reliability (Line 1)

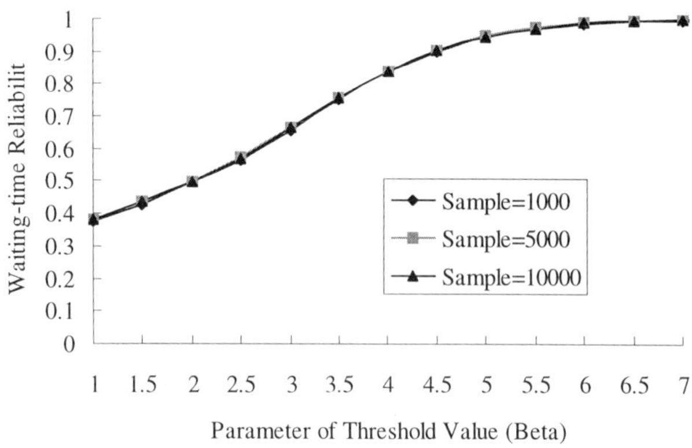

Figure 4 Effect of Sample sizes on the Waiting-time Reliability (Line 1, Station A)

Reliability Assessment

The reliability of the aforementioned transit service (denoted as base case) is assessed first. Compared with other lines, Line 1 suffers lower schedule reliability and waiting-time reliability, shown in Figure 5 and Figure 6 respectively. Since Line 1 offers an express service from A to B and the passengers who want to take this line are obviously beyond its transport capacity. As a result, some of passengers will wait more time for the next arriving vehicle, and furthermore, the actual dwelling times of buses serving Line 1 will be longer than the scheduled ones and thus Line 1 will has a less frequency than expected, which in turn leads to a more waiting time for passengers. In order to improve the service reliability of Line 1, three plans have been suggested as follows:

- Plan 1: slacking the schedule by increasing the nominal dwelling time at each stops from 2 minute to 2.5 minute without changing the frequency, which implies more vehicles are needed;
- Plan 2: increasing the capacity of each vehicle serving Line 1, which implies larger vehicle will be used;
- Plan 3: improving the service offered by Line 3: shorter running time (L_3^1, L_3^2: from 5 min. to 3 min.) and higher frequency (from 6 vehs/hr to 10 vehs/hr).

The corresponding reliability measures of each plan are presented in Figure 5 and Figure 6. It is illustrated that, with more slackness in the operation schedule as Plan 1, because the transit has more luxury to adjust its operation once the later-arrival happens, the schedule reliability will be significantly improved. However, regarding the waiting-time reliability, because the unreliability is mainly attributed to the unexpected overload delays, the slackness will not lead to much improvement of waiting-time reliability.

Plan 2 intends to improve the reliability of Line 1 by increasing its capacity. However, it is surprising to find that the objective cannot be achieved. As shown in Figure 5 and 6, the waiting-time reliability is not much different with that of base case while the schedule reliability is even deteriorated. The increase of the capacity of Line 1 encourages more passengers to take this line. This induced or transferred passenger flow further results in a longer dwelling time and a less frequency, and consequently worsens the schedule reliability, illustrated in Figure 5. Meanwhile, although the overload delays are reduced due to the increase of the capacity, the reduction is obviously offset by the increase of the first component of the waiting time as Eq. (3) due to the less frequency. Therefore, the waiting-time reliability presents no much change in Figure 6. If we further improve the capacity of Line 1 so that it can accommodate the total travel demand between A and B, the reliability of Line 1 will be improved. However, such plan is obviously unfavorable due to the limited budget.

Different with the above alternative, Plan 3 aims to improve the reliabilities of Line 1 by offering better service of Line 3. It can be seen from Figure 5 and 6 that since some of passengers will transfer to the route of the segment of Line 2 (L_2^1) and Line 3, the schedule and waiting-time reliability of Line 1 are improved respectively. This observation underscores the importance of improving the reliability of transit service from a systematic point of view. Systematic improvement plan should play a significant role in improving transit service reliability.

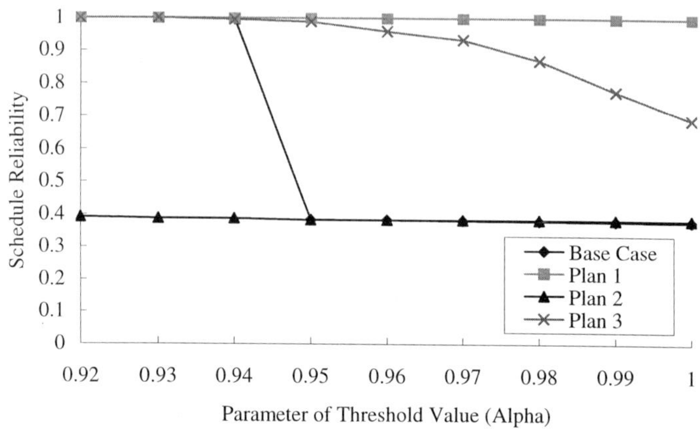

Figure 5 Schedule Reliabilities of Different Alternatives (Line 1)

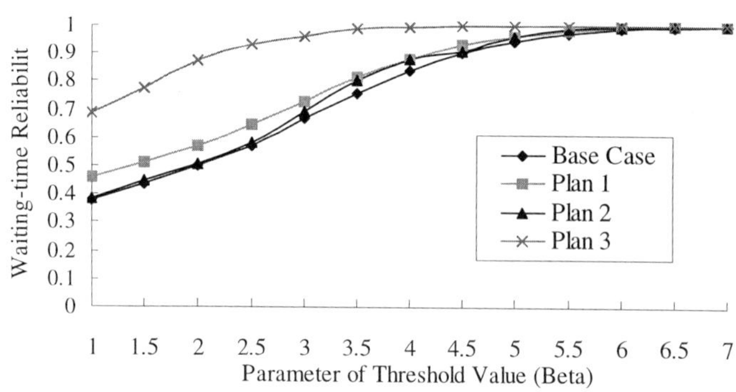

Figure 6 Waiting-Time Reliabilities of Different Alternatives (Line 1, Station A)

CONCLUSIONS

In this paper, we have proposed a Monte Carlo simulation procedure as a tool for assessing the transit service reliability. Two measures, namely the schedule reliability and the waiting-time reliability have been defined and evaluated with taking into account the interaction between the network performance and passengers' travel choice behaviors. The research will continue to investigate the applicability of the proposed procedure to the real transit network together with the model calibration and validation.

ACKNOWLEDGEMENTS

The second author would like to express his gratitude to the Institute of Systems Science of Japan for providing a grant to the Hong Kong Polytechnic University (A/C H-ZF21), which partially supported the work described in this paper.

REFERENCES

Asakura, Y. (1999). Evaluation of network reliability using stochastic user equilibrium. *Journal of Advanced Transportation*, **33,** 147-158.

Asakura, Y. (2000). Stochastic network design problem. *Paper Presented on Matsuyama Workshop on Transport Network Analysis*, Matsuyama, Japan.

Bell, M. G. H. (1995). Stochastic user equilibrium assignment in networks with queues. *Transportation Research*, **29B**, 125-137.

Bell, M.G.H. (1999) Measuring network reliability: a game theoretic approach. *Journal of Advanced Transportation*, **33,** 1999, 135-146.

Bonsall, P. (2000). Travellers' response to uncertainty. In. *Reliability of Transport Networks* (Bell, M.G.H. and Cassir, C.), 1-10. RSP, England.

Carey, M. (1999). Ex ante heuristic measures of schedule reliability. *Transportation Research*, **33B**, 473-494.

Chen, A., Yang, H. and Tang, W.H. A (1999). capacity related reliability for transportation networks." *Journal of Advanced Transportation*, **33,** 183-200.

Chriqui, C. and Robillard, P. (1975) Common bus lines. *Transportation Science*, **9**, 115-121.

De Cea, J. and Fernández, E. (1993). Transit assignment for congested public transport system: an equilibrium model. *Transportation Science*, **27**, 133-147.

Frima, M, Edardsson, B. and Gaerling, T. (1998) Perceived service quality attributes in public transport: inferences from complaints and negative critical incidents. *Journal of Public Transportation*, **2**, 67-89.

Fujii, S. and Kitamura, R. (2000). Rethinking of drivers' departure time choice under uncertainty. *Paper Presented on Matsuyama Workshop on Transport Network Analysis*, Matsuyama, Japan.

Hallowell, S.F. and Harker, P.T. (1998). Predicting on-time performance in scheduled railroad operations: methodology and application to train scheduling. *Transportation Research*, **32A**, 279-296.

Iida, Y. (1999). Bsic concepts and future directions of road network reliability analysis. *Journal of Advanced Transportation*, **33,** 125-134.

Lam, W. and Morrall, J. (1982) Bus passenger walking distances and waiting times: a summer-winter comparison. *Transportation Quarterly*, **36**, 407-421.

Lam, W. H. K., Gao, Z. Y., Chan, K. S. and Yang, H. (1999a). A stochastic user equilibrium assignment model for congested transit networks. *Transportation Research*, **33B**, 1-18.

Lam, W. H. K. Chueng, C. Y. and Lam, C. F. (1999b). A study of crowding effects at Hong Kong Light Rail Transit stations. *Transportation Research*, **33A**, 401-415.

Lam, W. H. K. and Xu, G. (1999). A traffic flow simulator for network reliability assessment. *Journal of Advanced Transportation*, **33,** 159-182.

Lam, W. H. K., Zhou, J. and Sheng, Z.H. (2000). A capacity restraint transit assignment with elastic line frequency. Re-submitted to *Transportation Research*, **B.**

Taniguchi, E., Yamada, T. and Tamagawa, D. (2000). Probabilistic routing and scheduling of urban pick-up/delivery trucks with variable travel times. In. *Reliability of Transport Networks* (Bell, M.G.H. and Cassir, C.), 73-90. RSP, England.

Uno, N. and Iida, Y. (2000). A study on traveler information system to improve travel time reliability. *Paper Presented on Matsuyama Workshop on Transport Network Analysis*, Matsuyama, Japan.

Wirasinghe, S. C. and Szplett, D. (1984). An investigation of passenger interchange and train standing time at LRT stations: (ii) estimation of standing time. *Journal of Advanced Transportation*, **18,** 13-24.

Wu, J. H., Florian, M. and Marcotte, P. (1994). Transit equilibrium assignment: a model and solution algorithms. *Transportation Science*, **28,** 193-203.

Yin, Y. and Ieda, H. (2000). Assessing performance reliability of road networks under nonrecurrent congestion. *Transportation Research Record*, **1771**, 148-155.

Yin, Y. and Ieda, H. (2001a) Does providing information to drivers improve road network reliability? *Paper presented on 9th World Conference of Transport Research*, Seoul, Korea.

Yin, Y. and Ieda, H. (2001b). An optimal improvement scheme for network reliability. *Transportation Research Record* (**accepted for publication**).

8

Modeling of Traveler's Uncertainty and Anxiety

Shinya Kikuchi, Department of Civil and Environmental Engineering, University of Delaware, Delaware, USA
Henk J. van Zuylen, Delft University of Technology, Delft, the Netherlands

Abstract

How an individual makes a travel decision under uncertain conditions has been one of the critical issues in designing the information that is delivered by Intelligent Transportation Systems (ITS). This has been a difficult problem because a suitable mathematical framework that deals with the interaction among uncertainty, information, and traveler's attitude toward uncertainty is not available. This paper introduces a possibility theory framework, and demonstrates how this framework represents the uncertainty perceived by the traveler, and calculates the feasibility of achieving the travel objective under different degrees of specificity of information. We present this framework in the setting of selecting the time of departure in the face of a not well-defined desired arrival time and estimated travel time. Feasibility of arrival is measured by the possibility and necessity measures; these measures represent two views, optimistic or conservative, respectively. Anxiety that is associated with the departure time, whether to leave now or not is modeled by Yager's anxiety measures. The anxiety measure considers the conflict of impelling forces between "to leave now" and "not to leave yet"; possibility and necessity measures for these outcomes represent these forces. Thus, along the time axis of possible departure time, possibility and necessity measures of arrival and non-arrival are computed and accordingly, the anxiety measure associated with each departure time. The range of time within which a traveler decides to leave is indicated by the anxiety measure. The size of the range is sensitive to the specificity of information. The more specific the information, the smaller the range of time. While the purpose of this paper is to introduce the mathematical framework useful for the analysis of a traveler's decision under uncertainty, the analysis raises an interesting issue of the paradoxical effects of information accuracy also.

INTRODUCTION

The reality of decision-making under uncertainty is that the decision-maker usually has only vague ideas about the outcomes because the available information is not complete. In most cases, available information is not the true sense of probability "distribution"; rather it is a single value or a range of values, expressed and perceived in natural language. For example, in the case of a travel decision, the estimated travel time is most often given by a natural language expression, such as "about 30 min." The conventional statistical framework has limited utility when it comes to dealing with a decision-making process involving perceptive natural language based information.

Types of uncertainty being analyzed with respect to human travel behavior have two layers; one related to the uncertainty that the analyst (or modeler) faces when examining the pattern of decisions made by a group of travelers; and the other, the uncertainty that individual travelers face given the incomplete information. The former is attributed to the lack of data and the lack of the analyst's knowledge about the behavioral pattern of all the travelers. This aspect of uncertainty can be dealt with by the conventional statistical approach using the available data. The stochastic choice model is a typical example. The latter, however, is attributable to the lack of information, and, sometimes, the lack of clear goal of the individual traveler. This type of uncertainty is more complicated to deal with than the former one. This paper focuses on this type of uncertainty, and introduces possibility theory for analyzing the effects of information.

When it comes to analyzing decisions of individuals, two analysis frames exist. They are the descriptive and prescriptive approaches. The descriptive approach studies the pattern how individuals make decisions. The prescriptive approach, on the other hand, studies how one should make a decision. A typical example of the latter is maximization of expected outcome, an often-recommended decision strategy under uncertainty. In this paper, we deal with the former, the descriptive decision model, by formulating a modeling framework of how information is perceived and processed by individuals in their travel decisions. It provides the basis for the prescriptive analysis.

DEALING WITH INTERVALS OR RANGES WHEN ANALYSING TRAVEL SCHEDULING DECISIONS

Imagine a case of establishing an itinerary of a trip. Before deciding on the time of departure one examines the desired arrival time, its allowable deviation, and the estimated travel time. Perhaps, in the mind of the traveler, information on the desired arrival time and the estimated travel time is perceived as a value with an interval. The concept of time intervals or ranges was introduced in some of the time–window based scheduling algorithms first. Among them are Jaw et al.(1986), Kikuchi and Rhee (1989), Kikuchi and Donnelly (1992), and Solomon and Desrosier (1988). A troublesome aspect of dealing with intervals is that the size of the interval increases as different intervals are added or subtracted, and, as a result, the arithmetic operations can become intractable.

For example, when two numbers with an interval [3, 5] and [6, 8] are added, then the answer is [9, 13]. When [3, 5] is subtracted from [9, 13], then the value is [4, 10]; this interval is greater than [6, 8] above. In other words, when intervals are introduced, A+B=C does not mean C-B=A. In the case of departure time determination, it is computed by {desired arrival time$^{(1)}$ - travel time = departure time$^{(1)}$}, initially. However, the arrival time computed based on this departure time, {departure time$^{(1)}$ + travel time =arrival time$^{(2)}$}, is not the same as the original desired arrival time; namely arrival time $^{(2)}$ ≠ desired arrival time $^{(1)}$. This means that departure time $^{(1)}$ may not guarantee the arrival at the desired arrival time $^{(1)}$.

In the above problem, two ranges of departure time exist, one that allows *possible* arrival (but does not always guarantee arrival) at the destination within the desired range, and the other, that allows *necessary (guaranteed)* arrival within the desired range. The latter is a subset of the former range. Depending on the sizes of the range of the estimated travel time and the desired arrival time, however, the latter range may not be available. It is available only as long as the range of the desired arrival time is larger than that of the estimated travel time. Otherwise, all the departure times within the range yield *possibility* of arrival within the desired time but do not guarantee it (they may yield an arrival outside the desired range.) The above illustrates the complexity when dealing with ranges in travel time analysis.

MATHEMATICAL FRAMEWORK FOR DEALING WITH UNCERTAINTY CAUSED BY INTERVAL EXPRESSIONS

Most natural language based expressions, such as "late," "early," or "approximately." are associated with a range of values whose boundaries are not well defined. They are, however, the essential information when making travel decisions. We need a mathematical framework that captures the interplay between information and decision. Such a framework will be useful for understanding the traveller's satisfaction with the itinerary and the quality of information. The framework, however, should be capable of handling the following:

(1) Representing a range with both the hard and the soft boundaries and performing arithmetic operations of ranges;

(2) Satisfying the basic axioms of measure theory, in which the traditional treatment of uncertainty, and probability theory, is subsumed;

(3) Providing the measure for the feasibility of travel (or arriving or not arriving at the desired time); and,

(4) Modeling the traveler's satisfaction and anxiety relative to the quality of information.

Range Representation by a Fuzzy Set

In order to deal with the ranges with hard and soft boundaries, we introduce the fuzzy set. A fuzzy set is characterized by the membership function. A membership function of concept A (in our case, the concepts such as the approximate travel time or desired arrival time), $h_A(x)$, $x \in X$, where X is the universal set, indicates the degree that a value (x) is compatible with the notion of A.

Readers are reminded that the probability density function and the membership function of a fuzzy set are fundamentally different. The former refers to the expectation of occurrence of the random events; the latter is the representation of a set with soft boundaries. We believe that a notion, such as the "desired" time, is best represented by a fuzzy set, not by the probability distribution. We refer to various textbooks (e.g. Klir, et al. [1997], Klir and Folger [1988], Yager and Filev [1994], Zimmermann [1996]) for detailed explanations of the membership function and its utility.

Figure 1 illustrates the basic difference in the structure between the membership function and the probability density function. The membership function of a fuzzy set is constructed based on "stacking up" the ranges, each range representing the nested set. The probability density function, on the other hand, is constructed by the weight of evidence supporting each exclusive subset.

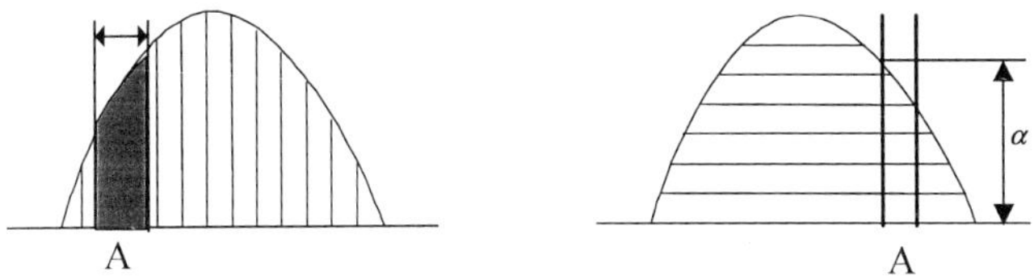

Event (Alternatives) | Event (Alternatives)

Prob(A)=Shaded area | *Poss(A)=α*

Probability Density Distribution | Possibility Distribution

Figure 1 Differences in Probability and Possibility Distributions

Once the notion of an approximate time is represented by a fuzzy set, then addition and subtraction of the approximate times can be performed by the extension principle, which allows the functional operations of fuzzy sets when the input to the function is a fuzzy number. The operations of the extension principle show how fuzziness propagates as more operations of fuzzy numbers are performed. Kaufmann and Gupta (1984) present various operations of approximate numbers.

Three Basic Measures of Uncertainty

The nature of uncertainty with regard to the truth of a proposition is closely related to how evidence supports the proposition. If evidence supports the proposition completely then the proposition is considered true, and no uncertainty exists. If there is no evidence, then the proposition is false. When evidence supports the proposition partially, then uncertainty exists. Figure 2 presents schematically the interplay between evidence and proposition for the following three cases:

<u>When the individual pieces of evidence point to the well-defined mutually exclusive sets</u>, as seen in Figure 2a, then the approach to measure uncertainty is probability theory. It weighs the amount evidence available for each set. Thus, the probability measure is additive, $P(A \cup B) = P(A) + P(B)$, when $A \cap B = \varnothing$. This additive characteristic allows various operations of probabilities without no and consistent with the classical .

<u>When the individual pieces of evidence point to nested sets,</u> as seen in Figure 2b, then the truth of the proposition cannot be measured easily. Two measures are considered. One weighs the evidence that exclusively points toward the proposition. The other weighs all the evidence including the one that even partially supports the proposition. The measure based on the former

view is called the necessity measure, and the latter view is called the possibility measure. The necessity measure represents the conservative view, and the possibility measure represents the optimistic view. The mathematical theory that formalizes uncertainty in this case is possibility theory.

The type of evidence that is relevant to this paper is the fuzzy set. For example, if the available information is "travel time is *approximately* 30min.", and the proposition is "arrive within 40 min.", then the truth is measured by possibility and necessity. This scenario can be expanded to include the case of a fuzzy proposition, such as the possibility that one "arrives within *approximately* 40 min.," given that the "travel time is *approximately* 30min." in the above context.

Shackle (1969), Cohen (1970), Klir (1985), and others have argued that estimations and the notion of "desire" are usually made in terms of possibility, not probability, and hence, possibilistic representation is appropriate. This view comes also from the structure of possibility distribution in which amalgamation of ranges, rather than the frequency of particular events, is considered, as shown in Figure 2.

When evidence points to a mix of nested sets and also to well-defined sets, as seen in Figure 2c, uncertainty is formalized by Dempser-Shafer theory. Under this theory, the truth of a proposition is still measured by weighting two kinds of evidence, one that points to the proposition exclusively, and the other that points to the proposition and also its complement. These two measures of truth are called belief and plausibility measures, respectively. Given the three frameworks below, we believe that possibility theory is useful in dealing with uncertainty related to a range of numbers, which arises in many travel-related decisions. Possibility theory can handle the truth of a proposition in the face of vague evidence or information. Hence, the remaining part of this paper uses possibility theory to deal with uncertainty related to travel decisions.

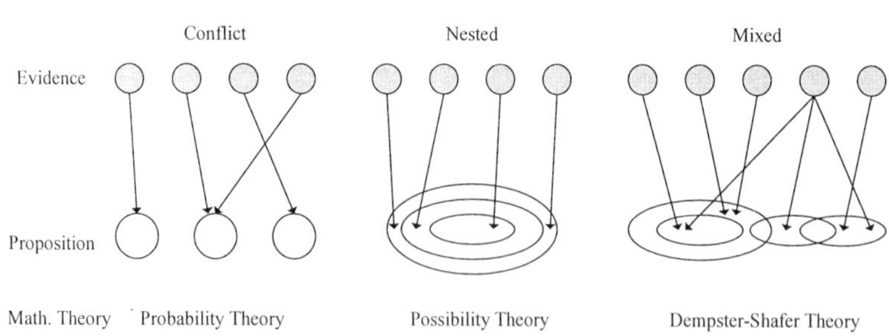

Figure 2 Three Patterns of Evidence – Proposition Relation and Corresponding Mathematical Frameworks

Measures of Feasibility of a Travel Plan: Possibility and Necessity Measures

A basic question that one asks when making a travel plan is "If I leave now, will I arrive at the destination at the desired time?" or "In order to arrive at the desired time, what time should I leave?" The answers to these questions are found in the confluence of the estimated arrival time and the desired arrival time. In essence, it is to find the truth-value of a proposition, "arrival at the desired time" in the face of the information of the estimated travel time.

If the estimated arrival time is represented by a fuzzy set, the membership function of this fuzzy set induces the possibility distribution, $\pi(x)$, (Zadeh [1978]); hence, $h_A(x)$ and $\pi(x)$ are numerically equal: $h_A(x)=\pi(x)$. We now consider $\pi(x)$ as the mathematical representation of evidence, a counterpart of probability density function for the random events. The match between the possibility distribution of the estimated arrival time and the fuzzy set for the desired arrival time indicates the possibility that one arrives within the desired arrival time..

In the following we describe the theory of possibility briefly. Given evidence E and a proposition A, the possibility measure, Poss(A), considers all non-negative evidence and measures the truth.

$$\text{Poss}(A)=1 \text{ if } A \cap E \neq \emptyset\text{; and Poss}(A)=0 \text{ otherwise.} \quad (1)$$

The necessity measure, Nec(A), considers only positive evidence, or the evidence that exclusively points to A, and measures the truth by

$$\text{Nec}(A) =1 \text{ only if } E\subseteq A\text{; and Nec}(A)=0 \text{ otherwise.} \quad (2)$$

These measures are axiomatically defined as

$$\text{Poss}(A\cup B) = \text{Max } \{\text{Poss}(A), \text{Poss}(B)\}, \text{ and Nec}(A\cap B)=\text{Min } \{\text{Nec}(A), \text{Nec}(B)\} \quad (3)$$

Further, Poss(A) and Nec(A) are related by a dual relationship:

$$\text{Poss}(A)= 1- \text{Nec (not A) or } \quad \text{Nec }(A) = 1-\text{Poss(not A)} \quad (4)$$

In other words, possibility of A counts all evidence except for "necessarily not A"; conversely, necessity of A means "impossibility of not A."

When the given possibility distribution of a shape in Figure 1, $\pi(x)$, and for a crisp set A (or a hard bound set), the measures are

$$\text{Poss}(A) = \text{Max } \pi(x) \qquad \text{for } x\in A,$$

$$\text{Nec (A)} = \text{Min } (1-\pi(x)) \quad \text{for } x \notin A. \tag{5}$$

When A is a fuzzy set whose membership function is expressed by $h_A(x)$, and when the evidence is possibility distribution, $\pi(x)$,

$$\text{Poss(A)} = \text{Max Min } [\pi(x), h_A(x)] \qquad \text{for } x \in X$$
$$\text{Nec (A)} = \text{Min Max } [(1-\pi(x)), h_A(x)] \quad \text{for } x \in X \tag{6}$$

Incidentally, if we assume that the estimated arrival time follows a probability density distribution, then the probability of arriving at the desired time, A, can be derived only for the case that A is a hard range. It is the area of the probability density function corresponding to set A. However, if A is a fuzzy set, such as A= " *approximately* 3p.m.", which is often the case of human travel decision, then probability theory cannot measure the truth of such a proposition. This aspect is discussed in detail in Kikuchi and Pursula (1998).

Figure 3 illustrates the possibility and necessity measures of "arriving" and "not arriving" at a desired time. The membership function of the desired arrival time has a triangular shape for the sake of simplicity of explanation. The estimated arrival time is represented by a fuzzy set whose membership function has a triangular shape. The figure shows how to compute the possibility and necessity measures of "arriving within the desired time" and "not arriving within the desired time." The height of the intersection of two distributions at point Z represents possibility measure. This is consistent with Eq(6). In addition, the duality between possibility and necessity measures, agrees with Eq(4).

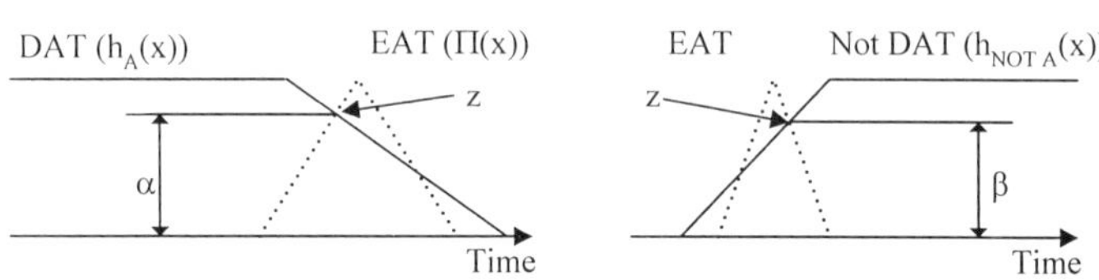

Pos (A) = α
Nec (A) = 1-β

Pos (NA) = β
Nec (NA) = 1-α

DAT: Desired Arrival Time: $h_A(x)$
EAT: Estimated Arrival Time: $\pi(x)$

DAT and EAT are a fuzzy set and a possibility distribution, respectively, and they are expressed using a triangular-shaped membership function, $h_A(x)$ and $\pi(x)$

Figure 3 Calculation of Possibility and Necessity of Arrival (A) and Non-Arrival (NA)

Measure of Anxiety When Selecting Departure Time

When the information on the estimated arrival time and the desired arrival time is not specific, then deciding on the time of departure is associated with a feeling of anxiety or uneasiness. The intensity of anxiety depends on the degree of specificity of the information. The less uncertain the information is, the greater the anxiety becomes. How anxiety manifests itself in relation to the specificity of information is the aspect that the conventional choice model has not been able to incorporate, because its mathematics does not question the quality of information. This section attempts to formalize the measure that represents decision difficulty or anxiety in the face of vague information.

At time t, decision to "to leave" or "not to leave yet" is controlled by the support for the proposition "arrive within the desired time range" if leaving at time t. If there is sufficient evidence that one will arrive within the desired time, then he will not leave yet. If there is little evidence that one will arrive within the desired arrival time, then he will leave. The possibility and necessity measures provide the amount of evidential support for these alternatives in relation to the quality of information about the desired arrival time and estimated travel time, as illustrated above.

The anxiety or decision difficulty can be symbolized by the conflict between the supports for different decision options. If both the estimated travel time and the desired arrival time are not specific, then the supports for both options are low; then, the traveler experiences anxiety as to which to choose, "to leave" or "not to leave," If, on the other hand, only one option is clearly supported and the other option is not supported, then anxiety is minimal and the decision is simple.

We measure possibility and necessity for "arrive within the desired time " or "not arrive within the desired time" for a given departure time t, and consider these values as the supports (or impelling forces) for the alternatives. We then use the approach suggested by Yager(1982) to model anxiety associated with selecting the departure time. In this approach, anxiety is measured by the cardinality of the decision set, which is characterised by a fuzzy set. The elements of this fuzzy set are the decision alternatives (d_i). The grade of membership of each element (decision option) is the strength of support for each decision alternative (γ_i). Thus, the membership function of this decision set is written as (γ_1/d_1, γ_2/d_2, γ_3/d_3..). A greater value of cardinality of this fuzzy set means more supporting evidence for each alternative; thus selecting one alternative becomes difficult, or anxiety increases. The degree of anxiety is the highest, however, when there is no support for any of the alternatives. It is the lowest when only one action is supported and all other actions are not supported.

The value of Yager's anxiety measure lies between 0 and 1, and it is expressed as

$$A_x = 1 - \int_0^1 \frac{1}{|D_\alpha|^\rho} d\rho \qquad (7)$$

where A_x= degree of anxiety given information x;

D_α is an α-cut of the fuzzy set, which is actually a crisp set; a collection of elements whose membership grades are greater than α. $|D_\alpha|$ is the cardinality (the number of elements) of this decision set for the membership grade is α. This is the number of alternatives whose membership grades (or support) are greater than α. $|D_\alpha|$ is large, then A_x becomes smaller.

If there are only two choices, such as our case "leave" or "not leave", the above expression is reduced to the following. (The derivation of the above is shown in Appendix.)

$$A_x = 1 - \text{Max}(M_L, M_{NL}) + \tfrac{1}{2}[\text{Min}(M_L, M_{NL})] \qquad (8)$$

where M_L corresponds to the amount of support (or impelling force) for "to leave now", and M_{NL} corresponds to the amount of support (or impelling force) for "not to leave now."

Perincherry (1994) illustrates the anxiety as a function of M_L and M_{NL} as shown in Figure 4. Based on this figure, the anxiety measure takes the highest value (=1) when both measures, M_L, M_{NL} are equal to 0; in other words, no information supports either to "leave" or "not leave." This is the situation of dilemma. Anxiety is the lowest when information supports just one of the options and not the other.

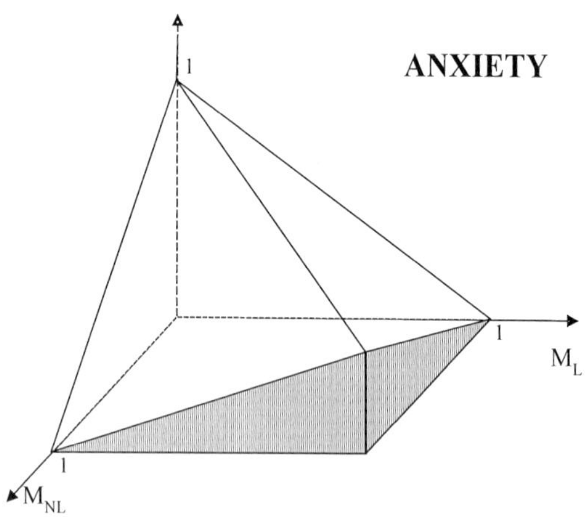

Figure 4 Anxiety Measures as a Function of Two Decision Alternatives

The values of M_L and M_{NL} at a given time t, vary depending on the attitude of the decision-maker. The optimistic person who likes to suppresses uncertainty may decide to "leave now" based only on the necessity measure of "arriving late". He decides "not to leave now" as long as there is possibility of arriving (if he leaves now). In other words, he decides to leave, only when absolutely clear that, he will be late unless leave now. In this discussion we use the optimistic person in the same sense as a procrastinator, who decides to take action only at the time when absolutely necessary to do so. In summary,

For the optimistic person,

Leave:
based on necessity of late arrival = Nec (not to arrive on time)= Nec(NA) = M_L
Not leave:
based on the possibility of arrival = Poss (to arrive on time) = Poss (A) = M_{NL} (9)

The conservative person, on the other hand, acknowledges the uncertainty and follows a safe strategy. He may decide "to leave now" based on the possibility of arriving late, and "not to leave now" based on the necessity of "arriving on time", or absolutely sure he will arrive. In summary,

For the conservative person,

Leave:
based on possibility of late arrive -= Poss (not to arrive on time)= Poss(NA)= M_L
Not leave:
based on necessity of arrival on time = Nec (to arrive on time)= Nec(A)= M_{NL} (10)

For each type of person, M_L and M_{NL} are evaluated, and anxiety is computed using Eq(8) along the axis of departure time. It must be noted that both the optimistic person and the conservative person desire to postpone the departure time as much as possible. However, the optimistic person wait until it becomes absolutely *necessary* to leave (otherwise he will be necessarily late); the conservative person, on the other hand, wait only until it becomes *possible* that he will be late.

NUMERICAL EXAMPLES AND DISCUSSION

The following values are used to demonstrate the level of anxiety associated with the selection of departure time. All the times are measured from a reference time point (t=0). The assumed values for the three basic times are

Estimated travel time = *approximately* 40 min. This is a fuzzy set, and its membership function is a triangular shape. The base of the triangle is 40 min ± c min. The value of c can take 3, 5, 10, and 15 min for the sensitivity analysis sake.

Estimated arrival time = t + Estimated arrival time. This is also a fuzzy set. Since this is used as the evidence of possible arrival time, this time is represented by a possibility distribution, which has a triangular shape.

Desired arrival time = *approximately* 80 min. This is a fuzzy set and its membership function is a semi-triangular shape; the membership grade is 1 at 80 min, and it decreases to the right and becomes zero at 80+d min. The value of d can take a value of 3, 5, 10, or 15 min in the sensitivity analysis.

We now compute anxiety for the optimistic person and also for the conservative person for selected departure time at t. It is computed for different combinations of the values of d and c, and also the value of t (departure time). c and d are the sizes of the ranges for the desired arrival time and estimated arrival time, respectively.

Figure 5 shows all possible combinations of the locations of the desired arrival time (DAT) and estimated arrival time (EAT) (both are expressed in a range). These combinations are the results of different departure times. The possibility and necessity of "arrival (A)" and "non-arrival (NA)" are calculated for each pattern of overlap between EAT and DAT according to the illustration in Figure 3.

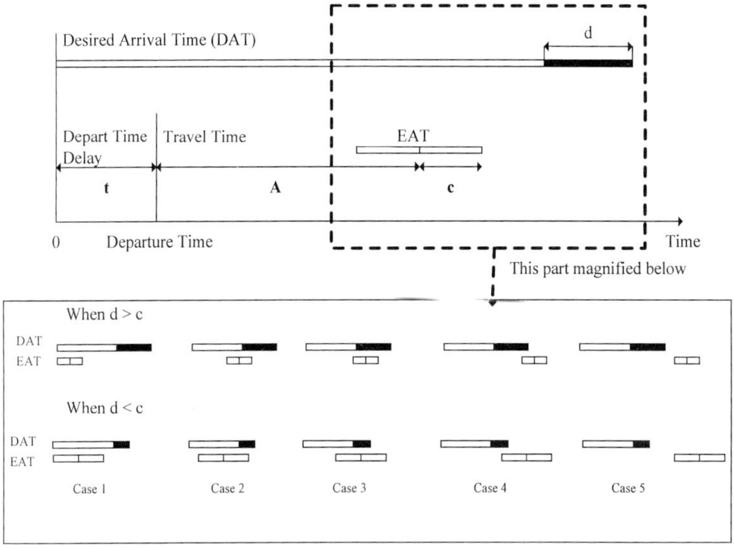

	CASES				
	1	**2**	**3**	**4**	**5**
Poss (A)	1	1	α	α	0
Nec (A)	1	1−β	1−β	0	0
Poss (NA)	0	β	β	1	1
Nec (NA)	0	0	1−α	1−α	1

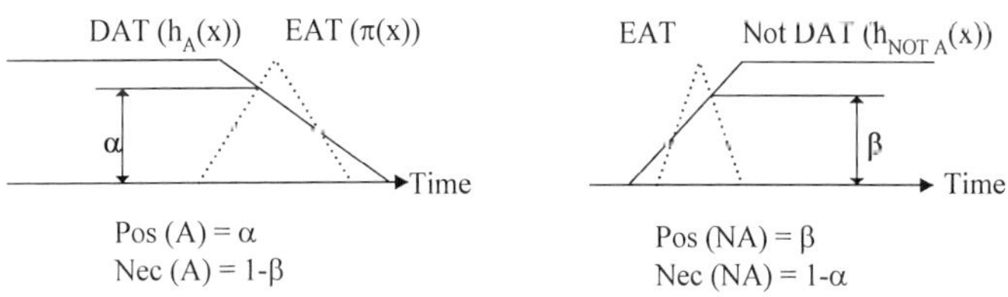

Figure 5 Patterns that Estimated Arrival Time matches Desired Arrival Time and Corresponding Possibility and Necessity Measures

Figure 6 shows the possibility and necessity measures of arrival and non-arrival for different departure time, t, and for different combinations of values of c and d. It is seen that generally Poss(A) and Nec(A) are large when the departure time is early (small t values). But, these values decrease as t increases. At the same time Poss (NA) and Nec (NA) increase as t shifts to the right.

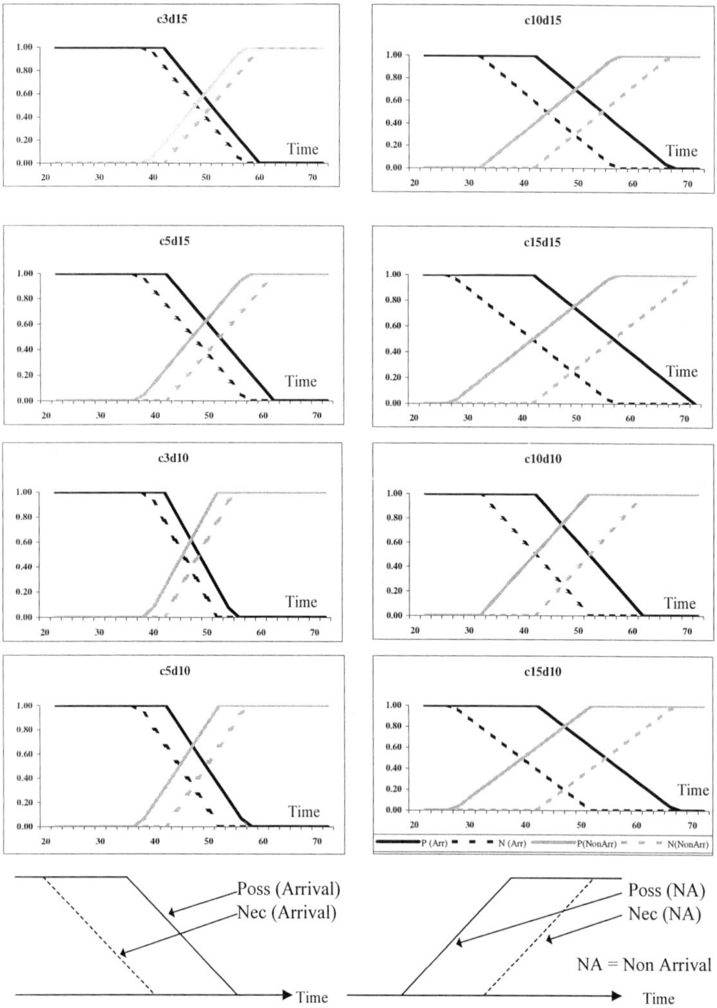

Figure 6 Possibility and Necessity Measures of Arrival and Non-Arrival for d=(10, 15) and C=(3, 5, 10, 15)

Given Poss(A), Nec(A), Poss(NA) and Nec (NA) at a given t, we now can calculate the traveler's anxiety levels according to Eq (9) and (10). Anxiety levels are shown for the two types of travelers, optimistic and conservative, and for different combinations of the values of d and c. Figures 7, 8, and 9, each shows the anxiety level for three sets of d and c values , (d=3min, and c= 3, 5, 10, and 15min), (d=10min, and c= 3, 5, 10, and 15min) and (d=15min, and c= 3, 5, 10, and 15min), respectively.

In each of these figures, the top two figures show the level of anxiety associated with the corresponding departure time for the optimistic and conservative person. It is seen that the anxiety occurs to the conservative person first along the time axis, and later to the optimistic person.

This is natural because the conservative person tends to leave earlier than the optimistic person in order to make sure that he arrives on time. The optimistic person, on the other hand, does not leave until when, otherwise, necessarily arrive late. The difference between the two types of persons is pronounced when the values of c and d are large; in other words, when the estimated travel time and desired arrival time are not specific. But, as the information becomes more specific (the values of c and d become smaller), the region of anxiety becomes smaller.

The bottom chart in each of Figures 7, 8, and 9 shows the time ranges where anxiety occurs by projecting the x-axis of the top two charts. It is seen that as the value of c decreases, in other words, as the estimated travel time becomes more specific, the range of anxiety decreases. When the value of d is smallest (d=3min in Figure 7), the range of anxiety is the smallest. This means that both the desired arrival time and the estimated travel time are specific, the travelers' anxiety with respect to the departure time selection decreases and eventually vanishes.

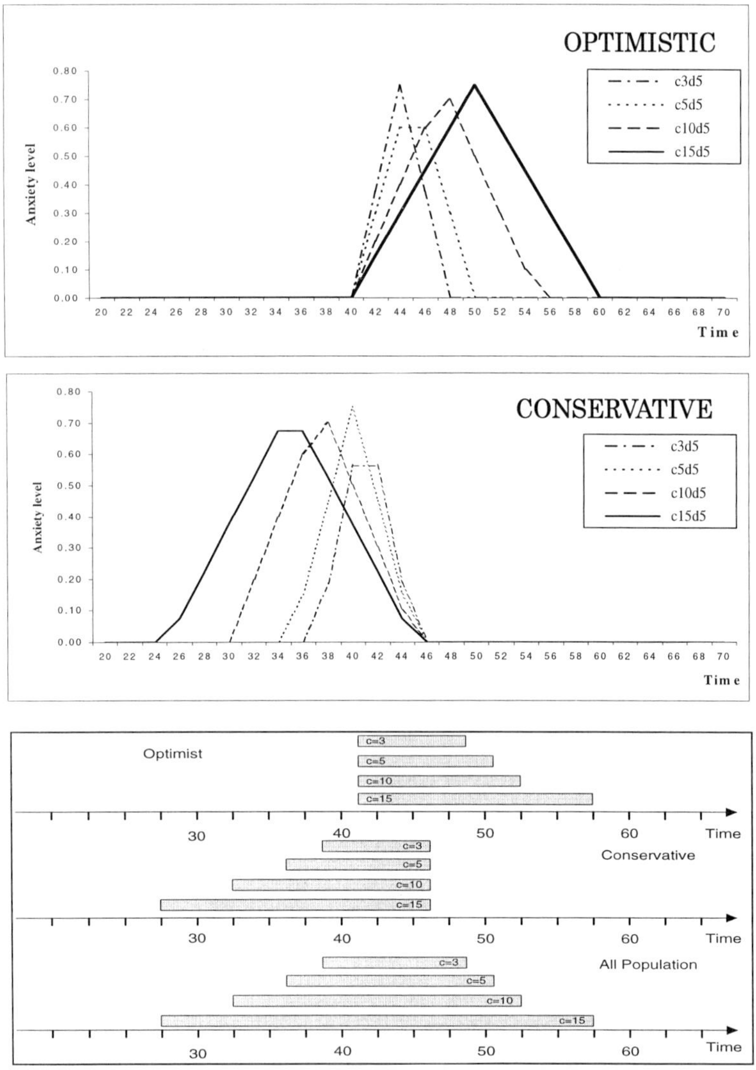

Figure 7 Anxiety levels of optimistic and conservative persons when d=5 and c=3, 5, 10 and 15 minutes

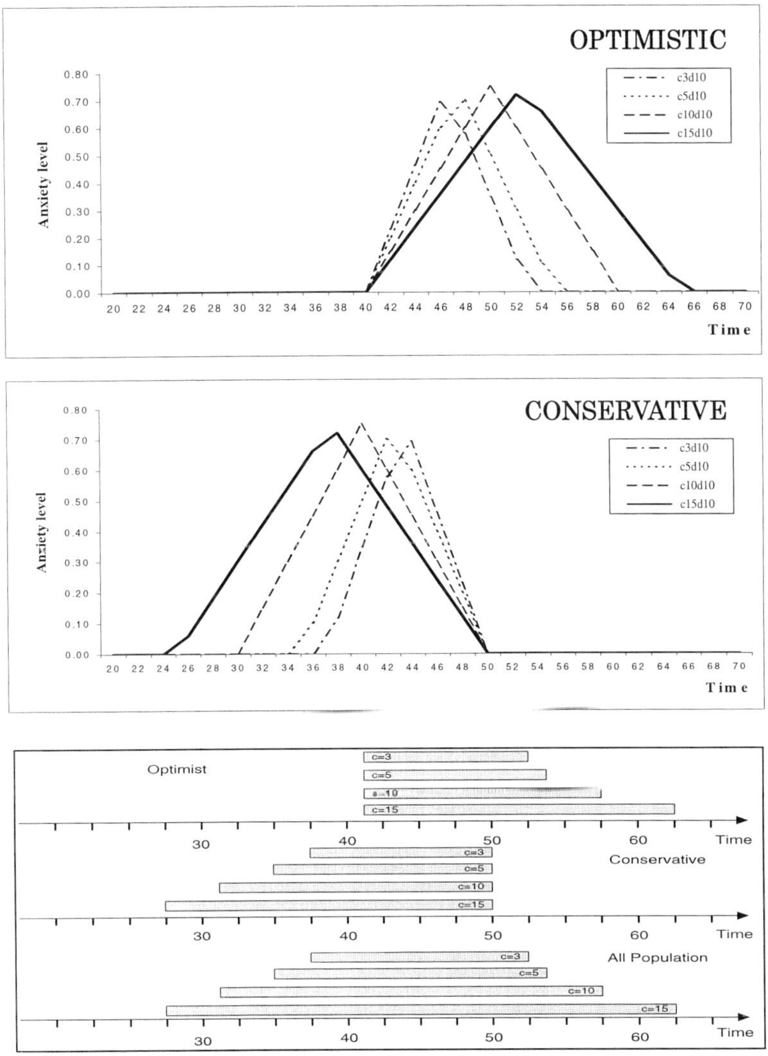

Figure 8 Anxiety levels of optimistic and conservative persons when d=10 and c=3, 5, 10 and 15 minutes.

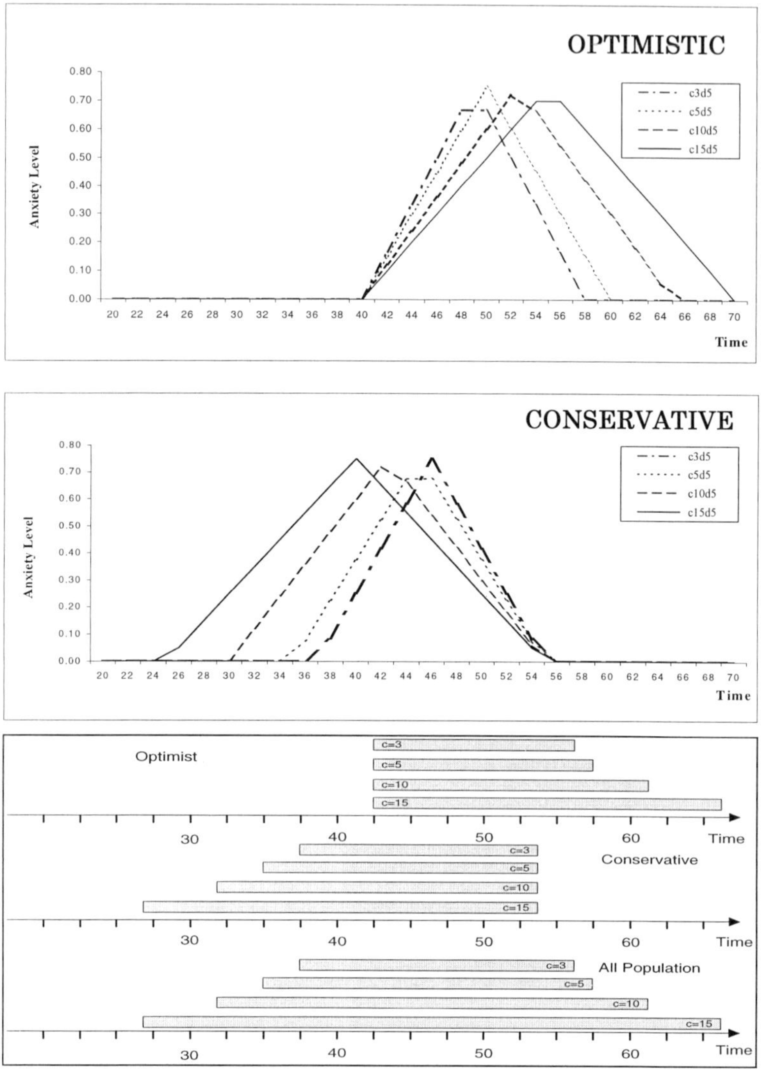

Figure 9 Anxiety levels of optimistic and conservative persons when d=15 and c=3, 5, 10 and 15

The two types of persons, optimistic and conservative, are the extreme cases of attitude. Most travelers fall between these two types. Thus, the anxiety ranges shown in the bottom charts of Figures 6, 7 and 8 should be the range where most persons experience anxiety. Let us examine the above findings. As the information becomes more specific to the traveler, anxiety reduces, and the range of departure time narrows. Further, the difference between the optimistic and conservative persons becomes irrelevant; they both behave the same under the specific information. However, if the information is less specific, the difference between the two types of persons is pronounced, and the time-range where the two types of persons experience anxiety becomes different. We can safely assume that the rational traveler decides to depart during the region of anxiety, because outside this region he will arrive too early or too late to the desired arrival time. Hence, the anxiety measure can provide a mathematical framework that identifies the departure time range in the face of the incomplete information of the travel time and desired arrival time.

A question may arise about the consequence of refinement of information, which is generally the aim of ITS. Assume that travelers share a road section from point A to B, where A is a sort of merge point of a set of branches from different origins, and B is the common destination. All travelers are supposed to wish to arrive at the destination at the same time (say, the common work start time). Assume also that very accurate information on the travel time is provided for the travel from A to B. In this case, the departure time from A can be defined; hence all travelers plan to arrive at A at the same time. Because of the exact information, all travelers behave the same, and the attitude difference among them is diminished. As a result, traffic volume between A and B will be concentrated within a small time interval. This is not necessarily a desirable phenomenon. The greater the concentration of traffic volume, the greater level of congestion, which subsequently, increases the travel time and also makes the prediction of travel time less reliable.

On the other hand, if the information on travel time between A and B is not too specific, then the optimistic person and conservative person have a difference in selecting the departure time, and, as a result, the departure times of the travelers spread. This creates a more or less even traffic volume during the time period, and the congestion level is also lessened and more reliable and shorter travel time is achieved. This paradoxical situation in which the more specific the information becomes, the less reliable travel conditions, may become a subject which requires further study.

CONCLUSIONS

A seemingly intractable problem of the effects of uncertainty on traveler's decision pattern can now be formalized, and the interplay between the quality of information and decision anxiety can be analyzed in the context of possibility theory. The feasibility of the decision objective is measured by the truth-value of the proposition in light of evidence. This framework incorporates two views, possibility and necessity measures, or the optimistic and conservative views, on the truth of the proposition. These two views emerge naturally in the face of uncertainty.

The possibility theory framework is found useful when modeling departure time selection under the incomplete information on the estimated travel time and desired arrival time. The possibility and necessity measures of "to arrive" and "not arrive" (if leave now) provide the degrees of support for "to leave now." or "not to leave now."

By applying Yager's anxiety measure, which considers decision difficulty caused by the conflicting forces supporting different decision options, we can now identify the degree of anxiety (as to whether to arrive on time or not) with respect to the selection of the departure time. We find that the time periods in which the traveler experiences anxiety are different between the two attitudes. The conservative person experiences anxiety first, and later along the time axis, the optimistic persons experiences anxiety. The size of the range depends on the specificity of the information. The more specific the information, the less the size of the range becomes; in other words, the time within which departure time is selected is narrowed. These ranges have turned out to be the logical time periods within which each type of traveler leaves in light of the incomplete information.

Thus, the proposed framework can serve as a tool to design the travel information, which is the focus of today's ITS efforts. This paper has dealt with the development of the mathematical framework. The forthcoming task is to examine the usefulness of the model in view of the actual data on human travel behavior. The values of the model parameters, such as the membership functions of the travel time, need to be calibrated. This process may be facilitated by equating the range of actual departure time of the travelers to the time period identified by the proposed framework in this analysis, or alternatively, by a survey in which the individuals are asked about the departure time under different hypothetical scenarios of information accuracy. Another task that needs to be preformed is more philosophical. It is to examine the potentially paradoxical situation, in which the better the information becomes, the worse the travel condition may become due to the concentration of traveler at a particular time period. Such a study will reveal that the utility of information to a traveler ultimately depends on his attitude towards handling uncertainty. This may suggest that, for ITS development, various forms of information in terms of quality, expression, and format, as well as the specificity, need to be considered.

ACKNOWLEDGEMENTS

The authors express thanks to Mr. Peter Pilko for preparing the calculation and graphics.

Model of Anxiety

Notation

d_1 and d_2: alternatives 1 and 2

$a1$ and $a2$: degree of support for alternatives 1 and 2, respectively.

A: anxiety measure [0, 1.0]

Anxiety

$$A = 1 - \int_0^1 \frac{1}{\left|\widetilde{D}\right|^{\rho}} d\rho$$

In the case of 2 alternatives

$$A = 1 - \int_0^1 \frac{1}{\left|\widetilde{D}\right|^{\rho}} d\rho$$

where $\widetilde{D} = a_1 / d_1 + a_2 / d_2$

If $a_1 > a_2$

$$\begin{cases} A = 1 - \left[\int_0^{a_2} \frac{1}{2} d\rho + \int_{a_2}^{a_1} \frac{1}{1} d\rho \right] \\ = 1 - \left[\frac{1}{2} a_2 + 1(a_1 - a_2) \right] \\ = 1 - \left[\frac{1}{2} a_2 + a_1 - a_2 \right] \\ = 1 + \frac{1}{2} a_2 - a_1 \end{cases}$$

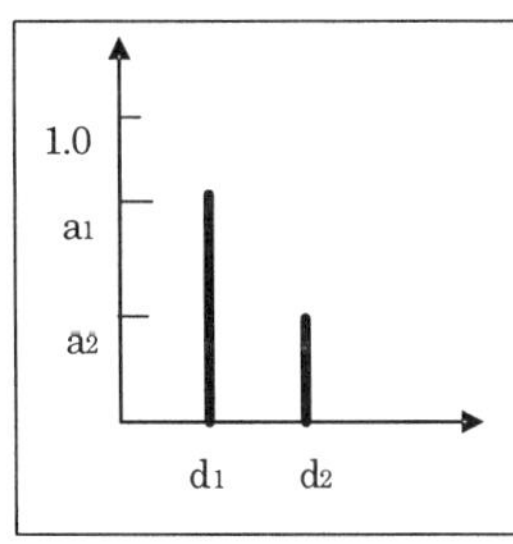

If $a_2 > a_1$

$$\begin{cases} A = 1 - \left[\int_0^{a_1} \frac{1}{2} d\rho + \int_{a_1}^{a_2} \frac{1}{1} d\rho \right] \\ = 1 - \left[\frac{1}{2} a_1 + a_2 - a_1 \right] \\ = 1 + \frac{1}{2} a_1 - a_2 \end{cases}$$

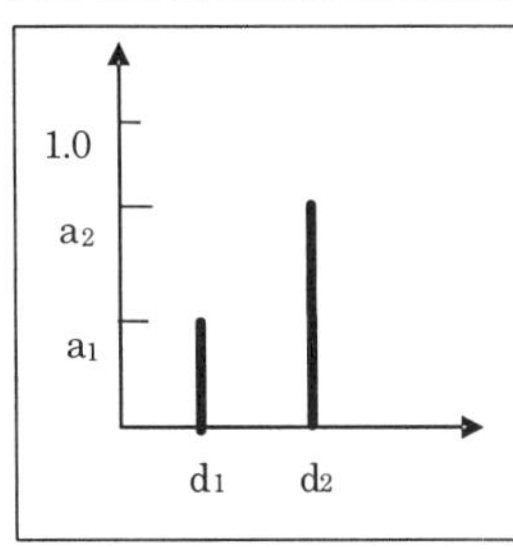

$$\therefore A = 1 - Max[a_1, a_2] + \frac{1}{2} Min[a_1, a_2]$$

REFERENCES

Cohen, I.J.(1970), The Implications of Induction, Methuen, London.

Jaw, J.J., A. R. Odoni, H.N. Psaraftis and N.H.M. Wilson (1986) "A Heuristic Algorithm for the Multi-Vehicle Advance Request Dial-A-Ride Problem With Time Windows", *Transportation Research B*, **20B(3)**, pp 243-257.

Kaufmann, A and Gupta, M.M. (1984) Introduction to Fuzzy Arithmetic, Van Nostrand Reinhold Company, New York.

Kikuchi, S. and Pursula, M., (1998) Treatment of Uncertainty in Study of Transportation: Fuzzy Set Theory and Evidence Theory, *Journal of Transportation Engineering, American Society of Civil Engineers*, **Vol. 124**, No. 1, pp1-8.

Kikuchi, S. and J.H. Rhee (1989) "Scheduling Method for Demand Response Transportation System," *Journal of Transportation Engineering*, **Vol. 115**, No. 6, pp 630-645.

Kikuchi, S. and Donnelly, R.A.(1992), "Scheduling Demand Responsive Transportation Vehicles Using Fuzzy Set Theory" *ASCE Journal of Transportation.*

Klir, G. J., (1985), Architecture of System problem Solving, Plenum Press, N.Y.

Klir, G.J. , St. Clair, U.H., and Yuah, B.(1997), Fuzzy Set Theory: Foundation and Applications, Prentice Hall, Upper Saddle River, NJ.

Klir, G. and T. Folger (1988), Fuzzy Sets, Uncertainty and Information, Prentice Hall, Englewood Cliffs, New Jersey.

Perincherry, V.(1994), Analysis of Large Scale Systems, Ph.D. Dissertation, University of Delaware, Newark, Delaware.

Shackle, G. L. S. (1969), Decision, Order and Time in Human Affairs, Cambridge University Press, London.

Solomon, M.M. and J. Desrosier (1988) "Survey Paper: Time-window Constrained Routing and Scheduling Problems," *Transportation Science*, **22(1)**.

Yager, R, R. and Filev, D. P.(1994), Essentials of Fuzzy Modelling and Control, John Wiley and Sons, Inc. New York.

Yager, R. R.(1982) , Measuring Tranquillity and Anxiety in Decision Making: An Application of Fuzzy Sets, *International Journal of General Systems*, **8**, pp139-146.

Zadeh, L. A. (1978), "Fuzzy Sets as a Theory of Possibility," *Fuzzy Sets and Systems*, **Vol. 1**, pp 3-28.

Zimmermann, H.J.(1996), Fuzzy Set Theory and Its Applications, Third Edition, Kluwer Academic Publishers, Norwell, Massachusetts.

9

Traveller's Behaviour under Uncertain Conditions

Henk J. van Zuylen, Delft University of Technology, The Netherlands
Shinya Kikuchi, Department of Civil and Environmental Engineering, University of Delaware, Delaware, USA

Abstract

Travel behaviour depends on the travellers' perception of the trip characteristics. The uncertainty of the journey time is one of the determinants of the choice. Studies of decision behaviour in uncertain conditions show the limited validity of the common assumptions in traffic models. Decision making under uncertainty has to be considered as a contingent process, depending on the objectives and conditions with which the choice is made. Expected utility is one of the many possible criteria used by people to decide. There is a discrepancy between the objective and subjective value of this concept. This is due to the bias with respect to the perception of very likely and very unlikely events. In many cases the expected utility is a less applicable objective, e.g. because people want to arrive before a certain deadline and maximize the probability to achieve that. The fact that probabilities of travel times have to be learned from experiences introduces a dynamic character of choice behaviour.

Most of the reported properties of decision-making under uncertainty still have to be verified for travel behaviour.

INTRODUCTION

The choices made by travellers with respect to their destinations, travel mode, departure time and routes, are believed to be governed by a set of simple incentives, such as travel costs, travel time, comfort and their combination. In economic terms, the traveller chooses the trip that maximizes the utility and minimizes the disutility. Disutility is assumed to be a (linear) combination of time and out-of-pocket costs. When choices have to be made under uncertain conditions, it is assumed that one chooses the alternative with a maximum expected utility assuming that the utilities of different alternatives have a stochastic character. However, since people seem to be biased in estimating the expected outcomes of their choice, the expected utility is not a realistic indicator for choice behaviour. Furthermore, because people learn from their experiences, their travel behaviour is only partly consistent over time, and hence, the equilibrium assumption of most economic theories is not applicable. Equilibrium means that the state of the system (e.g. the flows in a transport network) does not change over time if the external conditions remain the same. Of course, several systems exist which show periodic behaviour but most economic systems are assumed to converge to a stable state. Adaptive behaviour of travellers may even lead to periodic or quasi-periodic dynamics, from which no stable state emerges.

A final mechanism that influences decision-making in uncertainty is that people differ in their preferences and attitudes, depending on the decision conditions. These variations in choice behaviour are more than just a statistical phenomenon. It is a result of contingent decision-making, i.e. decision making that depends on the objectives of and conditions for the traveller, and the degree of uncertainty (or available information). Therefore, the choices made are determined by the expected outcome and its variance and the decision-maker's objectives and attitude.

In this paper, we review some recent psychological studies on decision-making under uncertain conditions and identify characteristics that are not always consistent with the assumptions of the probabilistic behavioural modelling. The approach of the reviewed studies is applied to route choice in networks with uncertain travel time and inaccurate travel information.

THE PROBLEM

People make choices in the confluence of the decision objective and the available information on the alternatives. When it comes to travel decisions, one usually makes choices under an uncertain environment because the events during the trip, in particular the travel time, are in the future and not known. The traditional approach that treats uncertainty in such situation has been to use probability theory assuming that the occurrences of future events are due to randomness. Random utility theory considers that one chooses the alternative with a maximum expected utility.

Various studies show that human behaviour under uncertainty does not exactly match the premise of the random utility theory, both logically and experimentally. Among the possible reasons for the mismatch are: (1) uncertainty is imbedded in the perceptions both of the objectives and of the available information about the alternatives, i.e. individuals make their choice about which they do not have certain knowledge with respect to the alternatives and the consequences of their choice for their objectives and, on the other hand, the modellers do not have certain knowledge about what the objectives of the individuals are; (2) each decision maker is different in the perception and interpretation of incomplete information; and (3) decisions made under uncertainty are not always rational in economic terms.

All these reasons are applicable for the decision making in travelling. The second reason, the differences in perception, is dealt with in the random utility theory for route choice (Domincich and McFadden 1975). For the first and second reason no alternative choice models have been developed yet.

In the traditional approach, the modeller analyses the choice phenomena of the subjects objectively and models the behaviour of the individuals from the statistical standpoint. In this case, the behaviour of individuals may appear to be the random phenomenon, and the uncertainty comes from the modeller's inability to predict the behaviour of the research subjects, or human subjects. During the choice process, the individual humans face the uncertainty due to the lack of information on the alternatives and due to the lack of clear decision criteria. In this case, the individual does not make decisions randomly; rather it is believed that each individual in each situation possesses some unique mechanism to cope with uncertainty. Perception of information and objectives when one makes travel decision - which in most cases, is not a critical decision of one's life - is believed to be an interval or a range and the decision process is the approximate reasoning process. This aspect of dealing with uncertainty is not reflected in the traditional probabilistic choice model so far.

Modelling decision patterns under uncertainty, especially under different types of information, is very important in the development of the Intelligent Transportation Systems, whose effectiveness depends on the interactions among information and decision. We feel that there is no reason that probability theory be the only framework that can deal with human choice modelling. Because probability theory can deal with only one type of information, probability distribution, and follows the rigorous mathematics, its utility is limited when dealing with an interval or approximate values or natural language based information. Different modelling frameworks for different types of information should be considered, if appropriate. There is another way of treating uncertainty if it is generated due to the imprecise or vague information perceived by the decision-maker. Fuzzy sets and possibility theory can be superior alternative paradigms to deal with uncertainty (Kikuchi and van Zuylen 2001).

We explore in this section some recent findings on the human choice behaviour - that often defies the principle of maximum expected utility - and examine the discrepancy between the theory and the reality.

Some Recent Studies of Decision Making under Uncertain Conditions

One who has to make decisions under uncertain situations needs a strategy to cope with it.

A common approach in transport models is to assume that travellers apply the strategy of weighted choice, in which unlikely outcomes have less influence on a decision than outcomes with a high probability. The assumption that people are able to take decisions based on expected utility of a choice appears to be invalid, due to a bias in the perception of probabilities and the limited ability of humans to have a good understanding of probabilities if partial information is available (Wu and Gonzales 1998).

Decision-makers seem to behave in an irrational way, if we assume that they behave as utility optimisers. In reality, maximum expected utility, even with the bias as mentioned before, is not always the objective of the decision-maker. Depending on the objectives of the decision-maker, (s)he may show a preference for one of the following choices: a large variance in expected outcomes, a small variance, or a high expected outcome (Rode et al. 1999). The strategy of the decision maker with respect to information about the uncertain choice depends on the perception of the choice situation (Lipshitz and Strauss 1997). One of the issues related to choice is the learning behaviour of decision-makers under uncertain situations. The adaptation of behaviour depends on experience, but the learning strategy seems also to change after more experiences (Barkan et al. 1998).

In the following sections, we review the recent research results on decision-making under uncertain conditions mentioned above and try to generalize the results to decision making of travellers.

Contingent Decision Making under Uncertainty

Depending on how decision-makers perceive the uncertainty about their decisions, they have different strategies. Lipshitz and Strauss (1997) investigated decision making in uncertainty and found that the following strategies were applied:

1. Reducing uncertainty, e.g. by collecting additional information, extrapolation (assumption based reasoning), mental simulation (scenario building), reducing the time horizon, selecting one's interpretation of ambiguous information,
2. Acknowledging uncertainty, e.g. by choosing options with clear outcomes, choosing a flexible strategy, prepare responses for possible outcomes, avoid irreversible actions,
3. Suppressing uncertainty, by denial of uncertainty, rationalizing and symbolic actions,
4. Weighted pros and cons.

The choice of strategy applied by the decision-makers depends on their interpretation of the situation or the nature of available information. Lipshitz and Strauss distinguish the following possible interpretations:

1. Lack of information,
2. Inadequate understanding of the information,
3. Conflict in choices or information.

The strategies preferred by decision-makers depend on their perception of the situation:

- In the case of a lack of information: assumption based reasoning,
- If they think that they have inadequate understanding: reduction of uncertainty,
- If there are conflicts: weighted pros and cons.

These results of Lipshitz and Strauss are especially interesting in the assessment of the value of travel information. For instance, if a traveller thinks that he can get no reliable information about the arrival time of the next bus, he will assume that the bus will arrive according to the time schedule. If a driver believes that no reliable information is available about the congestion on a road, he will assume that the condition will be the same as the day before. Information provision for these travellers first has to convince them of the validity and applicability of the information, otherwise these people will stick to their assumptions.

If a traveller believes that he does not have sufficient information or that he does not understand the information, he will ask for clarification or additional information. For such travellers, the value of travel information will be high, especially if the additional information will help them to reduce their uncertainty.

Often travellers receive information that is multi-dimensional: some aspects of a choice are positive, some others negative. For example the following choices may exist: choice 1 gives a low expected travel time, but the probability that yields a very short travel time is rather small, choice 2 gives a longer expected travel time, but probability that yields travel time within a given time interval is rather high. The choice depends on the weight given to the expected travel time and to the probability of a short trip. For travellers in this situation the information is valuable, especially if also information is given about probabilities of travel times.

The Biased Probability Perception

Wu and Gonzales (1998) analysed the perception of probabilities. It is known that people overestimate the importance of small probabilities and underestimate the effect of changing probabilities in the middle range. Allais (1979) describes this as a paradox. He observed that the assumption that people maximize their expected utility is not consistent with their behaviour in reality. For example, the gambling behaviour is not consistent with the strategy that the gamblers maximize their expected utility. In the following we consider three gambling situations. Assume that people had the choice between two options, R and S. Each of these options had a probability of payoffs.

Situation 1: R = prob.= 0.05 for payoff \$240, expected outcome \$12, chosen by 38%
S = prob.= 0.07 of payoff \$200, expected outcome \$14, chosen by 62%

Situation 2: R =prob. 0.05 for payoff \$240, 0.30 for \$200, expected outcome \$72, chosen by 65%
S = prob. 0.37 for payoff \$200, expected outcome \$74, chosen by 35%

Situation 3: R =prob. 0.05 for payoff \$240, 0.90 for \$200, expected outcome \$192, chosen by 39%
S = prob. 0.97 for payoff \$200, expected outcome \$194, chosen by 61%

The choice behaviour in situation 2 does not seem very logical, since the difference of 7% (37 – 30%) for payoff \$200 is valued less than the probability of 5% for a payoff of \$240 (Wu and Gonzales 1998).

Tversky and Kahneman (1992) explain the discrepancy by the assumption that people do not use the probability and the possible outcomes, but a weighted probability. The weighted probability is an S-shaped function of the objective probability as shown as Figure 1.

Changes of the probability near the extremes (0 and 1) have a larger impact on the choice behaviour than changes in the middle range of the probability For instance, the change of the probability of a profit (or loss) from 0 to 0.05 is valued higher than the change of probability from 0.40 to 0.45. If the assumption of the expected utility is valid, the change of probability has an equal consequence on the outcome. In practice people will overestimate the change from 0 to 0.05 and underestimate the change of 0.40 to 0.45. It is similar to the law of Weber, which states that the perceptions of changes in a stimulus are inverse proportional to the original value of the stimulus (Miller 1962).

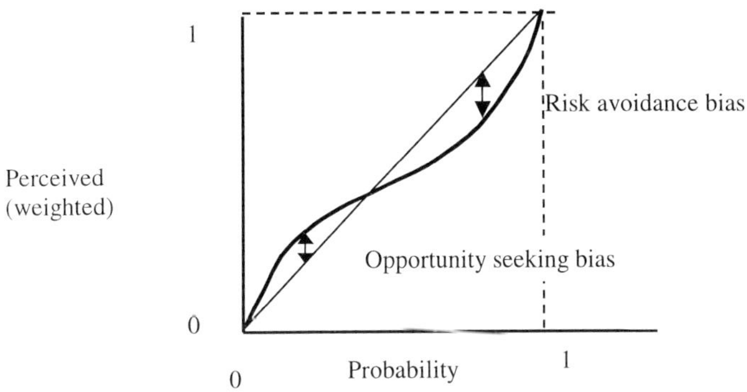

Figure 1 The non-linear relationship between the probability and the perceived probability (Tversky and Kahneman 1992)

Tversky and Kahneman (1992) found an asymmetry in the perception distortion: the risk-avoiding perception, i.e. the overestimation of the risk in the case of nearly certainty, is larger than the opportunity seeking tendency, the overestimation of low probabilities.

For travel behaviour this finding may mean that the expected utility theory as originally developed by Von Neumann and Morgenstern (1944) has a limited validity. The assumption that people maximize their expected utility has to be modified by the fact that the utility is not a simple additive and multiplicative quantity (Bonsall 2000), e.g. the consequence of a 50% probability to have 6 minutes delay is not equivalent to the probability of 1% to have 5 hours delay and the value assigned to such probabilities is normally context sensitive.

Contingent Decision-making: the Dependence of Choices on Objectives.

Rode et al. (1999) surveyed the choice behaviour of students using a hypothetical problem. They verified the hypothesis that people have a contingent strategy in dealing with uncertainty. They found that the following factors determine the strategy:

1. The need,
2. The expected (mean) outcome,
3. The variance of the expected outcome.

The strategy applied by the students appears to be aimed at maximizing the probability of getting what they need. If there are two choices available with the following probabilities, one with small variance but a high expected value, the other large variance with a lower expected value, the strategy followed was to choose the option with the highest probability to satisfy their needs, i.e. option 2 in figure 2, even if the expected outcome was lower than the alternative option. :

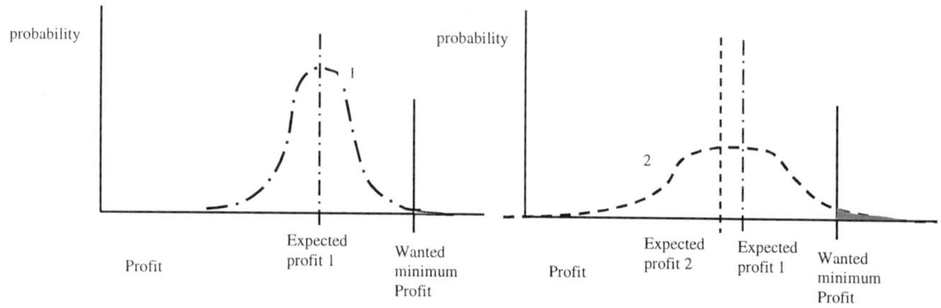

Figure 2 The probability distribution of profit for two choices: choice 1 with a low variance and choice 2 with a higher variance

In practice the interpretation of this result means that if people have a strong desire to get high profit, they become 'risk seeking', i.e. they will take an option with the highest chance of a high profit, even though the risk of a low profit is also high due to the high ambiguity of the situation. If their desire for high profit is low, they will prefer a choice with a less uncertainty in outcome and a profit that has high certainty but possibly a lower expected benefit.

If we put this problem in the context of route choice problem, it would mean that drivers who are in a hurry, would try a route that might be the quickest, but could also be very congested. If they are less hurried, they will choose a route with a longer expected travel time but with a small variance. This hypothesis is not within the scope of the survey done by Rode et al. but it is worthwhile to investigate this kind of travel behaviour. It seems irrational if we consider expected outcomes only, but it is completely rational if we would give the hurried travellers another objective than the normal travellers. In other words, the traditional probabilistic utility model does not take the decision-maker's decision objective or background into account.

Katsikopoulos et al. (2000) found similar outcomes. They executed a survey that showed that the behaviour of participants depends on their goals: to reduce the maximum loss they could suffer or to maximize the possible gain. With the goal to obtain a maximum gain (arrive as early as possible), people tend to become risk-aversive (they do not choose a route with a large variability of travel time). If they want to avoid being late, they are prepared to take more risks.

Katsikopoulos et al. found that the influence of variability of travel time on the choice behaviour is higher if participants have a low working load, e.g. if they can make their choice without being bothered by a (driving) task. If the same information is presented while the participants have to drive in a simulator, the influence of the variability of the travel time appears to be much less.

Learning Behaviour in Uncertain Conditions

Barkan et al. (1998) studied risk-taking behaviour under uncertain situations, especially the change in decision-making caused by learning effects. The scope of their study was accident prevention: operators had the choice to take actions that would prevent the negative consequences of an incident. The occurrence of the incident was not certain in advance but warnings are given that the situation might become critical. The costs of the measures were high but if the operator did not take measures and the situation became critical, the damage was much larger than the costs of the measures.

Accidents are known to happen if people take unsafe actions in risky situations. Two (binary) sequential judgement and decisions must be considered:

1. The judgement that a situation is safe or risky,
2. The decision to take measures to prevent possible damage,

The decision not to take measures under risky conditions does not always result in a critical situation with damage. Even if the situation is judged as risky, it is still possible that nothing will happen and even if the indicator gives no reason to take measures, the situation may become critical. The judgement that a situation is safe or risky is made on the basis of the observation of some criterion that is assumed to be an indicator of unsafety. E.g. in a factory it might be the measurement of pressure or temperature that indicate the possibility that overheating or overpressure will occur.

Figure 3 shows an example of the probability curve for a critical situation, as a function of the strength of the criterion (the temperature, pressure or warning in the examples above). For every value of the criterion there is a probability P that the situation will become critical and $1 - P$ that the situation will remain normal. The strategy of the decision maker is to take measures to deal with a possible critical situation as soon as the criterion is higher than a threshold value. It is possible that a critical situation arises even if the criterion is lower than the threshold value. This is a *missed opportunity*: the critical situation occurs while no measures have been taken. If the situation remains safe, we can call this situation a *correct recognition* of the signal given by the criterion. If the criterion is higher than the threshold, it is possible that the critical situation will occur. If the correct decision is taken to deal with the critical situation, we call this a *hit*. Finally it is possible that the criterion is higher than the threshold and still the situation would remain safe, we have a *false alarm*.

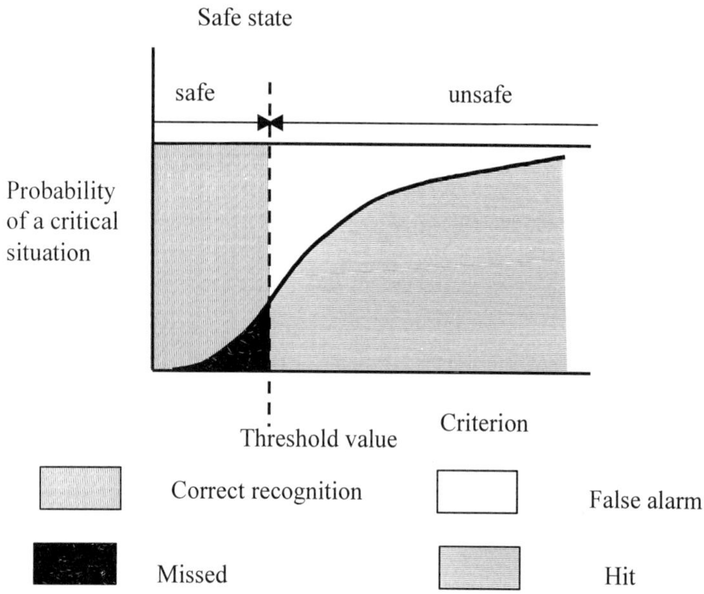

Figure 3 The probability for the four different situations: judgement safe/risky based on a criterion and the real situation safe/risky.

The optimal value of the threshold value depends of course of the costs of an unprepared critical situation and the costs of the measures. If the measures are very expensive and the costs of a critical state are rather low, one may take more risks and choose a higher value of the threshold.

The results of this study might be generalized to decision making in travel. For travellers we may think, for example, of the following choice situation represented in figure 4. Travellers waiting for a train get a warning in the station that an accident has happened on the track such that not trains can pass. The train operator tells that alternative transport is not available yet and that the journey along this route will take about two hours longer. The recommendation is given to take an alternative route, which takes one hour longer. However, it is not unlikely that the disturbance along the track will be removed and the track is ready at the moment that the train arrives at the spot of the incident.

The traveller has the choice to take the shortest route, with a risk to have two hours delay and the possibility to travel with no delay, or to take the alternative route with the certainty of one hour longer travel time. The decision depends, of course, on the objectives and constraints of the traveller, as described in the previous sections. If the traveller has to be exactly in time and there is no difference in being one hour or two hours late, the choice is different from the case where the traveller wants to arrive home as early as possible.

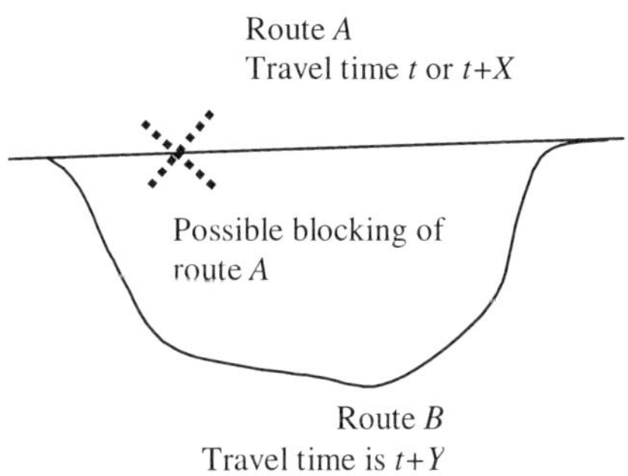

Figure 4 If the warning given is strong enough, travellers will decide to take the safe route *B* instead of the quick route *A*

A payoff matrix can be defined in which for each situation the costs for the traveller is be calculated, depending on the situation and the decision made by the traveller. This matrix is as given in table 1.

Table 1 Payoff matrix for the decision to take the longer route *B* after a warning that route A is blocked. The probability that *A* is blocked given a warning with strength *c* is *P*(*c*)

		State	
		Risky state: *A* is blocked, probability = $P(c)$	Safe state: *A* is accessible, probability = $1 - P(c)$
Decision	Route *B*	Hit: $P(c)\ Y$	False alarm: $(1 - P(c))\ Y$
	Route *A*	Missed: $P(c)X$	Correct recognition: 0

The strategy of the decision-maker might be to choose the threshold value such that this maximizes his expected utility. If the probability of getting the warning *c* is given by $p(c)$ and the decision to choose route *B* is taken when the warning is stronger than the threshold value *T*, the optimisation problem is

$$\mathrm{Min}_T \left\{ \int_0^T p(c)P(c)X dc + \int_T^\infty p(c)Y dc \right\} \qquad (1)$$

As an example we could look at a discrete case where the detour over route *B* costs $Y = 60$ minutes, but the delay at *A* if *A* is blocked is $X = 120$ minutes. We assume that 4 possible warnings are given with strength 1, 2, 3 or 4. The probability that link *A* is blocked is given by the Table 2.

Table 2 The probabilitythat link A is blocked

Strength of warning	Probability *A* is blocked
1	0.1
2	0.3
3	0.7
4	0.9

The traveller has to decide whether he will choose route *B* at a warning of strength 4, ≥ 3, ≥ 2 or at all warnings.

The expected penalty is given by Table 3.

Table 3 The expected penalty

Cut-off	Expected delay on route *A*	Expected additional travel time	Expected penalty
1	0	60	60
2	p(1)* 0.1*120	{p(2) + p(3) + p(4)}*60	48
3	{p(1)* 0.1 + p(2)*0.3}*120	{p(3) + p(4)}*60	42
4	{p(1)* 0.1 + p(2)*0.3 +p(3)*0.7}*120	p(4)*60	48

If we assume that the warnings are all equally probable, p(i) = 0.25 {i= 1,..4}, the optimal cut-off value would be 3. It is clear that this threshold is different when we assume that people have the objective to maximize the probability to arrive at their destination at a predetermined time. If they have to arrive at the scheduled time t, the threshold T above which they will choose route *B* will be very high. However, if their deadline for arriving is $t + Y$, they will always take the safe route *B*. If people have the bias in the observation as Allais showed, the threshold T may change, especially when the threshold calculated with eq. (1) is close to the low and high boundary of the range, e.g. if the difference $X - Y$ is very small or very large.

Even if people have experience with warnings and are able to estimate the probability $p(c)$ that a certain level will be issued, they have insufficient information about the probability that route *A* is blocked, given warning c. This means that an optimisation behaviour as defined with eq.

(1) is not realistic for ordinary people. In general, they have no idea about the probability $P(c)$ that route A is blocked given warning c. The possibility to build up this knowledge is limited. In fact it is knowledge about the reliability of the warning system. Such knowledge has an empirical character and in most cases it is impossible to get such knowledge accurately and completely. The rest of this section explains what learning strategy people apply in such situations to get an acceptable alternative.

Barkan et al. identified several models for learning strategies for decision-making in uncertain situations. Two strategies, relevant in the context of travel behaviour, are represented in the two following dynamic models:

- Hill climbing strategy
  The decision-maker shifts the cut-off value. If the shift results in an instantaneous higher payoff, the shift is repeated for the next time; if the payoff is lowered, then, the cut-off is reset to its previous value. If the last shift does not make any difference, then the next shift is made in a random direction. This means that, if a warning about the travel time over route A is given with certain strength, the traveller may still neglect that warning. If he is lucky and can use route A without additional delay, he will neglect similar warnings in the future. If he will find route A blocked, he will follow the warnings in the future.

- Cut-off reinforcement model
  This model assumes that the decision-maker chooses a cut-off from a set of possible values. The choice for a certain value will be reinforced if the decision-makers get a payoff from this choice. The difference with the Hill Climbing strategy, which is a systematic search for an optimum, is that the Cut-off reinforcement can select cut-off values in an arbitrary sequence.

The experiment reported by Barkan et al. shows that the learning behaviour changes in the course of time. It has the following stages:

1. Initial learning towards sub-optimal risky behaviour,
2. Slow learning towards a safer cut-off,
3. Decreased learning speed with decreased probability of a penalty,
4. The sensitivity of the choice of the cut-off changes with the experience and the probability of penalty.

The dynamic models with a cut-off value that changes with the experience of the decision maker, are successful to describe observed behaviour. The choices made by the participants of the survey and the sensitivities of the choice appear to depend on the payoff experience.

For travellers such a systematic survey has not been held yet. However, one can observe that warnings for fog, snow or other bad weather conditions in the beginning of the autumn or winter are often neglected and more accidents occur on the first days of the season with bad weather conditions while on later days drivers seem to be more prudent. The learning behaviours as found by Barkan are certainly not unlikely for travellers.

Discussion

For the modelling of the behaviour of travellers in uncertain conditions with unreliable and incomplete information we may first of all conclude that objective probabilities of the occurrence of certain travel times may not be adequate parameters for a descriptive model. The bias for low and high probabilities makes it necessary to transform the objective probabilities to experienced opportunities or risks and to take into account whether people are risk avoiding or opportunity seeking.

The difference in objectives of travellers may also give a distinction in their choice behaviour. The penalty that the traveller will experience after a late arrival may be proportional to the lateness or be constant (e.g. when a traveller misses the connection to a plane). Such conditions have influence on the choice strategy. Furthermore, the learning behaviour of the traveller makes it necessary to use a dynamic model for the choice behaviour.

The present choice models used in traffic studies are approximations of the actual behaviour of travellers. It is interesting to investigate, for example, whether and when the use of more accurate models for user behaviour gives a significant different macroscopic result of route choice and traffic assignment at a macroscopic level.

The fact that uncertainty about travel conditions emerges due to unique, irreproducible conditions, makes it necessary to reconsider the probabilistic approach followed in this paper. How can one justify the existence of a probability distribution for situations that occur only once? There are no possibilities to calibrate the distribution or to validate its assumed form. The probability functions are unobservable concepts. It would therefore, be good to model the uncertainty concept in terms of a mathematical framework that considers vagueness of decision maker's perception of the given situation and offers the measure of uncertainty in a possibilistic sense, rather than a rigid probability framework (Kikuchi and van Zuylen, 2001).

Conclusion

The behaviour of decision-makers under uncertainty often deviates from the rational optimising behaviour supposed by economists and other system modellers. A survey of the literature shows that people are not irrational in their behaviour, but have a contingent strategy: their behaviour depends on the conditions, the available knowledge, their experience and their abilities to cope with incomplete and inaccurate information. The assumption of an optimisation of expected outcomes does not have get support from laboratory surveys. This means that there is uncertainty among modellers about the question how people decide in uncertain situations. Knowledge about decision-making under uncertain conditions is necessary to optimise travel information, especially for unique, non-recurring conditions.

The maximization of the utilisation of transport infrastructure has made the transport system very sensitive for disturbances. In The Netherlands a quarter of all congestion at the freeways is caused by incidents. The robustness of the road infrastructure can be improved by a well designed strategy how to deal with travellers who have only a limited knowledge about the situation, while the information that can be provided is often insufficient, inaccurate and unreliable to make optimal choices possible.

The bias in perception is especially important for infrequently occurring situations. For these conditions traffic information has to be tailored to the bias in the probability perception of the travellers.

The following research tasks have to be done are, therefore,

- to verify the validity of the behaviour under uncertainty as observed in the experiments described in this chapter, in travel conditions, preferably for naturalistic decision making;
- to reformulate the conditions for decision making under uncertainty such that probabilities are only used when the probabilities are observable concepts and to use other concepts such as fuzzy sets otherwise;
- to determine the requirements that a more realistic model for travel behaviour imposes on route choice and assignment models.

REFERENCES

Allais, M. (1979) The so-called Allais Paradox and rational decision making under uncertainty, in: Allais, M. and O. Hagen (eds) *Expected utility hypotheses and the Allais paradox,* Reidel, Dordrecht, The Netherlands.

Barkan, R., D. Zohar, I. Erev (1998) Accidents and decision making under uncertainty: a comparison of four models. *Organizational behaviour and human decision processes,* 74 (2)pp. 118 – 144

Bonsall, P. (2000) Travellers' response to uncertainty. In M.G.H.Bell and Chris Cassir (eds) *Reliability of transport Networks.* Baldock UK, Research Studies Press Ltd.

Domincich, T.A. and D. McFadden (1975) *Urban Travel Demand: A Behavioral Analysis*. North-Holland, Amsterdam.

Katsikopoulos, K.V., Y. Duse-Anthony, D.L. Fisher and S.A. Duffy (2000) The framing of drivers' route choices when travel time information is provided under varying degrees of cognitive load. *Human Factors* 42 No 3, pp 470 – 481.

Kikuchi, S. and H.J. van Zuylen (2001) Modelling of Traveler's Uncertainty and Anxiety. In these proceedings.

Lipshitz R. and O. Strauss (1997) Coping with uncertainty: a naturalistic decision-making analysis. *Organizational behaviour and human decision processes* 69 (2) pp 149 – 163.

Miller, G.A. (1966) *Psychology, the science of mental life*; chapter 6. Penguin books.

WU, George, R. Gonzalez (1998) Common consequence conditions in decision making under risk. *Journal of Risk and Uncertainty* 16 pp.115 – 139.

Rode, C., L. Cosmides, W. Hell, J.Toody (1999) When and why do people avoid unknown probabilities in decisions under uncertainty? Testing some predictions from optimal foraging theory. *Cognition* 72, pp 269 – 304

Tversky, A. and D. Kahneman (1992) Advances in Prospect Theory, cumulative representation of uncertainty. *Journal of Risk and Uncertainty* 5, pp297 – 323

von Neumann, J. and O. Morgenstern (1944). *Theory of games and economic behavior*. Princeton (NJ), Princeton University Press.

10

MODE CHOICE MODEL WITH TRAVEL TIME RELIABILITY AND COMMUTERS' TRAVEL BEHAVIOR BEFORE/AFTER A MAJOR PUBLIC TRANSPORTATION SERVICE CLOSURE

Hiroshi Wakabayashi, Faculty of Urban Science, Meijo University, Kani, Japan
Katsuhiko Asaoka, Osaka Port Transport System Co., Ltd., Osaka, Japan
Yasunori Iida, Department of Civil Engineering, Kyoto University, Kyoto, Japan
Hiroyuki Kameda, Disaster Prevention Research Institute, Kyoto University, Kyoto, Japan

ABSTRACT

In this study, a mode choice model explicitly considering travel time reliability is developed. This model quantifies travelers' attitudes towards travel time variability as well as average travel time. Data were collected from the morning commuters who have two or three alternative modes including some public transportation and private vehicles. The survey period includes both a normal period where all the transportation modes were available and an abnormal period where the main major public transportation service was closed. The model is applied to practical commuters' decision making, and one of the findings in the mode choice model is that they pay relatively large attention to the travel time variability. In this model, travel time variability is dealt with as the possibilities that the commuters arrive before or after their job starting time separately. The best-fit model indicates that the commuters pay more attention to early arrival and less to late arrival in the normal period. In the abnormal period, however, their attention shifts drastically to late arrival. This suggests that the commuters behave optimistically in the normal period and pessimistically in the abnormal period.

INTRODUCTION

The conventional mode choice models generally employ travel costs such as the travel time, waiting time and fare as the explanatory variables. These travel costs are, however, generally average cost, including average travel time. Actually, the travel time fluctuates with day-to-day dynamics, and mode choice should be considered including travel time reliability. For the advanced transportation systems, the consideration of reliability is also strongly important.

This study aims at the evaluation of how commuters estimate their travel time reliability. For this purpose, a discrete mode choice model explicitly considering travel time reliability is developed. Here, the magnitude of commuters' attitude towards the travel time reliability adding to the average travel time is measured. Second, this model is applied to the actual commuting mode choice in order to find out the travel behavior in the normal period. In fact, many commuters choose their mode with much consideration for travel time variability. Thus they may choose a mode whose average travel time is larger but whose variability is smaller. Since the survey period includes the major public transportation's service closure, the mode choice behavior change is also quantified before and during the closure. The long-term purpose of the study is to establish the crisis management of urban transportation systems in the case of a disaster such as an earthquake.

Recent studies (Kazimi *et al.*, 2000, Ghosh, 2001, Lam and Small, 2001) suggest that travelers are interested in not only travel time saving but also in a reduction of travel time variability (Chen, 2001). As Lam and Small (2001) pointed out, the value of reliability has received much less attention so far. Kazimi *et al.* (2000), Ghosh (2001), and Lam and Small (2001) evaluate willingness-to-pay to reduce travel time variability as well as that to reduce travel time. Their mode choice models are, however, lane choice models on freeway, namely between the normal free lane and HOT (High Occupancy / Toll) lane. Although the travel time reliability is an emerging important issue as they pointed out, few mode choice models between public transportation and private vehicle have been proposed. Wakabayashi *et al.* (1998, 1999) proposed mode choice model explicitly considering travel time reliability between public transportation and private vehicle.

Let us consider, for example, a commuter who has two or more alternatives of travel modes to his/her working place. The rail transit system usually operates very punctually, and he/she can estimate his/her travel time accurately. The highway system, however, is usually congested, then the travel time of vehicle varies larger and schedule delay may occur. If the average travel times of both transportation systems are same, the traveler tends to choose the highly reliable transportation system, especially when he/she has a strictly restricted coupling time. Even if the average travel time of vehicle is less than that of rail transit, some people tend to choose the rail transit because if they use vehicle, they may fail to be on time for job starting time. The conventional mode choice models with average variables cannot consider such a phenomena. Thus the mode choice model considering travel time reliability is very important. A better understanding of this phenomenon aids the urban transportation system management strategies.

In this paper, a discrete mode choice model including travel time reliability is proposed. Variables that indicate travel time variability, *i.e.* travel time reliability, are added to the conventional average travel cost variables. Here, the travel time reliability is a result of travel time fluctuation. Hall (1983) adopted a hypothesis that travel time fluctuation follows normal distribution. In our study, however, it was directly observed as commuters' perception without consuming any shape of travel time distribution. The data were collected through a cross sectional survey. Commuters were asked their perceptions about the maximum commuting time and its frequency, the minimum commuting time and its frequency and others that they had ever experienced, adding to the conventional travel cost factors for each travel mode. Three models are tested. In the proposed model, the difference between maximum travel time and average travel time, and the difference between minimum travel time and average travel time were used as indicators for travel time reliability. They are named upper difference of travel time (UDTT) and lower difference of travel time (LDTT) in the paper, respectively.

The focus area is the Port-Town in Osaka Southern Bay Area, City of Osaka in Japan. It is a commuter town of City of Osaka with approximately a hundred hectare and 32,000 population and the number of households is about 10,000. The Port-Town is a reclaimed island and is isolated from the main body of the City of Osaka. Therefore, the transportation modes are limited when they go out of the town. There are three transportation systems between the island and the main land. One is the LRT (Light Rapid Transit) system of public rail service. Others are public bus service and personal vehicle of road transportation system. In Autumn 1993, there happened an accident at the terminal station of LRT, and the LRT transportation system was closed for more than one month, and the commuters by LRT were forced to change their transportation mode to bus service or personal vehicle after some trials for their most appropriate alternatives. These commuters' transportation choices were surveyed before and during LRT closure after the accident.

Parameters of the discrete choice model were estimated for the normal period and the abnormal period, and are compared. For normal period, for example, the commuters feel the magnitudes of average travel time and travel time reliability as follows: The ratio of "average travel time" : "upper difference of travel time" : "lower difference of travel time" equals approximately 1: 0.2: (-1). This suggests that commuters tend to expect their travel time to be the minimum travel time they have experienced when they leave home everyday for normal period. The ratio changes, however, drastically during the abnormal period. The magnitude of "lower difference of travel time" became very small, and that of "upper difference of travel time" increased. The comparison of the travel time attributes makes their contribution clear, which aids future transportation system's planning and its crisis management in case of a disaster.

First, this paper discusses the background of the research in the introduction. Second, the definition of reliability is described and the significance of reliability of transportation systems is presented where a state-of-the-art is briefly shown. Third section describes the case study area and the accident that caused the major transportation service halt. In the fourth section, mode choice model for assessing travel time reliability is proposed, after several models are tested using the actual data for normal period. Finally, the change in mode choice behavior before and after the temporal closure of major transportation service is presented. The commuters' attitude towards travel time reliability is discussed for the normal and abnormal periods.

RELIABILITY: THE DEFINITION AND SIGNIFICANCE

There are many ways for evaluating reliability of transportation system. A highly reliable highway network provides sure and stable traffic service by offering drivers alternative routes even when some part of the network is unavailable. In the network analysis for highway system, there are two major considerations: network connectivity reliability and travel time reliability.

According to the mathematical definition, reliability is the probability of a device performing its purpose adequately for the period of time intended under the operating conditions encountered (Barlow and Proschan, 1965). In a more general sense, reliability is the probability that a system or a unit performs a given function for a certain time period. When a system composed of many units is described by means of graph theory with a set of links and nodes, the system reliability is defined as the probability that the input node and the output node are connected. This is called terminal reliability.

Similarly, the highway network reliability can be treated as the probability of connectivity between two given nodes. Network reliability analysis, however, requires in general a great amount of computational work and it increases exponentially with the size of the network. In addition, the computational effort is very complicated. Thus, network reliability analysis, especially for large systems, has been said difficult so far. Recently, an efficient and practical method has been developed by Iida and Wakabayashi (1989) even for a large size network. In their works, the terminal reliability of highway network is defined as the probability that two given nodes are connected for certain traffic service levels for given time periods. An assessment for improving highway network reliability before/after introducing traffic management alternatives were reported (Wakabayashi and Iida, 1994). Du and Nicholson

(1993) took another approach for the degradable network which is degraded by the events such as earthquakes and floods, etc. Wakabayashi (1996) also reported the reliability assessment both for the damaged network hit by the 1995 Kobe earthquake and its future revival network.

Travel time reliability is another kind of index. The travel time reliability is defined as the probability that travel time between two given nodes is guaranteed within a given travel time. It can be also treated as the maximum travel time to arrive at the destination with given probability. Assuming that the travel time reliability is attributed to the stochastic traffic variation, improvement of travel time reliability under the traffic management which conducts less fluctuation in traffic flow was demonstrated by Wakabayashi and Iida (1993).

The other reliability has been proposed by Ferrari (1988). He considers maintaining speed on a motorway using traffic control, where the reliability is defined as the probability that the reduction in speed does not exceed the given criteria.

In the mode choice model, the consideration of reliability can be found as follows:

Paine *et al.* (1967) reported that reliability-related attitudes scored higher in importance than the traditional predictors of travel choice, average travel time, and travel cost. Prashker (1977) described that "...no demand model has been estimated with a measure of the reliability of a travel mode included". Despite Prashker's effort, there still had been little empirical works directed to an understanding of the effect of reliability on the actual decisions of the traveler (Jackson and Jucker, 1982). Jackson and Jucker (1982) pointed out that "one complication is determining an appropriate measure of reliability. Attitudinal studies have included various aspects of reliability..."

Abkowitz (1981) tested the impact of transit service reliability on work-travel, where reliability was treated as loss function of early-arrival and late-arrival. Hendrickson and Plank (1984) reported mode choice and departure time choice model, where variance of transit waiting time was considered. However, they did not consider travel time variability explicitly.

Jackson and Jucker (1982) tested a mean variance model that is a linearly combined model of the expected (mean) travel time and the variance of travel time, proposed by Brastow and Jucker (1977), using stated preference data. The risk aversion coefficient, *that is*, the ratio of the variance to the mean travel time, λ, was reported to vary from zero to 2.25. These results may be useful, but difficult in understanding the travelers' perception.

Recent studies motivate from the conversion of HOV (High Occupancy Vehicle) lane to HOT (High Occupancy / Toll) lane on freeways. Solo drivers can use the HOT lanes by paying a toll that varies every certain time period to maintain a free flowing traffic, while carpoolers continue to use the lanes for free. Ghosh (2001) suggests that travelers are interested in not only travel time saving but also reduction of travel time variability. Kazimi *et al.* (2000) estimate the willingness-to-pay in money terms not only to reduce commute time but also to reduce variability of commute time using reveled preference data and traffic flow data obtained from loop detectors embedded. Lam and Small (2001) also estimate both value of time (VOT) and value of reliability (VOR), and evaluate the ratio (that varies between man and woman) to the average wage rate.

Wakabayashi *et al.* (1998, 1999) proposed a mode choice model explicitly considering travel time reliability between public transportation and private vehicle. In this study, travel time reliability is treated as the maximum travel time and the minimum travel time besides average travel time that have been experienced by the commuters. The maximum travel time and the minimum travel time were directly asked to respondents, who have several mode choice alternatives and have experienced them.

In the mode choice study, better understanding of travelers' behavior towards the reliability of transportation system should greatly aid the transportation system management and planning strategies. Further research for reliability control method in the urban transportation system is also important.

Study Area and Data

The focus area is named Port-Town in Osaka Southern Bay Area. It is western part of City of Osaka in Japan. It is a bed-room town of City of Osaka with about a hundred hectare and 32,000 populations and the number of households is about 10,000. Approximately 9,000 workers commute out of this town. The distance to CBD is about 17km. The travel time by LRT and subway is approximately 35 minutes. Travel time by vehicle varies from 20 to 60 minutes depending on a usage of expressway and on traffic congestion.

The Port-Town is a reclaimed island and then is isolated from the main body of City of Osaka. Therefore, the transportation modes to go out of the town are limited. There are three transportation systems between the island and the main land. One is the LRT (Light Rapid Transit) system of public rail service, named *Newtram* of *Osaka Municipal Transportation Bureau*. Others are public bus service and personal vehicle as highway transportation system. Thus there are four routes between the Port-Town and the main land as shown in Table 1. The highway route choice is also limited, since there are only two routes to go outside of the town. One route is a free highway, and the other is an expressway route charged fare of \500 (approximately $4.00).

Newtram has 6.6km route length with eight stations including end terminals. One of the end terminals locates in the Port-Town, and the other locates in the main land connecting with the subway named *Yotsubashi line* of *Osaka Municipal Transportation Bureau.* Newtram is a remote-operated system; there is no boarding crew in the LRT train. The travel time is 15 minutes between the end terminals, and it is operated with 2.25 minutes headway for peak period. The train runs on elevated guideway. Each train is electrically powered and consists of four cars equipped air filled robber tires. 294 passengers can be carried in a normal condition, and more in a peak period. Total passengers are approximately 60,000 a day.

In Autumn 1993, there happened an accident at the terminal station of Newtram and the service was temporally closed for more than one month. All passengers using Newtram were forced to change their transportation mode to bus service or personal vehicle after some trials for their most appropriate alternative. Alternative bus service with 60 buses was introduced, and it carried approximately 50,000 passengers through more congested highways. Table 2 summarizes the consequences of mode choice change before and during LRT service closure.

Table 1 Four Routes from Port-Town to C.B.D. in City of Osaka

No.	Route	Note
Route A	Port-Town ⟶ Suminoe Park ⟶ C.B.D. (NewTram) (Subway)	61,761passengers per day [1992]
Route B	Port-Town ---➤ Asashio-bashi ⟶ C.B.D. (Bus) (Subway)	5,587 passengers per day [1992]
Route C	Port-Town --------------------➤ C.B.D. (Hanshin Expressway [via Minato-Ohashi])	
Route D	Port-Town --------------------➤ C.B.D. (Highway[via southern route])	

Legend: ⟶ : railway route ---➤ : highway route

Table 2 Mode Changes Before and During the LRT Service Closure

Route No	Normal Period		Abnormal Period	Number of Passengers
1	Vehicle	→	Vehicle	88 (47)
2	Vehicle	→	Alternative Bus Service	0
3	Vehicle	→	Usual Bus Service	0
4	Newtram	→	Vehicle	20
5	Newtram	→	Alternative Bus Service	254
6	Newtram	→	Usual Bus Service	46
7	Newtram	→	Bus Service(unknown)	18
8	Usual Bus	→	Vehicle	3
9	Usual Bus	→	Alternative Bus Service	4
10	Usual Bus	→	Usual Bus Service	38
11	Usual Bus	→	Bus Service(unknown)	3
Total				474 (47)

( ):Captive Vehicle Traveler

These commuters' transportation choices were surveyed before and after the accident. Two major research goals were set: to develop a mode choice model explicitly considering travel time reliability and to investigate commuters' attitude change before/after the accident.

Commuters were asked their perceptions about the minimum commuting time and its frequency, the maximum commuting time and its frequency and others that they had ever experienced, adding to the conventional average travel costs for each travel modes. At the same time, questions about their constraints of work starting time, desired arrival time and departure time were asked.

The questionnaire was delivered to each condominium or apartment, and collected by mail. The number of delivered questionnaire was 7,462. Although it took much time to answer the questionnaire, 1,034 samples were collected. We screened these samples and eliminate those with lack of significant answer, those without rationality for mode choice, those with limited mode choices. Limited mode choice commuters are those who have single commuting mode. And we also use the samples within 50km of commuting distance only. Furthermore, samples considered to be captive in mode choice are also eliminated. These captive commuters are those who are constrained to use vehicle again on his/her whole trip chain after the commuting. Eventually 515 and 427 samples are analyzed for pre-accident investigation and post-accident investigation, respectively. All of these samples have commuting alternatives with public transportation and private vehicles.

MODE CHOICE MODEL INCLUDING TRAVEL TIME RELIABILITY

Mode Choice Model Candidates

Most of the focused travelers in this study are morning commuters with arrival time constraints. We assume they behave to minimize risk to be late on the arrival time and consequently decide to choose their travel modes using their perception about their travel time and travel time variability they have ever experienced. Table 3 summarizes Average Travel Time (AVTT), Lower Difference Travel Time (LDTT), Upper Difference Travel Time (UDTT), their standard deviations, and the actual numbers of each mode choice,

where,

$$\text{LDTT} = \text{AVTT} - \text{MNTT}, \tag{1}$$

$$\text{UDTT} = \text{MATT} - \text{AVTT}, \text{ and;} \tag{2}$$

MNTT is the Minimum Travel Time, and MATT is the Maximum Travel Time that the commuters have ever experienced.

According to Table 3, the fastest mode in the average travel time is vehicle, *i.e.* 40.515 min and the second fastest is *Newtram* (LRT) *i.e.* 59.435 min, and the third fastest is bus, *i.e.* 62.241 min. Actually, however, the most popular mode is *Newtram* (LRT), the second is bus and the third is vehicle. The fastest mode is not always chosen by commuters. This shows that the mode choice is not clearly explained only with the average travel time, which is indicated in the standard deviations in Table 3. Since UDTT for vehicle, *i.e.* 17.673 and the standard deviation 22.285 are much higher than those of public transportation, the mode choice depends on travel time reliability.

Table 3 Average and Variability of Each Mode Travel Time

	Mode	Normal Period		Abnormal Period (Newtram Servive Closure)	
		Average (min.)	Std. Deviation (min.)	Average (min.)	Std. Deviation (min.)
1.AVTT	Vehicle	40.515	20.217	46.752	22.560
	Newtram	59.435	18.270	-	-
	Bus	62.241	22.975	69.857	22.318
2.LDTT	Vehicle	8.606	9.027	9.862	8.150
	Newtram	6.715	5.895	-	-
	Bus	5.907	6.305	9.813	9.049
3.UDTT	Vehicle	17.673	22.285	15.269	11.506
	Newtram	7.072	7.560	-	-
	Bus	12.146	8.984	11.642	10.089
Number of Users	Vehicle	48 (9.3%)		64 (15.0%)	
	Newtram	407 (79.0%)		-	
	Bus	60 (11.7%)		363 (85.0%)	
Number of Available Samples		515 (100.0%)		427 (100.0%)	

To analyze commuters' mode choice attitude, logit models are estimated. Variables that indicate travel time reliability are added to the conventional average travel cost variables. Here, the travel time reliability is result of travel time variability. Hall (1983) assumed a normal distribution as travel time variability. In our study, however, it was directly observed as commuters' perception without Hall type model. The data were collected through a cross sectional survey mentioned above.

Before the Newtram service closure, since there were three modes available, Multinomial Logit model is developed. The probability P_{in} that an individual n selects available mode i is assumed to be the logit form:

$$P_{in} = \exp(V_{in}) / \sum_{i} \exp(V_{in}). \tag{3}$$

For each mode choice, the deterministic term of utility function, V_{in}, is specified to be:

$$V_{in} = \theta_1 X_{1in} + \theta_2 X_{2in} + \theta_3 X_{3in} + \theta_4 Y_{41n} + \theta_5 Z_{51n} + \theta_6 Z_{62n} + \theta_7 Z_{71n} + \theta_8 Z_{82n}, \tag{4}$$

where variables are defined in Table 4 for individual n, and mode 1, 2 and 3 of i indicates vehicle, *Newtram* and bus, respectively.

Our purpose is finding out the coefficients that give maximum $L(\theta^*)$. Logarithmic likelihood function $L(\theta)$ is ;

$$L(\theta) = \ln(\prod_n \prod_i P_{in}^{\delta_{in}}) = \sum_n \sum_i \delta_{in} \ln(P_{in}), \qquad (5)$$

where, when individual n uses mode i, δ in=1, and when individual n does not use mode i, δ in=0.

During the service closure, since there were only two modes, *i.e.* bus service and individual vehicle, Binary Logit model is developed. In the course of developing mode choice model, there are several ways to consider three indicators of travel time such as average travel time, minimum travel time and maximum travel time the commuters had ever experienced.

Three types of models for considering travel time reliability are tested as follows:

Model 1: Utility function consists of average travel time (AVTT), Minimum Travel Time (MNTT), and Maximum Travel Time (MATT), directly.

Model 2: Utility function consists of average travel time (AVTT), Lower Difference Travel Time (LDTT) and Upper Difference Travel Time (UDTT),

where,

$$\text{LDTT} = \text{AVTT} - \text{MNTT}, \qquad (6)$$

$$\text{UDTT} = \text{MATT} - \text{AVTT}. \qquad (7)$$

Model 3: Utility function consists of average travel time (AVTT), Ratio of Lower Difference (RLD) and Ratio of Upper Difference (RUD),

where,

$$\text{RLD} = (\text{AVTT} - \text{MNTT})/\text{AVTT}, \qquad (8)$$

$$\text{RUD} = (\text{MATT} - \text{AVTT})/\text{AVTT}. \qquad (9)$$

The results are shown in Table 4. For each model, a negative coefficient for AVTT is expected. For each coefficient related to minimum travel time, a positive sign is anticipated to reflect the increasing utility associated with smaller travel time. For each coefficient related to maximum travel time, a negative sign is expected.

Model 1 has a simple structure in its utility function, namely, that the Minimum Travel Time (MNTT) and Maximum Travel Time (MATT) are simply added to average travel time (AVTT). Although the variables in utility function of logit model should be independent each other, MNTT and MATT are anticipated to have strong correlation with AVTT because MNTT and MATT are the values of average travel time (AVTT) and its variability. This correlation may cause no rationality in coefficient sign and deterioration in t-statistic for AVTT in Model 1

Table 4 Estimated Coefficients for Three Models

			Model 1	Model 2	Model 3
[Comman variables]					
X_{1in}	1	AVTT	**0.0263** (1.614)	**-0.1058** (-9.551)	**-0.0877** (-8.996)
X_{2in}	2	MNTT & related var.	**-0.1141** (-6.282)	**0.1141** (6.282)	**0.0658** (6.293)
X_{3in}	3	MATT & ralated var.	**-0.0180** (-2.003)	**-0.0180** (-2.003)	**-0.0248** (-1.070)
[Alternative specific variable: Vehicle]					
Y_{41n}	4	Use expressway	**2.9237** (6.655)	**2.9237** (6.655)	**2.9106** (6.614)
[Socio-economic variables]					
Z_{51n}	5	Owing license:Vehicle	**-5.1599** (-4.815)	**-5.1599** (-4.815)	**-5.4829** (-5.065)
Z_{62n}	6	Owing license:Newtram	**0.9302** (3.303)	**0.9302** (3.303)	**0.9503** (3.361)
Z_{71n}	7	Owing Vehicle:Vehicle	**-1.5620** (-2.361)	**-1.5620** (-2.361)	**-1.4499** (-2.226)
Z_{82n}	8	Owing Vehicle:Newtram	**0.6352** (2.519)	**0.6352** (2.519)	**0.6350** (2.514)

Model 2 introduces the travel time variability separately from the average travel time (AVTT). Here, the travel time variability is treated as lower variation and upper variation separately as described in Eqs. (6) and (7). Thus the correlation of average travel time to other two variables is expected to be removed as mentioned below. Table 5 indicates the correlation among variables of AVTT, UDTT and LDTT used in Model 2. The largest is 0.606 between AVTT and LDTT for vehicle, and other correlation coefficient are small. This suggests that these values are appropriate for constructing the mode choice model. In addition, Model 2 is rational in its signs of coefficients and is good in t-statistic.

Model 3 introduces the standardized (*i.e.* non-dimensional) variables in its utility function using similar explaining variables as Model 2. However, t-statistic for RUD is not significant, thus the model cannot consider the travel time variability both for upper and lower directions. In these discussions, Model 1 and 3 are inappropriate and Model 2 is adopted in this study.

Table 5 Correlation among AVTT, UDTT and LDTT in Model 2

	Vehicle			Newtram			Bus		
	AVTT	LDTT	UDTT	AVTT	LDTT	UDTT	AVTT	LDTT	UDTT
AVTT	1.000			1.000			1.000		
LDTT	0.606	1.000		0.419	1.000		0.303	1.000	
UDTT	0.084	0.026	1.000	0.195	-0.011	1.000	0.160	-0.183	1.000

Commuters' Attitude for Mode Choice Behavior

Comparing the coefficients permits understanding of commuters' attitude for mode choice behavior. From the results of Model 2 (Table 4), the coefficients for average travel time (AVTT) and Lower Difference Travel Time (LDTT) (*i.e.* the difference between minimum travel time and average travel time; positive index) have similar value. This shows that the LDTT has the similar magnitude on choosing the mode as the AVTT. Whereas commuters take LDTT into much consideration, they take the UDTT (Upper Difference Travel Time, *i.e.* the difference between maximum travel time and average travel time; positive index) into less consideration. The ratio of AVTT, LDTT and UDTT is approximately 1: -1: 0.2 from Table 4. Thus the ratio of absolute values for AVTT, LDTT and UDTT is 1: 1: 0.2.

This indicates that commuters tend to expect the minimum travel time they have ever experienced when they leave home every day, and to expect little to be delayed. This is one of the interesting findings of this study. It is not clear, however, whether the less consideration for UDTT is a result from their minimizing the risk or Hall (1983) type disutility function for being late on the coupling time or not. It needs further researches, for example, explaining the behavior by nested logit model.

CHANGE IN MODE CHOICE BEHAVIOR AFTER THE MAJOR TRANSPORTATION SERVICE CLOSURE

This section quantifies the commuters' mode choice behavior changes and discusses their mode choice attitude changes after the major transportation service closure. As is stated in the previous section, one train of the Newtram crashed into the terminal station in Autumn 1993 and the Newtram transportation service was temporally closed for more than one and half month. Many commuters lost their major and usual transportation mode after the accident. During this period, they were forced to change their commuting modes after some trials for their most appropriate alternative.

In pre-accident period, the percentages commuters using vehicle, Newtram and bus usage are 9.3%, 79.0%, and 11.7%, respectively as is indicated in Table 3. After the service closure of Newtram, the percentages of commuters using vehicle and bus changed to 15.0% and 85.0%, respectively with increasing vehicle's usage. The average travel time for vehicle and bus increased by six to seven minutes as is also shown in Table 3.

The mode choice model, which has the same utility function as Model 2 for normal period, is estimated for this abnormal period. The coefficients for both normal period and abnormal period are summarized in Table 6. The mode choice model for abnormal period also has the rationality in its signs of coefficients and goodness in t-statistic excluding LDTT. It is not allowed to compare the coefficients between normal period and abnormal period because the estimates of coefficients are only relative values for each individual model. Thus we compare the coefficient within each mode choice model separately and then discuss the mode choice attitude change between pre-accident and post-accident period.

Table 7 indicates the ratio of estimated coefficients before and after the service closure. Commuters' attitude changed drastically after the Newtram service closure: the ratio of LDTT to AVTT is 1.079 in absolute value in the normal period decreased to 0.145 in the abnormal period. The t-statistic value that was 6.282 in the normal period decreases to -0.284 that is not 5% significant. On the other hand, the ratio of UDTT to AVTT increases from 0.171 in the normal period to 0.900 in the abnormal period. The t-statistics also increased. These result indicate that the commuters who expect the reduction in travel time in the normal condition changed to consider seriously about the potential schedule delay in the abnormal condition. The commuters who behave optimistically in the normal period changed to behave pessimistically in the abnormal period.

Table 6 Estimated Models' Coefficients Before/After Service Closure

	Normal Period	**Abnormal Period**
[Common variables]		
AVTT(min)	-0.1058 (-9.551)	-0.0322 (-3.851)
LDTT(min)	0.1141 (6.282)	-0.0047 (-0.284)
UDTT(min)	-0.0180 (-2.003)	-0.0290 (-2.195)
[Alternative specific variables :Vehicle]		
Use expressway	2.9237 (6.655)	1.5953 (5.616)
[Socio-economic variables]		
Owing license:Vehicle	-5.1599 (-4.815)	-2.3589 (-4.867)
Owing license:Newtram	0.9302 (3.303)	- -
Owing vehicle:Vehicle	-1.5620 (-2.361)	-1.6988 (-4.209)
Owing vehicle:Newtram	0.6352 (2.519)	- -
L(0)	-565.7853	-295.9739
L(θ *)	-225.7242	-130.4199
Ratio of likelihood	0.6010	0.5594
Hit-ratio[%]	82.14=423/515	86.89=371/427
Share(%)	Vehicle Newtram Bus	Vehicle Bus
Predict	7.4 87.8 4.9	7.0 93.0
Actual	9.3 79.0 11.7	15.0 85.0

Table 7 Change of Ratio of Estimated Coefficients Before/After Service Closure

	Normal Period	Abnormal Period
[parameter for LDTT] / [parameter for AVTT]	-1.079*	0.145
[parameter for UDTT] / [parameter for AVTT]	0.171*	0.900*

* Coefficient is significant at 5% level.

This is an important point to manage the malfunctioning transportation system. Traveler wants to know his/her potential schedule delay rather than potential early arrival. When the transportation system becomes ill-condition, the strategy for system management and information provision may be required to be changed. For example, the detour route information, its expected travel time and its potential delay are important. In the near future, it may be insufficient to provide travelers with the average or expected travel time in advanced transportation system. Travelers may be eager to know the degree of variability of travel time, particularly when their trips are crucial.

Dummy Variables in Mode Choice Model

It is expected that the utility function with the dummy variables explains the actual mode choice better than without the dummy variables, since the dummy variables' role is to explain the unidentified attributions (including parking). Thus we expect the contribution of each variable becomes clearer, and the convergence may also be improved in the course of model estimation. The results are shown in Table 8, compared with the mode choice models without the dummy variables. The models without the dummy variables are the same as in Table 6.

In case of model with dummy variables for normal period, the problem was found in strong correlation between 5th variable and 9th dummy variables in the course of Newton-Raphson method. Although tendencies in the magnitude among AVTT, LDTT and UDTT are similar to the results from Model 2, the t-statistic deteriorates for UDTT coefficient. The similar results are obtained for the model for abnormal period. Either for normal period or for abnormal period, introducing dummy variables does not improve the explanation of actual mode choice very much. Thus we decided to leave these models with dummy variables.

Table 8 Estimated Coefficients for Models with/without Dummy Variables

	Normal period		Abnormal period	
	with dummy var.	without dummy var.	with dummy var.	without dummy var.
[Common variables]				
1 AVTT(min)	-0.1087	-0.1058	-0.0328	-0.0322
	(-9.637)	(-9.551)	(-3.807)	(-3.851)
2 LDTT(min)	0.1178	0.1141	-0.0043	-0.0047
	(6.425)	(6.282)	(-0.261)	(-0.284)
3 UDTT(min)	-0.0146	-0.0180	-0.0284	-0.0290
	(-1.684)	(-2.003)	(-2.129)	(-2.195)
[Alternative specific variables :Vehicle]				
4 Use expressway	3.0305	2.9237	1.6454	1.5953
	(6.713)	(6.655)	(4.952)	(5.616)
[Socio-economic variables]				
5 Owing license:Vehicle	-25.2590	-5.1599	-2.2667	-2.3589
	(-44.680)	(-4.815)	(-4.007)	(-4.867)
6 Owing license:Newtram	0.0566	0.9302	-	-
	(0.166)	(3.303)		
7 Owing Vehicle:Vehicle	-2.2090	-1.5620	-1.6384	-1.6988
	(-3.175)	(-2.361)	(-3.658)	(-4.209)
8 Owing Vehicle:Newtram	-0.1357	0.6352	-	-
	(-0.428)	(2.519)		
[Dummy variables]				
9 Dummy:Vehicle	21.0580	-	-0.2719	-
	(37.249)		(-0.294)	
10 Dummy:Newtram	2.3563	-	-	-
	(3.790)			
No. of Iterations	26	6	6	6
L(**0**)	-565.7853	-565.7853	-295.9739	-295.9739
L(θ *)	-218.2088	225.7242	-130.3742	-130.4199
Ratio of likelihood	0.6143	0.6010	0.5595	0.5594
Share(%)	Veh. NT Bus	Veh. NT Bus	Veh. Bus	Veh. Bus
Predict	7.0 88.7 4.3	7.4 87.8 4.9	6.8 93.2	7.0 93.0
Actual	9.3 79.0 11.7	9.3 79.0 11.7	15.0 85.0	15.0 85.0
Data	515		427	

CONCLUSION

This paper proposed a mode choice model explicitly considering travel time reliability. Travel time reliability was explained in the utility function as terms for maximum and minimum travel time attributes adding to the term for the traditional average travel time attribute. The best-fit model has good t-statistics for the explanation variables and the coefficients of the variables indicate the magnitude of commuters' attitude for the corresponding attributes of the transportation mode. Thus, this formulation of the utility function enables the direct measurement of the travelers' attitude toward travel time reliability. These discussions demonstrate that the morning commuters take the travel time variation into much consideration as well as the average travel time in their mode choice decision-makings.

First, the model was estimated for the normal period when all transportation systems were normally operated. In the normal period, the effects of AVTT and LDTT on choice behavior are approximately same. However, UDTT has approximately 20% of magnitude against AVTT. This shows that the morning commuters tend to expect their travel time to be the minimum they have ever experienced when they leave home every day as much as they consider the average travel time. In addition, they pay 1/5 times attention for being late as much as for being early and being average on time.

However, once the major transportation service closed, magnitudes among AVTT, LDTT and UDTT drastically changed. Magnitude of LDTT decreases and that of UDTT increases. The coefficient of UDTT increases nearly same as that of AVTT. This shows that the commuters who expect reduction in travel time in the normal condition change to consider seriously the potential schedule delay in the abnormal condition. This suggests that the commuters behave optimistically in the normal period, and once some accident or incident occurs, they change to behave pessimistically.

The proposed method that directly measures the magnitude of travel time reliability is an alternative method, whereas penalty disutility function approaches have been made by many authors. The further research is needed for the relationship between these two approaches.

REFERENCES

Abkowitz, M. D.(1981). Understanding the Effect of Transit Service Reliability on Work-Travel Behavior. Transportation Research Record 794, 33-41.

Barlow, R. E. and Proschan, F.(1965). Mathematical Theory of Reliability. John Wiley & Sons, Inc., New York.

Brastow, W. C. and Jucker, J. V.(1977). Use of a Mean-Variance Criterion for Traffic Assignment. Unpublished manuscript, Department of Industrial Engineering, Stanford University (referred in Jackson and Jucker, 1982)

Chen, A, Ji, Z. and Recker, W. (2001). Travel Time Reliability With Risk-Sensitive Travelers. Presented at The 1st International Symposium on Transportation Network Reliability, Extended Abstract, 61-64.

Du, Z. P. and Nicholson, A. J.(1993). Degradable Transportation Systems Performance, Sensitivity and Reliability Analysis, Research Report 93-8, Department of Civil Engineering, University of Canterbury, Christchurch, New Zealand.

Ferrari, P. (1988). The Reliability of the Motorway Transport System. Transportation Research-B, Vol.22B, No.4, 291-310.

Ghosh, A (2001). To Pay or Not To Pay: Commuters' Mode Choice Under Real Time Congestion Pricing. Paper presented to the 80th Annual Meeting, Transportation Research Board. Available on Preprint CD-ROM. Transportation Research Board, Washington, DC.

Hall, R. W. (1983). Travel Outcome and Performance: The Effect of Uncertainty on Accessibility, Transportation. Research-B, Vol.17B, No.4, 275-290.

Hendrickson, C. and Plank, E.(1984). The Flexibility of Departure Times for Work Trips. Transportation. Research-A, Vol.18A, No.1, 25-36.

Iida, Y. and Wakabayashi, H.(1989). An Approximation Method of Terminal Reliability of Road Network Using Partial Minimal Path and Cut Sets, Transport Policy, Management and Technology Towards 2001 (Selected Proceedings of The Fifth World Conference on Transport Research), Vol.4, 367-380, Western Periodicals, Co..

Jackson, W. B. and Jucker, J. V.(1982). An Empirical Study of Travel Time Variability and Travel Choice Behavior. Transportation Science, 460-475.

Kazimi, C., Brownstone, D., Ghosh, A., Golob, T. F. and van Amelsfort, D. (2000). Willingness-to-pay to Reduce Commute Time and Its Variance: Evidence from the San Diego I-15 Congestion Pricing Project. Paper presented to the 79th Annual Meeting, Transportation Research Board. Available on Preprint CD-ROM. Transportation Research Board, Washington, DC.

Lam, T. C. and Small, K.A. (2001). The Value of Time and Reliability: Measurement from A Value Pricing Experiment. Transportation Research Part E 37 (2001), 231-251.

Paine, F. T., Nash, A. N., Hille, S. J. and Brunner, G. A.(1967). Consumer Conceived Attributes of Transportation: An Attitude Study. Clearinghouse Edition, University of Maryland.

Wakabayashi, H.(1996). Reliability Assessment and Importance Analysis of Highway Network: A Case Study of the 1995 Kobe Earthquake. Proceedings of the 1st Conference of Hong Kong Society for Transportation Studies, (Hong Kong, December 7, 1996), 155-169, The Hong Kong University of Science and Technology.

Wakabayashi, H. and Iida, Y.(1993). Improvement of Terminal Reliability and Travel Time Reliability under Traffic Management. Pacific Rim TransTech Conference Proceedings Volume 1, Advanced Technologies, ASCE 3rd International Conference on Applications of Advanced Technologies in Transportation Engineering. 211-217, American Society of Civil Engineers.

Wakabayashi, H. and Iida, Y.(1995). Improvement of Road Network Reliability with Traffic Management. in B. Liu and J.M. Blosseville (eds.), Transportation Systems: Theory and Applications of Advanced Technology. Pergamon Elsevier Science, United Kingdom, 603-608.

Wakabayashi, H. and Asaoka, K. (1998). Mode Choice Model Considering Travel Time Reliability Between Vehicle and Rail Transit in Commuting Travel: Case Study-Before/After Public Transportation Service Closure, Presented at the 8th World Conference on Transport Research, Antwerp Belgium.

Wakabayashi, H., Asaoka, K., Kameda, H. and Iida, Y. (1999). Effect of Travel Time Reliability on Mode Choice Behavior of Commuters and Travel Behavior Change Before/After Public Transportation Service Closure, Journal of Infrastructure Planning and Management, No.632/IV-45, 29-40 (Abstract in English).

11

AN ANALYSIS OF POTENTIAL OF PROVIDING INFORMATION ON TRAFFIC ACCIDENT TO ENHANCE TRAVEL TIME RELIABILITY

Nobuhiro Uno, Department of Civil Engineering, Kyoto University, Kyoto, Japan
Yasunori Iida, Department of Civil Engineering, Kyoto University, Kyoto, Japan
Seiichiro Kawaratani, OMRON Corporation, Osaka, Japan
Masumi Suganuma, Department of Civil Engineering, Kyoto University, Kyoto, Japan

ABSTRACT

This study was performed to examine the relationship between dynamic traffic information and driver's route choice behavior when non-recurrent events cause traffic jams. To enhance the effectiveness and applicability of the information system as a traffic control measure, it is important to investigate the influence of the software aspects of a dynamic traffic information system; these software aspects include information content and the timing of information delivery. This study was undertaken to investigate whether providing drivers with dynamic traffic information might relieve traffic jams and improve travel time reliability. A questionnaire survey was conducted to obtain data regarding the respondents' attitudes to detours and their stated preference (SP) of route choice behavior. These data are used to analyze the characteristics of drivers who show negative and positive attitudes to detours, including driver information usage and knowledge about the alternative routes. Also, this study analyzes the relation between the way to provide drivers with information and driver's decision-making on route using the SP data. Especially, the analytical attentions are paid to the influences of both timing and contents of information provided upon the respondent' route choices.

INTRODUCTION

Traditionally, the road network performance has been evaluated in terms of absolute or average travel time. It is anticipated that advances in information technology will reduce uncertainties in the route choice decision-making process. In addition, intensive progresses in socioeconomic activities may lead to an increase in the value placed on peoples' time. Thus, it is becoming more important to provide travelers with stable road transportation services. Accordingly, road transport services should be evaluated from the viewpoint of network reliability (Iida, 1999).

With the rapid progress and penetration of ITS (Intelligent Transportation System) technologies, the concept of providing travelers with dynamic information is now regarded as a practical means of traffic control. The Advanced Traveler Information System (ATIS), an ITS component, provides drivers with real-time information and is aimed at affecting their travel behavior to obtain a better distribution of routes used, thereby promoting the efficient use of road networks. This study was undertaken to examine the potential effectiveness of providing drivers with information on traffic accidents, and related jams, as a means of traffic control for urban road networks. As the traffic demand for urban road networks is generally massive, and the number of alternative routes available is limited, a traffic accident on a road section may result in extremely inefficient traffic conditions over an entire road network. Thus, a traffic accident may markedly reduce travel time reliability. Accordingly, providing drivers with suitable information immediately after an accident offers an important way of preventing both inefficient traffic conditions and the deterioration of travel time reliability.

Many researchers, however, have reported the negative effects of information provision on traffic conditions (Arnott *et al.*, 1991; Ben-Akiva *et al.*, 1991). If numerous drivers in a network depend on provided information in determining their behavior, it is likely that the providing information regarding a less crowded route will lead, in turn, to an excessive concentration of drivers on that route. Thus, traffic conditions may become worse than would have been the case had no information been provided. Such adverse effects may occur when the provider of information does not explicitly consider the relation between travelers' behavior and the information.

To enhance the effectiveness of the dynamic information system, and ensure its applicability as a new means of controlling the impact of non-recurrent events such as traffic accidents, it is necessary to investigate the relationship between traveler response to information and the software aspects of the information system, which include the content, timing, and accuracy *etc.* of the information provided (Iida *et al.*, 1999). However, it is difficult to establish an economical and effective method to monitor both the traffic information provided, and driver response to the information, under real conditions. In this study, a stated preference (SP) survey was adopted to collect data on driver response to the traffic information provided. The SP survey technique has often been used in the field of transportation research, especially in cases where traditional RP survey methods cannot provide the full range of information required for analysis (Abdel-Aty *et al.*, 1997).

SP surveys are expected to provide data on driver response to a wide range of information provision methods in a systematic manner (Bonsall and Palmer, 1999). Some researchers have successfully applied SP surveys to obtain driver response to information (Khattak *et al.*, 1993; Hato *et al.*, 1995; Wardman *et al.*, 1997; Abdel-Aty *et al.*, 1997; Lai *et al.*, 2000). However, SP surveys are sometimes criticized for oversimplifying the situation that travelers face, and for being removed from the real decision-making process. However, SP surveys are believed to be useful for identifying potential problems caused by inappropriate methods of providing information to travelers. In addition, analyses using SP data are effective for identifying the functional form of models to represent travelers' decision-making and the relative influences of various factors upon their decision-making (Horowitz, 1985).

As an initial step in the investigation of the effects of the software aspects of information systems on driver behavior, the following subjects are examined in this study. 1) In order to discuss the potential usefulness of traffic information following an accident, driver attitudes to detours were investigated through a questionnaire survey. Especially, an analytical attention is paid to the characteristics of drivers who show negative and positive attitudes to detours, including their information usage and knowledge about alternative routes. 2) The relationship between the way to provide drivers with information and driver route decision-making is examined using the stated preference (SP) survey method. Our analytical attention is paid to the influence of both the timing and the content of the information provided on the respondents' route choices.

POSSIBLE TRAFFIC CONDITIONS CAUSED BY AN ACCIDENT

Figure 1 shows an example that illustrates the possible impact of a traffic accident on traffic conditions. For clarity, the discussion below is based on the concept of static network equilibrium. The network is assumed to consist of one OD pair and two alternative routes, as shown in Figure 1. It is assumed that an accident occurs on route 1 and thereby the capacity of route 1 drops to 1000 (veh/hour).

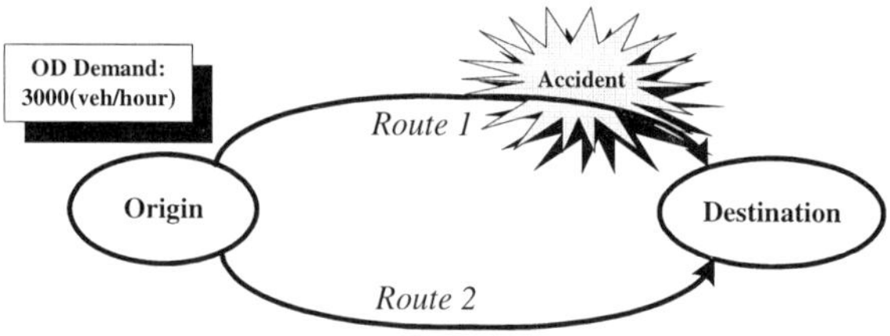

	Route 1	Route 2
Capacity with no accident	2000 (veh/hour)	2000 (veh/hour)
Distance	$L_1 = L_2 = L$ (same distance)	

Figure 1 Hypothetical Network to Illustrate the Possible Impact of a Traffic Accident

Figure 2 indicates the relationship between the flow assignment onto the network and the corresponding travel time. Point A in Figure 2 represents the network equilibrium state with no traffic accident. The accident on route 1 causes a marked change in traffic conditions. If drivers neither realize that an accident has occurred, nor switch their route, due to a lack of appropriate information, the traffic state may shift to points B_1 and B_2. The subscripts 1 and 2 indicate the routes. The traffic conditions of route 1, represented by point B_1, correspond to heavy congestion. Points D_1 and D_2 correspond to traffic conditions resulting from the overreaction of drivers to information regarding the accident. In this case, it is assumed that 1,000 drivers shift from route 1 to route 2. It is clear that points B and D represent extreme, but possible, traffic conditions, and these points highlight both the advantages and disadvantages of information provision.

From the viewpoint of efficient use of the entire road network, the traffic conditions represented by point C can be regarded as ideal conditions following an accident. The long-term goal of our studies is to establish a suitable method of information provision that can, in the case of an accident, bring about the traffic conditions represented by point C. The effects of information provision are strongly dependent on driver reaction to the information. To build an effective information system that will enhance travel time reliability, it is important to analyze the relation between driver route choice behavior and dynamic traffic information. Especially, the analytical attention should be paid to the software aspects of the information system, including the content and the timing of the information provided.

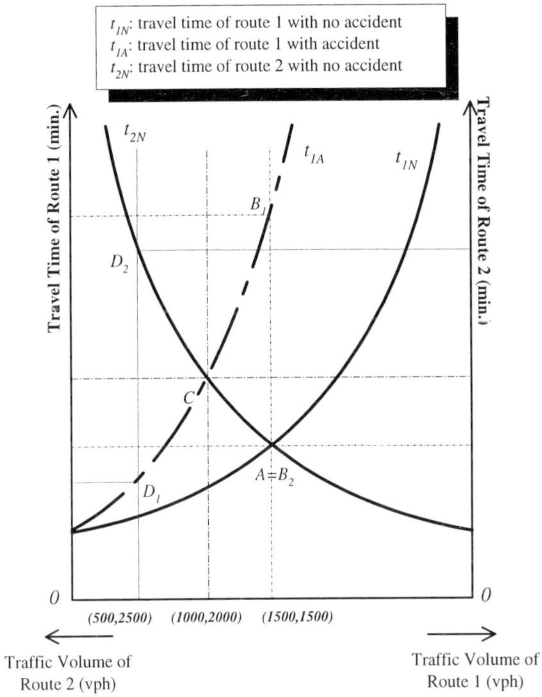

Figure 2 Relation between Flow Assignment and Travel Time

QUESTIONNAIRE SURVEY

A questionnaire survey was conducted among drivers using the Hanshin Expressway on October 21, 1999. The Hanshin Expressway has become important in the operation of daily life and the economic activity of the Kansai area, which is the second largest metropolitan area in Japan after Tokyo. The expressway network is currently 221.2 km in total length, and is traveled by about 920,000 vehicles, or about 1.4 million people, on average, per day. The Hanshin Expressway is a toll road system connecting the major cities of the Kansai area (Osaka, Kobe, Kyoto).

The questionnaire sheet was distributed to drivers at the tollbooths. After answering the questions, the respondents were required to return the questionnaire sheet by mail. The numbers of sheets distributed and returned were 14,000 and 2,063, respectively. Due to inadequate answers to some questions, the sample size actually analyzed in this study became 1,792.

The questionnaire used in this study includes the following items.

1) Respondent's demographic characteristics: gender, age, occupation, *etc.*
2) Items on the trip being undertaken by the respondent, including questions on the purpose of the trip, origin and destination, information used for route choice, *etc.*
3) Items on the respondent's usage of the Hanshin Expressway, including questions on frequency of use, reasons for preference of the Hanshin Expressway, *etc.*
4) Respondent's attitude toward diversion away from the route with a traffic accident.
5) Respondent's knowledge of alternative routes, including the number of alternative routes known, frequency of use, *etc.*
6) Stated preference on route choice following the provision of information regarding a traffic accident.

This study focuses mainly on items 4) and 6) from the above list. The design of the stated preference survey is explained in detail later.

PRELIMINARY ANALYSIS OF ROUTE SWITCHING BEHAVIOR

Situation Where Respondents Use an Alternative Route

This section describes our analysis of the respondents' attitudes toward a diversion away from the route on which an accident has occurred. Based on the results of the questionnaire survey, out analytical attention is paid to a situation in which respondents use an alternative route to the Hanshin Expressway. These situations are classified into the three categories shown in Table 1.

Table 1 Situation in Which Respondents Use an Alternative Route

Category	Number of Respondents	Percentage
(a) Only when traffic on the expressway has stopped completely because of an accident.	406	22.7%
(b) When heavy traffic congestion occurs on the expressway.	1233	68.8%
(c) Even when only slight traffic congestion occurs on the expressway.	153	8.5%
Total	1792	100.0%

The numbers of respondents belonging to categories (a), (b), and (c) are 406 (22.7%), 1233 (68.8%), and 153 (8.5%), respectively. The respondents who use the alternative route only when the traffic on the expressway has stopped completely, as a result of an accident, are defined as *'drivers negative towards detours'*. The respondents who switch from the expressway to the alternative route even when only slight traffic congestion develops on the expressway are defined as *'drivers positive towards detours'*. The majority of respondents belong to the *intermediate category*, that is, those between the drivers showing positive and negative attitudes to detours.

This study also analyzes the characteristics of drivers, with both negative and positive attitudes to detours, from the viewpoint of the respondents' knowledge about the road network and information usage. Figures 3 to 5 show the relationship between a situation in which respondents use an alternative route and various characteristics of the respondents themselves, including: 1) reference to traffic information (Figure 3), 2) the number of alternative routes known to the respondent (Figure 4), and 3) the frequency of use of alternative routes (Figure 5). As shown in Figure 3, the *negative* category tends to include more respondents who do not refer to traffic information than do the *positive* and *intermediate* categories.

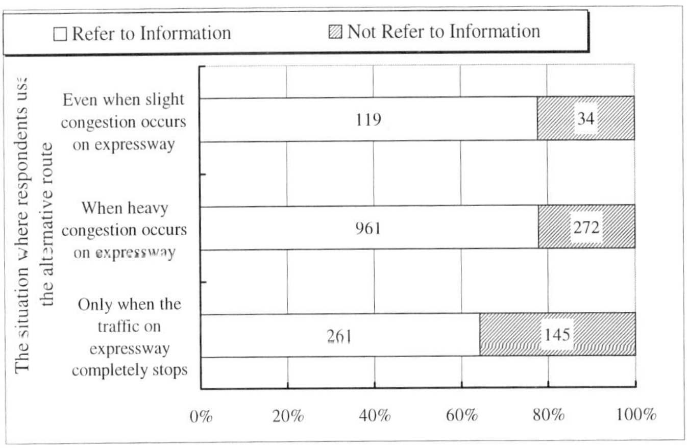

Figure 3 Situations in which Respondents Use an Alternative Route, and Reference to Information

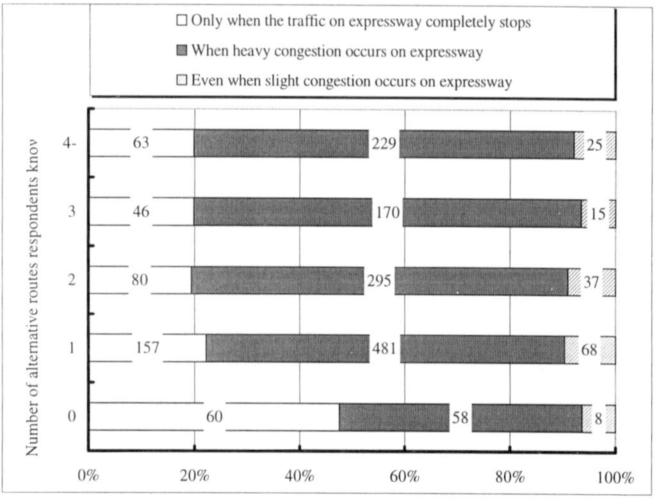

Figure 4 Situations in which Respondents Use Alternative Routes, and Number of Alternative Routes Known

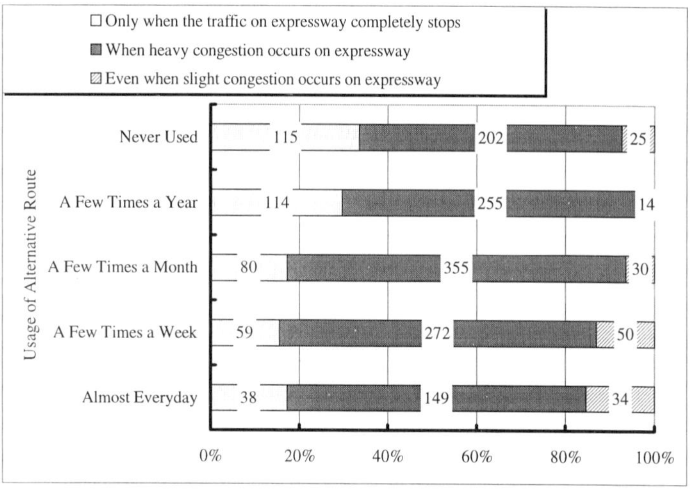

Figure 5 Situation in which Respondents Used Alternative Routes, and Frequency of Use of Alternative Routes

The number of alternative routes known to a respondent corresponds directly to the respondent's knowledge of the road network. The frequency of use of alternative routes can thus be regarded as an indirect index representing the respondent's knowledge of the road network. In general, drivers that use alternative routes frequently may be expected to have more knowledge about the network. As shown in Figure 4, the category with no knowledge of alternative routes includes more *drivers negative towards detours* than do the other categories. As shown in Figure 5, the categories with less usage of alternative routes include more *drivers negative towards detours.* Judging from the above findings, lack of knowledge of alternative routes appears to discourage drivers from using such alternatives. Thus, the provision of appropriate information, which enhances driver knowledge of the road network and traffic conditions, may encourage drivers to divert to alternative routes.

Reasons Why Respondents Did Not Divert to an Alternative Route

This subsection describes our analysis of the reason why drivers who are negative towards detours do not divert from the Hanshin Expressway to an alternative route. Accordingly, only the data given by the 406 respondents belonging to this category were analyzed in detail. The reasons why the *negative* respondents do not divert to an alternative route are classified into five categories, as shown in Table 2 below.

Table 2 Reasons Why Drivers Who Are Negative towards Detours Do Not Divert

Categories	Number of Respondents	Percentage
(a) The respondents believe that travel time on the Hanshin Expressway is shorter than that on alternative routes, even with the traffic congestion caused by an accident on the expressway.	281	69.2%
(b) The respondents are unsure of the severity of the traffic conditions on the alternative routes.	52	12.8%
(c) The respondents do not know of any alternative routes.	30	7.4%
(d) Traffic information can be obtained only after entering the Hanshin Expressway.	26	6.4%
(e) Others	17	4.2%
Total	406	100.0%

The numbers of respondents belonging to categories (a), (b), (c) and (d) are 281 (69.2%), 52 (12.8%), 30 (7.4%) and 26 (6.4%), respectively. Categories (b), (c) and (d) correspond to a lack of suitable guidance on, or relevant traffic information regarding, alternative routes. Table 2 suggests that improvements in the provision of information regarding traffic accidents and alternative routes might encourage drivers who are negative towards detours to take a diversion. Although the total percentage of categories (b), (c) and (d) relative to the total number of respondents (1792) is only 6%, a significant effect on traffic conditions may be expected if 5 or 6 % of all drivers changed their behavior.

Figure 6 shows the relationship between the reasons why respondents actually use the Hanshin Expressway and the reasons why drivers who are negative towards detours do not attempt to make a diversion. The former are classified into four categories:

(i) Likelihood that travel time of Hanshin Expressway is shorter than that of other routes.
(ii) Less fluctuation in travel time of Hanshin Expressway.
(iii) The Hanshin Expressway provides a simpler route to the destination.
(iv) Others

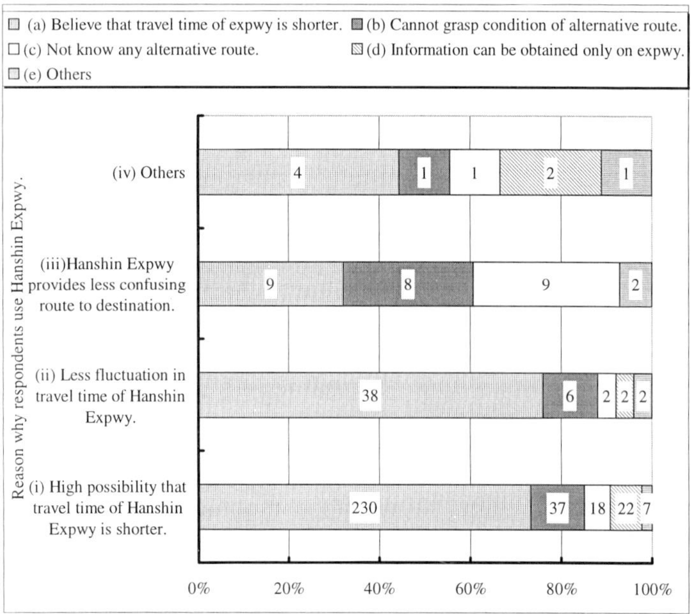

Figure 6 Reasons Why Respondents Actually Use the Hanshin Expressway, and Reasons Why Drivers who are Negative Towards Detours Do Not Divert to an Alternative Route

The respondents included in the intersections of (i) &(a) (shown in Table 2) and (ii) & (a) express a belief in the superiority of the Hanshin Expressway with regard to travel time over the alternative routes under both ordinary and extraordinary conditions. It may be difficult to encourage this kind of driver to divert to an alternative route by providing them with traffic information. However, if there is a large discrepancy between the respondent's subjective evaluation of the level of service provided by the Hanshin Expressway, and the actual conditions of this highway, the provision of suitable traffic information might encourage this kind of drivers to make a diversion.

ANALYSES OF ROUTE CHOICE BEHAVIOR

Outlines of SP Survey on Route Choice Behavior

A Stated Preference (SP) survey on route choice behavior was conducted to collect data for analysis of the relationship between information provided and respondents' decision-making. It is deemed especially important to analyze the influence of the software aspects of information provision on the respondents' route choice behavior. Among the many software aspects of information provision, this study focuses on the type and timing of the information provided.

There are six scenarios, differing as to the type and timing of the information provided.

Scenarios 1b and 1a : Respondents obtain information regarding a traffic accident and related jam on the Hanshin expressway.

Scenarios 2b and 2a : Respondents obtain information regarding a traffic accident, related traffic jams and expected travel time on the Hanshin expressway.

Scenarios 3b and 3a : Respondents obtain information regarding a traffic accident, related traffic jams, expected travel time on the Hanshin expressway, and information regarding travel times using alternative route.

The subscripts 'b' and 'a' indicate that the respondents obtain the above information 'before' and 'after' entering the Hanshin Expressway, respectively. The networks adopted in the SP survey are shown in Figure 7. In the SP survey on route choice, the respondents are assumed to travel from Ikeda City Hall to the Kitahama off-ramp in the central business district (CBD) of Osaka. Ikeda is located about 15 km north of Osaka and includes a high percentage of residential areas.

As shown in Figure 7, the SP survey assumes two types of network: a) the network for the respondents who obtain information after entering the Hanshin Expressway; and b) the network for the respondents who obtain information before entering the Hanshin Expressway. The former is used in Scenarios 1a, 2a and 3a, and the latter is adopted in Scenarios 1b, 2b and 3b.

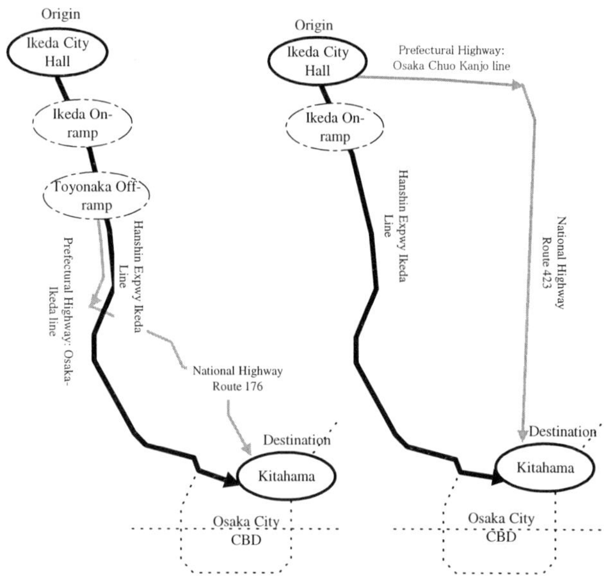

a) Provision of Information *after* Entering Expwy

b) Provision of Information *before* Entering Expwy

Figure 7 Network Used in SP Survey

In the SP survey, it is assumed that a traffic accident occurs in a road section of the Hanshin Expressway close to the CBD of Osaka. For both the Hanshin Expressway and the alternative route, we assume three levels of traffic congestion; *i.e.,* slight, medium, and heavy congestion. In addition, the traffic information provided hypothetically in the survey corresponds to the levels of congestion. Table 3 shows the assumed congestion level and the corresponding information used in Scenarios 1 and 2. In Scenarios 1 and 2, all respondents were required to give their route preference under three different conditions corresponding to the levels of congestion and the information shown in Table 3, respectively.

Table 3 Congestion Level and Information Provided for Scenarios 1 and 2

Congestion Level of Hanshin Expressway	Information Provided	
	Jam Length on Hanshin Expressway	Travel Time of Hanshin Expressway
Slight Congestion	2km	25 minutes
Medium Congestion	5km	40 minutes
Heavy Congestion	9km	60 minutes

Table 4(a) Congestion Level and Information Provided for Scenario 3a

Congestion Level		Alternative Route: Slight Congestion	Alternative Route: Medium Congestion	Alternative Route: Heavy Congestion
Hanshin Expressway	Slight Congestion	2km 25 minutes *35 minutes*	2km 25 minutes *45 minutes*	2km 25 minutes *55 minutes*
	Medium Congestion	5km 40 minutes *35 minutes*	5km 40 minutes *45 minutes*	5km 40 minutes *55 minutes*
	Heavy Congestion	9km 60 minutes *35 minutes*	9km 60 minutes *45 minutes*	9km 60 minutes *55 minutes*

Key (top to bottom in each cell): Jam Length of Expressway / Travel Time of Expressway / *Travel Time of Alternative Rt.*

Table 4(b) Congestion Level and Information Provided for Scenario 3b

Congestion Level		Alternative Route: Slight Congestion	Alternative Route: Medium Congestion	Alternative Route: Heavy Congestion
Hanshin Expressway	Slight Congestion	2km 25 minutes *30 minutes*	2km 25 minutes *40 minutes*	2km 25 minutes *45 minutes*
	Medium Congestion	5km 40 minutes *30 minutes*	5km 40 minutes *40 minutes*	5km 40 minutes *50 minutes*
	Heavy Congestion	9km 60 minutes *45 minutes*	9km 60 minutes *55 minutes*	9km 60 minutes *60 minutes*

Key (top to bottom in each cell): Jam Length of Expressway / Travel Time of Expressway / *Travel Time of Alternative Rt.*

Table 4 shows the combinations of congestion level on both the Hanshin Expressway and the alternative route, and information provided in Scenario 3. Each cell in Table 4, corresponding to the combination of congestion level between the Hanshin Expressway and the alternative route, contains three numbers representing the assumed information provided to the respondents. The first number from the top in each cell indicates the information about jam length on the Hanshin Expressway. The second and third numbers indicate the travel time information for the Hanshin Expressway and the alternative route, respectively. In Scenario 3, respondents were required to state their route preference in nine different possible situations corresponding to the levels of congestion and the information provided, as shown in Table 4.

We prepared two different questionnaire sheets: *Questionnaire A* and *Questionnaire B*; they differ with respect to the SP survey included. In the SP survey in *Questionnaire A*, the respondents were assumed to obtain the traffic information only after entering the Hanshin Expressway. That is, *Questionnaire A* included Scenarios 1a, 2a and 3a. In the SP survey of *Questionnaire B*, it was assumed that the traffic information was provided to the respondents before entering the Hanshin Expressway. Thus, *Questionnaire B* included Scenarios 1b, 2b and 3b. Each respondent received either *Questionnaire A* or *Questionnaire B* at the tollbooth of Hanshin Expressway.

Estimation of Route Choice Model

The stated preference on route choice is used to analyze the relationship between information regarding traffic accidents and route choice behavior according to the logit type route choice models. Based on the estimated parameters of these models, the influences of both the timing and the content of information provision on the respondents' route choice are analyzed. The results of this analysis will be discussed later. The estimated parameters of route choice models are shown in Table 5. In this study, three types of route choice model are built, corresponding to the three scenarios described above (Scenarios 1, 2 and 3).

As shown in Table 5, each model includes the information provided both before and after entering the Hanshin Expressway as explanatory variables to allow explicit analysis of the influence of the timing of information provision. For each route choice model, it is possible for us to directly compare the parameter related to the information before entering the expressway with that after entering the expressway. The route choice models also include a number of socioeconomic attributes (gender, age and occupation) and the frequency of use of alternative routes, as well as the respondents' attitude toward the detour as dummy explanatory variables. Table 5 shows only the statistically significant parameters. In this study, the LOGISTIC procedure of the Statistical Analysis System (SAS) was used to estimate the parameters of the route choice models (SAS Institute Inc., 1990). The LOGISTIC procedure was used to statistically test a number of variables and to select suitable explanatory variables in order to maximize the likelihood of the estimated route choice model. Accordingly, all the estimated parameters shown in Table 5 are statistically significant. Judging from the likelihood ratio index and the hit ratio, it seems valid to analyze and discuss the relationship between the traffic information provided and route choice behavior, based on the estimated parameters shown in Table 5.

Before discussing the relationship between the information provided and the respondents' route choice behavior, we analyze the impact of route switch attitude and behavior in the real environment. The estimated parameters of *low frequency of use of alternative route* have a negative sign for all scenarios. This implies that the respondents who are unfamiliar with an alternative route tend not to make route switches, as compared with those who are familiar with the alternatives.

Table 5 Estimated Parameters of the Logit Type Route Choice Model

Type of Information \ Scenarios		Scenario 1	Scenario 2	Scenario 3
Traffic Jam of Hanshin Expwy		provided	provided	provided
Travel Time of Hanshin Expwy			provided	provided
Travel Time of Alternative Route				provided
Specific Dummy Variables of Hanshin Expwy	Constant	3.544 *24.698*	3.887 *25.406*	2.191 *13.157*
	Gender: Female	0.716 *3.088*	- -	- -
	Age: 60 or higher	- -	- -	0.239 *2.480*
Specific Dummy Variables of Alternative Route	Occupation: Transportation & Communication	0.753 *2.658*	- -	- -
	High Frequency of Use of Alternative Route	- -	0.479 *2.448*	0.474 *4.107*
	Low Frequecy of Use of Alternative Route	-0.585 *-4.276*	-0.634 *-4.514*	-0.771 *-8.641*
	Negative Attitude toward Route Switch	-1.632 *-12.099*	-1.381 *-10.349*	-0.972 *-11.953*
	Positive Attitude toward Route Switch	0.961 *4.859*	1.022 *5.094*	0.812 *6.859*
Information Provided ***after*** Entering Expwy	Information on Traffic Jam of Hanshin Expwy	-0.543 *-23.798*	-0.558 *-24.060*	- -
	Information on Travel Time of Hanshin Expwy	 -	 -	-0.149 *-24.735*
	Information on Travel Time of Alternative Route	- -	- -	-0.1322 *-19.673*
Information Provided ***before*** Entering Expwy	Information on Traffic Jam of Hanshin Expwy	-0.593 *-25.123*	-0.608 *-25.435*	- -
	Information on Travel Time of Hanshin Expwy	- -	- -	-0.122 *-33.767*
	Information on Travel Time of Alternative Route	- -	- -	-0.089 *-19.017*
Sample Size		2763	2763	8289
Degree of Freedom		8	7	10
Likelihood Ratio Index (ρ^2)		0.391	0.406	0.462
Hit Ratio (%)		80.4	80.7	83.5

* Italic numbers indicate t-statistics for estimated parameters.

As all the estimated parameters related to information are statistically significant, the provision of information may affect respondents' route choice. However, the extent to which the information affects the respondents' decision-making might depend on their attitude toward the detour. There is a coincidence between respondents' attitude to detours and their route choice preference as determined in the SP survey. The negative sign of the parameter *negative attitude towards detours* indicated that the respondents belonging to this category show a greater tendency not to use alternative routes, as compared with those in the other categories. Also, the positive sign of the parameter *positive attitude toward detours* implies that the respondents in this category are more apt to use an alternative route, as compared with those in the other categories. Taking both these parameters, and the parameters of information regarding traffic jams of the expressway in Scenarios 1 and 2, into consideration, it appears that the provision of information may encourage more respondents belonging to the category *positive attitude towards detours* to take an alternative route than it does those in the *negative attitude towards detours* category.

Influence of the Timing of Information Delivery

In this subsection, we focus on the estimated parameters of information provision. This subsection focuses, in particular, on the influence of the timing of information delivery on driver route choice behavior.

In Scenarios 1 and 2, the estimated parameter of information regarding traffic jams of the Hanshin Expressway is significant, as determined from t-statistics. Although the respondents are provided with information regarding both traffic jams and travel time of the Hanshin Expressway in Scenario 2, they seem to attach more importance to information regarding traffic jams in route choice decision-making. The absolute values of parameters of information on the traffic jam provided before entering the expressway are about 9% higher than those after entering the expressway. In the SP survey, it is assumed that the respondents are required to pay 600 Yen as a toll for the expressway. Thus, it is possible that the toll, which the respondent had already paid, may reduce the influence of information on route switching. Accordingly, it is possible that providing drivers with information regarding traffic jams on the Hanshin Expressway prior to entering the expressway may have a stronger influence on their route choice than the provision of such information after entering the expressway.

In Scenario 3, in which the respondents obtained information regarding the traffic jam of the Hanshin Expressway, and information on travel times for both routes, the estimated parameters of information on travel time for both routes are shown by t-statistics to be significant. As shown in Table 5, the differences in estimated parameters of travel time information, provided before entering the expressway, between the expressway and the alternative route, are larger than those in the situation where the respondents obtain information after entering the expressway. The estimated parameter of travel time information on the alternative route is just 73% of that on the Hanshin Expressway, in the case where the respondents obtain information before entering the expressway.

To clearly explain the influence of the timing of information provision in Scenario 3, a simple example will be described here. Our analyses concentrate on the relationship between the estimated parameters related to information and the respondents' utility related to route choice only. First, the travel time of the Hanshin Expressway according to the information provided, is assumed to be 30 minutes, either before or after entering the expressway. From the viewpoint of the influence of the information on driver route choice, when the respondents obtain the information before entering the expressway, a travel time of the alternative route of 41 minutes become equivalent to the information for the Hanshin expressway, as noted above. That is, if the travel time of the alternative route, according to information provided, is less than 41 minutes, the chance that a driver would make use of the alternative route instead of the Hanshin Expressway, is greater. In the case where the respondents are assumed to obtain information after entering the expressway, a travel time of 34 minutes of the alternative route become equivalent to the information on the Hanshin expressway, as noted above. The drivers' use of the alternative route, rather than the Hanshin Expressway, on the basis of travel time information may be higher only when the travel time on the alternative route is less than 34 minutes. Accordingly, focusing on the influence of the timing of information delivery on driver route choice / switch behavior, providing drivers with the information before they enter the expressway tend to encourage them to use the alternative route. This effect does not occur to the same extent when the information is provided after entering the expressway.

Influence of the Content of Information Provided

This subsection aims to analyze the influence of the content of the information provided on driver route choice behavior. That is, it is necessary to compare and analyze route choice behavior in different scenarios. However, it is difficult to directly compare the estimated parameters of a logit model with those of other logit models, because the scale for measuring utility differs between models. Instead, the estimated choice rates for the Hanshin Expressway are used for the analysis of the relationship between the content of the information provided and driver route choice behavior. The sample enumeration method for the disaggregate model can be applied to estimate the choice rate for the Hanshin Expressway (Ben-Akiva, 1985).

Figure 8 shows the relationship between the rates of choice for the Hanshin Expressway and the content of the information provided. Regardless of differences in information content, the changes in choice rates in Scenario 1a are very similar to those of Scenario 2a. In Scenario 3a, where the respondents are assumed to obtain travel time information for both the Hanshin Expressway and the alternative route, the rate of choice for the Hanshin Expressway is dependent on the level of congestion on the alternative route; greater degrees of congestion on the alternative route are associated with a higher frequency of choice of the Hanshin Expressway. The changes in choice rate in Scenario 3a, with light congestion on the alternative route, are similar to those of Scenarios 1a and 2a. As described above, no information on the traffic conditions for the alternative route are provided to respondents in Scenarios 1a and 2a.

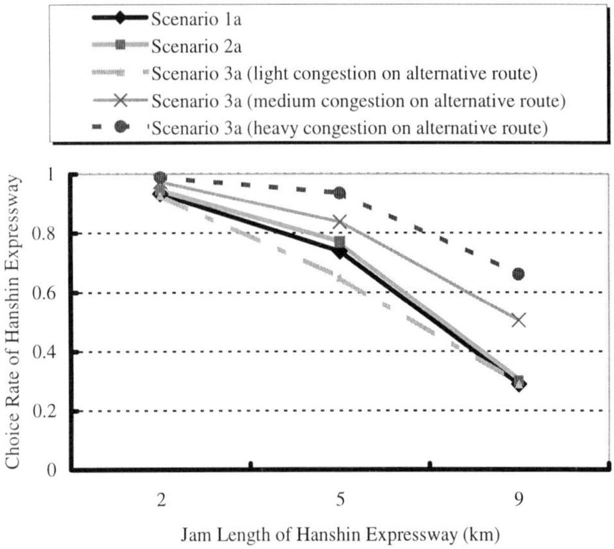

Figure 8 Relations between Expressway Choice Rate and Content of Information Provided

The results of the analyses described above suggest that some respondents might take a lack of information regarding the alternative route as an indicator of good traffic conditions. That is, when only information regarding the expressway is available, it is possible that drivers may overreact to the information provided, and that the diverted traffic volume might be greater than that which the alternative route can accept. The above discussion implies that it is necessary for traffic control systems to provide drivers with information regarding the alternative routes designated to accept the diverted traffic.

Concluding Remarks

To enhance the effectiveness of dynamic information systems, and to ensure their applicability as a new means of controlling the influence of non-recurrent events, in addition to enhancing travel time reliability, it is necessary to investigate the software aspects of information systems. These aspects include information content and the timing of information delivery. In this study, a questionnaire survey was conducted to investigate driver attitudes to detours and their responses to information provided. The findings that are useful for the design of the software aspects of systems for information provision can be summarized as follows:

1) A lack of knowledge regarding alternative routes may discourage drivers from using such routes; provision of appropriate information that enhances driver knowledge about the road network and traffic conditions may encourage diversion to alternative routes.
2) The estimated parameters of route choice models suggest that the timing of information delivery, and the content of the information provided on accidents and related traffic conditions, should be regarded as crucial factors affecting route choice behavior.
3) Providing drivers with information before entering the expressway tend to encourage them to use an alternative route, as compared with information that is provided after entering the expressway.
4) Where no traffic information regarding an alternative route is available, drivers may overreact to the information provided regarding the expressway.

In this study, we analyzed the potential effectiveness of traffic information provision in preventing both inefficient traffic conditions and the deterioration of travel time reliability, from the viewpoint of driver response to the information. However, to discuss the potential effectiveness of traffic information in a more precise manner, it is necessary to consider the relationship between drivers' decision-making, as affected by the information provided, and the traffic conditions on the network. When an accident occurs on a road network, the question of whether drivers should be encouraged to divert to an alternative route is dependent on both the severity of the accident and the level of congestion of the alternative route. From the viewpoint of network flow analysis, the optimal (or suitable) level of diversion to an alternative route should be analyzed and discussed using the traffic network simulation technique. We are currently planning future studies to perform such analyses.

ACKNOWLEDGMENTS

This study was supported by grants from the Ministry of Land, Infrastructure and Transport, Japan. The authors are grateful to the members of the Committee of Road Transport Informatics for Improving Efficiency (ROTIIE), part of the Japan Society of Civil Engineers, for their helpful suggestions. We are indebted to Hanshin Expressway Public Corporation and the Institute of Urban Transport Planning Co., Ltd. for conducting the questionnaire survey.

REFERENCES

Abdel-Aty, M.A., Kitamura, R. and Jovanis, P.P. (1997), 'Using Stated Preference Data for Studying the Effect of Advanced Traffic Information on Drivers' Route', Transportation Research -C **Vol.5C No.1** pp.39-50

Arnott, R., de Palma, A. and Lindsey, R. (1991), 'Does Providing Information to Drivers Reduce Traffic Congestion?', Transportation Research -A, **Vol.25A, No.5**, pp.309-318.

Ben-Akiva, M. (1985), 'Discrete Choice Analysis', MIT Press, pp.253-275.

Ben-Akiva, M., de Palma, A. and Kaysi, I. (1991), 'Dynamic Network Models and Drivers Information Systems', Transportation Research -A, **Vol.25A, No.5**, pp.251-266.

Bonsall, P.W. and Palmer, I.A. (1999), 'Route Choice in Response to Variable Message Signs: Factors Affecting Compliance', Behavioral and Network Impacts of Driver Information Systems, Ashgate, pp181-214.

Hato, E, Taniguchi, M. and Sugie, Y. (1995), 'Influence of Traffic Information on Drivers' Route Choice', Paper presented at the Seventh World Conference on Transport Research, Sydney.

Horowitz, J.L. (1985), 'Travel and Location Behavior: State of the Art and Research Opportunities', Transportation Research -A, **Vol.19A, No.5/6**, pp.441-453.

Iida, Y. (1999), 'Can Intelligent Transport System Help Solve Urban Traffic Problem?', '99 Shanghai International Symposium on Urban Transportation Proceedings, pp.16-22.

Iida, Y., Uno, N. and Yamada, T. (1999), 'Experimental Analysis of Effects of Travel Time Information on Dynamic Route Choice Behavior', Behavioral and Network Impacts of Driver Information Systems, Ashgate, pp215-239.

Khattak, A.J, Koppelman, F.S. and Schofer, J.L. (1993), 'Stated Preference for Investigating Commuters' Diversion Propensity', Transportation, **Vol. 20**, pp.107-127.

Lai, K. and Wong, W. (2000), 'SP Approach toward Driver Comprehension of Message Formats on VMS', Journal of Transportation Engineering, May/June, pp.221-227.

SAS Institute Inc. (1990), 'The LOGISTIC Procedure', SAS/STAT User's Guide, Vol.2, Version 6, 4th Edition, pp.1071-1126.

Wardman, M. R., Bonsall, P. W. and Shires, J. D. (1997), 'Stated Preference Analysis of Driver Route Choice Reaction to VMS Information', Transportation Research -C, **Vol.5C, No. 6**, pp.441-453.

12

Drivers' Cognition and Dynamics of Traffic-Condition Induced by Network Modification

Satoshi Fujii, Department of Civil Engineering, Tokyo Institute of Technology, Tokyo, Japan
Ryuichi Kitamura, Department of Civil Engineering Systems, Kyoto University, Kyoto, Japan

Abstract

In this paper, we discussed relations between drivers' cognition and the dynamics of traffic conditions. When we take into account drivers' cognition, many types of psychological impedance to behavioral change could be accounted for, including cognitive conservatism and habitual decision making. Based on psychological theories with respect to these impedance to behavioral change, we elucidated the process of changes in traffic conditions that might be induced by changes in road-network structure. As a result of the discussion, we concluded that traffic conditions are unlikely change as much as predicted by theories that assume a simple, and sometimes unique, relation between a traffic condition and a road-network structure, e.g., theories that assume user equilibrium.

Introduction

The dynamics of macroscopic traffic condition depends directly on how drivers' behaviors change. For example, even when a new road is constructed, traffic condition would not change at all unless travelers change their travel modes, destinations, routes or departure times. Traffic condition would change if and only if travelers' behaviors change. In turn, traveler' behaviors depend on their cognitions or mental representations of travel condition; in case of drivers, traffic condition. One of the most important mental representations of traffic condition for drivers' behaviors, would be *anticipated travel time*, that is, drivers' subjective travel time to a destination using a particular route. If a driver anticipates that the travel time along a route may decrease or increase for some reasons, e.g., changes in road-network structure, he may change his route. If many drivers change their routes due to such reasons, traffic conditions would change. We have to keep in mind that the dynamics of macroscopic traffic condition is dependent on change in drivers' behaviors, and the change in drivers' behaviors, in turn, is motivated by the change in drivers' cognition. It is also noted that the network reliability also rests on drivers' cognition; it is not guaranteed that the objectively defined reliability of a network coincides with its reliability as conceived by a driver. Thus, for the purpose of understanding the dynamics of traffic condition or network reliability, it is imperative that the drivers' cognition be explored.

In this paper, the dynamics of traffic condition induced by a change in road-networks, which would be one of the most straightforward determinants of the dynamics, will be discussed, while accounting for drivers' cognition. The discussion will be tripartite. First, we will review studies on anticipated travel time including one by the authors (Fujii & Kitamura, 2000), and discuss the cognitive conservatism that may limit effects on anticipated travel time of information on traffic condition. Second, we will review studies investigating drivers' information acquisition and discuss the reluctance drivers have toward behavioral change. Finally, we will discuss the process of traffic-condition changes induced by changes in rode-network structure, while accounting for drivers cognitive aspects including anticipated travel time and information acquisition. The concluding discussion will indicate it is unlikely that traffic conditions change as much as some network researchers and practitioners might believe.

Information Dominance Hypothesis and Experience Dominance Hypothesis

Studies on Anticipated Travel Times

Several studies have examined anticipated travel time empirically and theoretically (Bonsall, 1991; Chang & Mahmassani, 1998; Emmerink et al., 1993, 1994; Iida et al., 1992; Kobayashi, 1994; Kobayashi and Yasuno, 1998; Mahmassani & Chang, 1998; Nakayama et al., 1999; Polak & Oladeinde; 1999). There are yet areas that remain to be investigated. First, although these studies mainly investigated the effect of past driving experiences on the future anticipated travel time, the accuracy of subjective prediction of travel time has not been well investigated. As far as the authors know, Kobayashi and his colleague (Kobayashi, 1994; Kobayashi and Yasuno, 1998) investigated the accuracy of anticipated travel time, but it did not offer knowledge why and when such accuracy materializes. Second, the past studies investigated the anticipation of travel time when drivers use a road network repeatedly as, for example, commuting drivers do. However, the anticipation of travel time under unfamiliar condition has not been well understood. Third, some studies investigated the anticipation of travel time when travel time information is provided by roadside variable message signs or advanced traveler information systems (Bonsall, 1991; Emmerink et al., 1993, 1994; Iida et al., 1992), but they have not investigated the anticipation when information is acquired through other media (e.g. mass media, telephone services, or word-of-mouth).

To summarize, the followings have not been well investigated in the past studies:

- drivers' ability to accurately predict travel time
- anticipated travel time under unfamiliar condition
- the effect of qualitative information other than travel time information on anticipated travel time

Hypotheses

In order to understand these aspects of anticipated travel time that have not been well investigated in the past studies, the authors proposed the *Information Dominance Hypothesis* and the *Experience Dominance Hypothesis* with respect to formation of anticipated travel time, and tested these hypotheses using empirical data not from laboratory experiment, but from the real world (Fujii & Kitamura, 2000).

Before explaining these hypotheses, basic two concepts underlying these two hypotheses should be explained, i.e., the *state dependence* and the *information effect*. It has been demonstrated repeatedly that a behavioral or psychological state is influenced by the past state (Kitamura & Bunch, 1990; Heckman, 1981; Oullette & Wood, 1998). Therefore, we assume that an anticipated travel time affects anticipated travel times held in the future. This is labeled state dependence. On the other hand, the information effect is associated with the driver's ability to predict travel times. When the driver's predictive ability is high, anticipated travel time becomes correlated with actual travel time, because predictions match actual values more accurately as the predictive ability increases (Tversky & Kahneman, 1974). We labeled this as information effect, since predictive ability might be influenced by the amount and quality of information an individual has acquired.

Using these two basic concepts, the Information Dominance Hypothesis can be stated as follows;

Information effects become larger as the driver acquires more information on travel time, and dominate state-dependence effects.

This hypothesis predicts that if a driver acquires information on the travel time of a route from, e.g., telephone information services, radio traffic reports, or actual driving experiences on the route, the driver should be able to predict travel time more accurately than could a driver without such information. It is also predicted that if the driver is capable of anticipating travel times well based on the information he has acquired, it would not always be necessary to refer to anticipated travel times from the past.

On the other hand, the Experience Dominance Hypothesis states;

The influences of generic information on the anticipated travel time are weaker with actual driving experience than without it.

This hypothesis is based on the notion that the information provided by the media, e.g. TV, radio or telephone services, is, under the current technology, generic and not specific to each driver or each trip, whereas actual experience provides information specific to the driver, i.e. actual door-to-door travel time for his trip (the former type of information was called "generic information"). The hypothesis predicts that driving experience, especially the initial one, affects substantially the anticipated travel time and tends to negate the influences of generic information.

The authors have empirically tested the hypotheses using data from actual commuting drivers in a real road network, not from laboratory experiments (Fujii & Kitamura, 2000). In the followings section, the empirical test was briefly described.

Empirical Test

The data used for the test were collected in a multi-day survey which targeted commuters on the Hanshin Expressway Sakai Route, which is a toll road and was closed for maintenance works for 9 consecutive days from November 1 till 8, 1998. A total of 370 drivers who commute regularly on the Sakai Route, were recruited before the closure. The recruited drivers were requested to answer questions every day during the freeway closure (November 1 through November 8), and also on the day before the closure (October 31). The questions on the day before the closure were about the respondent's personal attributes and regular commuting behavior. The questions during the freeway closure were about the commuting behavior and information acquired on each day. On each day, respondents were requested to report the commuting travel mode, commuting route if automobile was used, anticipated travel time, and whether or not they obtained any information on traffic condition or travel time through mass media, telephone information services, or word-of-mouth. With respect to anticipated travel time, we asked actual starting time for commuting and anticipated arrival time at their work places that was formed before departure, and we calculated anticipated travel time using these two reported points of time afterward.

Table 1 Results of statistical test of the Information Dominance Hypothesis and the Experience Dominance Hypothesis

		1th-commuting day[a]		2nd-commuting day[a]		difference between 1th and 2nd commuting day the impact of experience[b]	
	parameter	coef.	t	coef.	t	coef.	t
State Dependence Effect[c]	*sd*	0.80	3.57****	0.22	2.91***	-0.58	-2.45**
	is(Mass-Media)	-0.78	-3.07***	-0.36	-1.55	0.42	1.20
	is(Telephone)	-1.39	-2.30**	0.39	0.59	1.78	2.00**
	is(Word-of-Mouth)	0.71	2.02**	-0.32	-1.92*	-1.03	-2.64**
Information Effect[d]	*pp*	0.18	1.10	0.57	6.47****	0.38	2.02**
	ip(Mass-Media)	0.63	2.64**	0.43	1.98**	-0.20	-0.63
	ip(Telephone)	1.00	2.26**	-0.28	-0.57	-1.28	-1.93*
	ip(Word-of-Mouth)	-0.53	-1.51	0.28	2.08**	0.81	2.15**

Sample size = 41 $\quad {}^{*}p<0.1$, ${}^{**}p<0.05$, ${}^{***}p<0.01$, ${}^{****}p<0.001$

[a] The coefficients and their t-statistics are estimated by LISREL 8.
[b] The coefficients and their t-statistics are calculated based on the estimation results of [a].
[c] The size of State Dependence Effect = *sd* + *is*(Mass-Media) + *is*(Telephone) + *is*(*Word-of-Mouth*)
[d] The size of Information Effect = *pp* + *ip*(Mass-Media) + *ip*(Telephone) + *ip*(*Word-of-Mouth*)

(This table was modified from the table reported in Fujii & Kitamura (2001))

Using these data, the authors found the following results from statistical analysis (see Table 1):

1) The state dependence for drivers who had information through mass media or telephone was significantly smaller than that for drivers who did not have such information.

2) The information effect, i.e. ability to accurately predict travel time, for drivers who had information through mass media or telephone was significantly larger than that for drivers who did not have such information.

3) The state dependence for drivers who had an experience to drive during the freeway closure was significantly smaller than that for drivers who had no experiences to drive during freeway closure.

4) The information effect for drivers who had an experience to drive during the freeway closure was significantly larger than that for drivers who had no experiences to drive during freeway closure.

These four results clearly supported the Information Dominance Hypothesis. With respect to the Experience Dominance Hypothesis, we obtained the following results.

5) When drivers did not have any driving experiences during the freeway closure, the information effect for drivers who had information through mass media or telephone was significantly larger than that for drivers who did not have information. On the other hand, for the drivers who had an experience to drive during the freeway closure, the effect of information was not significant. In other words, a driver's ability to accurately predict was enhanced by information acquisition through mass media and telephone, only when they had no experience.

6) Similarly, state dependence of anticipated travel time decreased by information acquisition through mass media and telephone, only when they had no driving experience.

These two results supported the Experience Dominance Hypothesis.

However, the results for word-of-mouth information were not always consistent with the Information Dominance Hypothesis and the Experience Dominance Hypothesis. What we found with respect to word-of-month were as followings:

7) When drivers did not have any experiences to drive during the freeway closure, the state dependence for drivers who had word-of-month information was significantly larger than that for drivers who did not have such information.

8) On the other hand, the information effect for drivers who had word-of-mouth information was not significantly different from that for drivers who did not have such information.

These findings were against the Information Dominance Hypothesis. Since the Experience Dominance Hypothesis was based on the presence of the effects assumed in the Information Dominance Hypothesis, the Experience Dominance Hypothesis was reasoned to be against the data as well. However, the following findings were in line with the Information Dominance Hypothesis:

9) When all the drivers had experiences to drive during the freeway closure, the state dependence for drivers who had word-of-month information was significantly smaller than that for drivers who did not have such information.

10) When all the drivers had experiences to drive during the freeway closure, the information effect for drivers who had word-of-mouth information was significantly larger than that for drivers who did not have such information.

In other words, the findings through 7) to 10) indicate that word-of-mouth information enhances information effect and decreases state dependence on the second commuting day during freeway closure, whereas it did not have any effects on information effect and state dependence in such directions on the first commuting day. One plausible explanation of these results is that information gained by word-of-mouth when everybody was ignorant of the traffic condition was not guaranteed to provide accurate information, whereas relatively more accurate information might be exchanged by word-of-mouth after some driving experience had been gained. The hypotheses would hold true only if the information being acquired has some level of accuracy.

To summarize, although information that did not have some level of accuracy, e.g. word-of-mouth information, did not have as much impact so as to be consistent with these two hypotheses, information that had some level of accuracy, e.g., information through mass media and telephone-based information services, had such impacts that were consistent with the two hypotheses.

COGNITIVE CONSERVATISM

Implications of the Hypotheses

The Information Dominance Hypothesis implies that information acquired by travelers about their trips influences their anticipated travel times. It has been demonstrated in the past studies that anticipated travel time was influenced by several types of information (Iida et al., 1992; Emmerink, et al., 1993, 1994; Bonsall, Pickup and Stathopoulos, 1991). However, the hypothesis states that acquired information enhances predictive ability and weakens state dependence. These cognitive, causal relations of information acquisition and anticipated travel time were not hypothesized and empirically investigated in the past studies. Therefore, the empirical evidence that supported the hypotheses has increased our knowledge about cognitive and psychological aspects of formation of anticipated travel times.

The Experience Dominance Hypothesis has the following two implications with respect to traffic information policies. The first is that it is not difficult to influence travel behaviors of drivers without any driving experiences, by simply providing information. This is because the hypothesis implies that the impacts of information on anticipated travel time are larger for drivers without any driving experiences than those with experiences. This, furthermore, implies that advanced traveler information systems are especially effective to, for example, rental car drivers or sightseeing drivers, since they might not have driving experience in the area.

The second implication of the hypothesis is that it is difficult to influence the travel behaviors of drivers who have driving experiences in the particular area by providing information. Such drivers might believe that they know the traffic condition very well even if their beliefs were not compatible with objective traffic condition. Accordingly, they might not trust generic information including that from traveler information systems, since they might have self-confidence on their experiences. This is because effects of experiences dominate effects of any other generic information, as the Experience Dominance Hypothesis implies. This implies that the impacts of advanced traveler information systems on traffic flow might be smaller than one might expect without the Experience Dominance Hypothesis.

Cognitive Conservatism

The Experience Dominance Hypothesis implies that drivers are cognitively conservative, since the hypothesis states that anticipated travel time did not largely change due to acquired generic information once it was formed based on driving experience. In fact, the *cognitive conservatism* is a well know cognitive phenomenon. For example, it has been demonstrated repeatedly that subjective probability changes due to additional information to smaller extent than Bayes' theory would predict (c.f. Anderson, 1990).

The authors also obtained the following statistical result which would directly illustrate drivers' cognitive conservatism (Fujii & Kitamura, 2000).

11) The state dependence effect from an anticipated travel time on the day before freeway closure to that on the first day during freeway closure, was statistically significant, although the freeway closure had large impacts on traffic condition and drivers should know the fact that the traffic condition had changed due to the closure.

However, we also found the following;

12) The state dependence effect from an anticipated travel time on the day before freeway closure to that on the first day during freeway closure, disappeared when they acquired mass-media information or telephone information.

The finding 11) indicates that drivers did have cognitive conservatism, but the finding 12) indicates that the cognitive conservatism may disappear, if they did not have any experience of driving in a new traffic environment, and if they have credible traffic information about that.

Overcoming Cognitive Conservatism

In order to modify traffic conditions by means of policies including information provision and changes in road-network structure, anticipated travel time should be influenced by the policies. However, as argued above, drivers seems to be conservative when modifying their anticipated travel time. Only the policies which overcome drivers' cognitive conservatism might be able to change their behaviors and the macroscopic traffic condition. The above findings from the authors' data indicate that the conditions for overcoming cognitive conservatism are;

a) drivers acquire information,
b) drivers believe the information is credible,
c) drivers believe the actual current traffic condition is different from what they used to know.

If all these three conditions were realized, policy instruments including traffic-information provision might change drivers' anticipated travel time and their behaviors. Unless drivers acquire any information, there is no hope of change, i.e., condition *a*) is the necessary condition. That condition *b*) is necessary, is indicated by the results that less credible information, i.e., word-of-mouth information on the first commuting day, did not have any impacts on anticipated travel time.

One may ask if condition *c*) must be *sufficient* condition, rather than a necessary condition, because rational drivers must change their anticipated travel time, when they know the objective traffic condition has been changed. However, the finding 11) demonstrated that drivers did *not* change their anticipated travel time even if they know the existence of change in objective traffic condition. Thus, *c*) is not a sufficient condition, but necessary condition.

HABIT AND INFORMATION ACQUISITION

Changes in Anticipated Travel Time and Changes in Road-Network Structure

The authors believe that the fact that condition *c)* is not a sufficient condition is important for theoretical understanding of dynamics of traffic condition in real road-networks. In real road-networks, the network structure changes gradually, e.g., adding new links or expanding road capacity. The freeway closure in the authors' study (Fujii & Kitamura, 2000) was also an example of structural change in road network. It would have been easier for drivers to know the fact of change itself, than to know traffic condition after the structural change. Therefore, many drivers knew only the fact of structural change, but were ignorant about traffic condition. If the condition *c*) was not a sufficient condition but just a necessary condition, many drivers would *not* change their anticipated travel time and their behavior, even if they knew the fact of structural change in a road network, e.g., road constructions or expansion of road capacity. Thus, traffic condition may not change as drastically as predicted without theories of drivers' cognition.

Of course, if the condition *a*) and *b*) were satisfied in addition to condition *c*), the drivers might change their anticipated travel time and their behaviors. Namely, if they actually acquired traffic information corresponding to the modified road-network structure in addition to know the fact of the existence of the modification of the road-network structure and they believed in the credibility of the information, they might modify their anticipated travel time. Therefore, although an information provision system is not sufficient for efficient use of road network, it would be still necessary especially after road-network modification.

Now, it is an important question for understanding the dynamics of traffic condition induced by modifications of the road-network structure, to know whether or not drivers do actually acquire information about the traffic condition. However, psychological studies have demonstrated that people tend *not* to acquire new information if they have developed a *habit,* and people easily develop habits.

Do Drivers Acquire Information?

One of the authors and colleagues proposed a model of formation of driving habits (Gärling et al., 2001). In the model, the decision making process of travel mode choice is hypothesized to become more automatic as the decision maker repeats driving. People who do not have experiences to repeatedly use a travel mode are hypothesized to choose based on their attitudes toward travel modes after they acquired information with respect to travel modes. However, they are hypothesized to acquire information less frequently and to refer to their own attitudes less weakly, as they repeat always choosing an alternative, e.g., automobile. Thus, the model predicts that people develop habits, that is automaticity in decision making, after repeated behaviors.

A survey and a laboratory experiment were conducted to test the model (Gärling, et al., 2001). In a survey study (n = 60), the statistical test using the data confirmed the hypothesized causal relations between attitude towards driving, self-reported frequency of driving choice, and degree of automaticity in decision making. For a laboratory experiment, 50 highschool students were recruited who did not have a driving license, who were expected not to have developed a driving habit at all. They were randomly assigned into two experimental groups each consisting of 20 subjects and one control group consisting of 10 subjects. They were requested to repeatedly choose between driving and walking. The subjects were isolated and seated in a sound-proof room facing a computer screen on which the choice tasks were presented.

In the first 16 choice tasks for one experimental group or 32 choice tasks for another experimental group, the attributes of driving always dominated those of walking (these tasks were called learning tasks). Therefore, almost all (i.e. approximately 95%) subjects in both experimental groups always chose driving. On the other hand, in the first 16 choice tasks for a control group, the attributes of driving did not always dominate those of walking. The control-group subjects therefore chose both driving and walking. The situation was thus produced that subjects in experimental groups choose driving repeatedly and subjects in a control group did not.

After these first 16 or 32 choice tasks, the subjects in all groups were requested to perform 8 choice tasks where the attributes of walking always dominated those of driving. Since the subjects were not explicitly told that the characteristics of choice tasks had been changed, some of them might not have noticed the fact unless they acquired all the information displayed in the screen. Therefore, the travel modes chosen by subjects indicated whether or not they acquired information. The subjects in control group chose walking at the rate of 97.5%. This indicated that most of them acquired information about the travel modes displayed in computer screen. However, the rate of choosing walking for an experimental group with 16 learning tasks was 59.4%, and that for an experimental group with 32 learning tasks was 45.0%. Namely, the subjects who were forced to choose to drive repeatedly did not acquire information even though the information was displayed in the screen *in front of them*, and this tendency to ignore displayed information increased with the number of times of repeated driving in the learning. The data thus confirmed the model proposed by the model by Gärling et al. (2001).

One may suspect that this result was just an artifact rather than a fact, since the result was from a laboratory experiment. However, the model was also tested by survey data as well as laboratory experiment. In addition, similar results were also obtained by Verplanken et al. (1997) and the authors and colleagues (Nakayama et al., 2002).

An implication of the experiment by Gärling et al. (2001) was that people might easily develop habits, since the experimental data indicated that 16 or 32 forced choices of driving made subjects ignore displayed information. Other evidence with respect to the strength of driving habit of actual drivers was reported in Fujii et al. (2001) that also used the survey data used in Fujii & Kitamura (2000). It was reported that 65.8% respondents of the survey commuted by automobile everyday, and 23.4% respondents commuted by automobile 80% of the time or more. Therefore, approximately 90% of respondents used automobile 80% of the time or more. The data indicated that most drivers seem to have developed a driving habit.

Changes in Traffic Condition

The authors have described some theoretical concepts with respect to drivers' cognition that may be necessary for understanding the dynamics of traffic condition induced by changes in road-network structure, including the Information Dominance Hypothesis, the Experience Dominance Hypothesis, cognitive conservatism, and habit. Using these concepts, the process of traffic-condition changes induced by changes in the road-network structure is described in Figure 1.

There are two types of changes in road-network structure, that is enhancement of service level and reduction of service level. The former includes construction of new routes and expansion of road capacity. The latter includes closure of existing routes and reduction of road capacity. The freeway closure investigated in Fujii & Kitamura (2000) was an example of reduction of service level in road-network. There is a difference between these two. In the case of a reduction of service level, some drivers are *forced* to change in their routes or travel mode; but any drivers are not forced to change in the case of an enhancement of service level. For example, all the drivers who routinely used Hanshin Expressway Sakai Route had to change their behaviors when the route was closed. In such a case, we do not need to understand drivers' cognition for understanding why and when drivers' behaviors would change, since they were just forced to.

However, in the case of an enhancement of service level, drivers would not be forced to use, for example, a new route. If they change their route, the behavioral change would be voluntary. Even in case of a reduction of service level, behavioral change of drivers who are not forced to change their behavior, would be also voluntary. Accordingly, we have to understand why and when such drivers' behaviors would change.

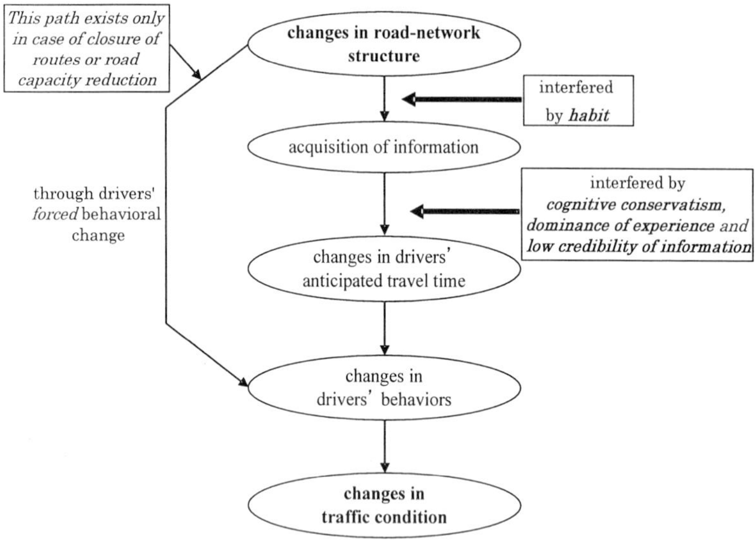

Figure 1 The process of changes in traffic condition induced by changes in road-network structure

As can be seen in Figure 1, there are two steps between changes in road-network structure and drivers' voluntary behavioral-changes. The first step is to acquire information about changes in road-network structure. Only drivers who know the existence of changes in road-network structure or traffic-condition changes induced by the other drivers' behavioral change, might change their behaviors. Otherwise, they would not have any motivation to change their behavior. However, as argued above, some drivers would not acquire any information due to habits.

The second step for behavioral change, is to modify their anticipated travel time. However, as the authors data illustrated (Fujii & Kitamura, 2000), even if drivers knew the existence of the freeway closure, they did not modify their anticipated travel time substantially if they did not have any information about traffic conditions. That is, drivers' cognitive conservatism may interfere with modification of anticipated travel time. Furthermore, even if drivers acquire information about new traffic conditions, they may not change their anticipated travel time if they already have actual driving experience in the new traffic condition, because actual experiences dominate the effects of generic information, as implied by the Experience Dominance Hypothesis. Finally, if the drivers do not believe in the credibility of the information, they might not change their anticipated travel time based on the information.

Thus, habit, cognitive conservatism, dominance of experiences over generic information, and low credibility of acquired information, could all impede behavioral changes which might be induced by changes in road-network structure.

Conclusion

In this paper, we discussed relations between drivers' cognition and the dynamics of traffic conditions, and elucidated the process of changes in traffic conditions that might be induced by changes in road-network structure. When we take into account drivers' cognition, many types of psychological impedance to behavioral change could be accounted for, including cognitive conservatism and habitual decision making. Needless to say, the authors do not insist at all that traffic conditions do not change in response to changes in road-network structure. What we do insist is that traffic conditions are unlikely change as much as predicted by theories that assume a simple, and sometimes unique, relation between a traffic condition and a road-network structure, e.g., theories that assume user equilibrium. For the purpose of understanding the dynamics of macroscopic traffic condition, rather than a static traffic condition, it would be obvious that we can not ignore the dynamics of microscopic drivers' cognition.

REFERENCES

Anderson, J. R. (1990) *Cognitive Psychology and Its Implications*, 3rd ed., W.H. Freeman, New York.

Bonsall, P.W., L. Pickup and A. Stathopoulos (1991) Measuring behavioural responses to road transport informatics, *Advanced Telematics in Road Transport* (2), pp. 1457-1487. Elsevier, Amsterdam.

Chang, G.L. and H.S. Mahmassani (1998) Travel Time Prediction and Departure Time Adjustment Behavior Dynamics in a Congested Traffic System, *Transportation Research*, **22**B, 217-232.

Emmerink, R.H.M., P. Axhausen, P. Nijkamp and P. Rietveld (1993). Effects of information in road transport networks with recurrent congestion, Discussion Paper, TI 93-229, Tinbergen Institute, Vrije Universiteit, Amsterdam.

Emmerink, R.H.M., P. Axhausen, P. Nijkamp and P. Rietveld (1994) Effects of information in road transport networks with recurrent congestion, Discussion Paper, TI 94-30, Tinbergen Institute, Vrije Universiteit, Amsterdam.

Mahmassani, H. S. and J. Rong-Chang (1998) Bounded Rationality in Commuter Decision Dynamics: Incorporating Trip Chaining in Departure Time and Route Switching Decisions, In *Theoretical Foundations of Travel Choice Modelling* (edited by T. Gärling, T. Laitila and K. Westin), New York, Elsevier, pp. 201-229.

Fujii, S. and T. Gärling and R. Kitamura (2001) Changes in drivers' perceptions and use of public transport during a freeway closure: Effects of temporary structural change on cooperation in a real-life social dilemma, *Environment and Behavior*, **33**(6), 796-808.

Fujii, S. and R. Kitamura (2000) Anticipated travel time, information acquisition and actual experience: The case of Hanshin expressway route closure, *Transportation Research Record*, **1725**, 79-85.

Gärling, T., S. Fujii and O. Boe (2001). Empirical tests of a model of determinants of script-based driving choice, *Transportation Research F: Traffic Psychology and Behavior,* **4**, 89-102.

Heckman, J.J. (1981). Heterogeneity and State Dependence. In *Studies in Labor Markets.* (S. Rosen, ed.), pp. 91-139. University of Chicago Press, Chicago, IL.

Iida, Y., T. Akiyama and T. Uchida: Experimental analysis of dynamic route choice behavior, *Transportation Research B*, **26**(1), 17-32, 1992.

Kitamura, R. and D.S. Bunch (1990). Heterogeneity and State Dependence in Household Car Ownership: A Panel Analysis Using Ordered-Response Probit Models with Error

Components. In: *Transportation and Traffic Theory*, (M. Koshi, ed.), pp. 477-496. Elsevier.

Kobayashi, K and T. Yasuno. (1998). Testing the non-neutrality hypothesis of traffic information: An experimental approach (in Japanese with English abstract), *Journal of Infrastructure Planning and Management*, Japan Society of Civil Engineers, **No. 567/IV-40**, pp. 49-60.

Kobayashi, K. (1994). Information, rational expectations, and network equilibria: An Analytical Perspective for Route Navigation Systems, *The Annuals of Regional Science*, **Vol. 28**, 493-500.

Nakayama, S., R. Kitamura and S. Fujii (1999) Drivers' Learning and Network Behavior: A Dynamic Analysis of the Driver-Network System as a Complex System, *Transportation Research Record*, **1676**, pp. 30- 36.

Nakayama, S., S. Fujii, R. Kitamura, N. Yamada (2002) An experimental analysis of delusion and reduction of alternatives in repeated route choice behavior, *Journal of Infrastructure Planning and Management* (in Japanese with English abstract), **No. 709/IV-56**, pp. 51-60.

Oullette, J. A., and W. Wood (1998). Habit and intention in everyday life: The Multiple Processes by Which Past Behavior Predicts Future Behavior. *Psychological Bulletin,* **Vol. 124**, pp. 54-74.

Polak, J. and F. Oladeinde (1999) An Empirical Model of Travelers' Day-to-Day Learning in the Presence of Uncertain Travel Time, presented at Transport Network Reliability Workshop, Newcastle, UK, July.

Ronis, D. L., J. F. Yates, and J. P. Kirscht (1989). Attitudes, decisions, and habits as determinants of repeated behavior. In: *Attitude Structure and Function* (A. R. Pratkanis, S. J. Breckler & A. G. Greenwald, eds.), pp. 213-239. Hillsdale, Erlbaum, NJ.

Tversky, A. and D. Kahneman (1974). Judgement under uncertainty: Heuristics and biases. *Science*, **185**, 1124-1131.

Verplanken, B., H. Aarts and A. van Knippenberg. (1997). Habit, information acquisition, and the process of making travel mode choices, *European Journal of Social Psychology*, **27**, 539-560.

13

A NORMATIVE ASSESSMENT OF TRANSPORT NETWORK RELIABILITY BASED ON GAME THEORY

Chris Cassir, Institute for Transport Research, DLR Berlin, Germany
Michael G H Bell, Centre for Transport Studies, Imperial College London, UK
Jan-Dirk Schmöcker, Transport Operations Research Group, Newcastle University, UK

ABSTRACT

This paper describes and extends the game theoretic approach to network vulnerability assessment. The basic idea is to set up a game between the network users who are trying to minimise their expected travel time by choice of route and a network tester who is trying to penalise the users most by degrading a link through capacity reduction leading to congestion. The method therefore finds the worst possible location for a link degradation, taking re-routing options into account (an upper, lower bound of impact). The original game identifies the weakest link for routes between an OD pair in the network. Two variations are introduced in this paper in order to determine the weak links for a specific origin or a specific destination and for the whole network. All three game variations are tested on a small network in Leicester and the results are presented.

INTRODUCTION

In systems engineering, reliability is defined as the probability that the system can perform its intended function satisfactorily during a given period of time. Reliability is an important quality as the costs associated with a system failure can be high. There has recently been some interest in applying reliability analysis to transportation networks, and the tragic events of September 11th have certainly increased the interest still further. In a crisis, caused by natural disasters, terrorist attacks or any other reason, the remaining transportation network is subject to severe pressure. It is therefore important to assess the weak points in a network. Although disasters of this scale are fortunately relatively rare (except in disaster prone areas), other small size (but higher frequency) events can also have a considerable impact on the transportation network. Nicholson and Du (1997) mention that variations in demand as well as supply can lead to a significant increase in travel cost. This paper considers changes in the network supply side, i.e. link closures or link degradation. Daily experience shows that accidents, building sites or vehicles blocking the road can delay traffic greatly. In many cases not only is the traffic on the degraded link affected by the event, but also the traffic in the surrounding area. Network users are therefore confronted with an uncertainty about which is the best route to their specific destination. Travellers who are familiar with the area might use their knowledge and take alternative routes, if they think that these routes are less congested than their normal, preferred routes.

The variability and uncertainty surrounding these day-to-day traffic conditions has led to the definition of some probabilistic reliability indicators for planning purposes. These indicators should reflect the robustness of the network to cope with adverse events such as those mentioned above. Research into *terminal reliability* (see Wakabayashi and Iida, 1992) reaches furthest back and is concerned with the question whether it is possible to reach the destination after an adverse event. The performance indicator that has been most widely used is *travel time reliability* (see Asakura, 1996). For a more detailed review see Lam and Zhang (2000) or Nicholson *et al.* (2002). The latter review points out that it is important to consider the level of information available to the user when reliability is "measured". As will be described later in more detail, the approach considered in this paper assumes users who have information about the probabilities of link degradations and therefore adjust their route strategy accordingly. If the users had less (more) information this might increase (decrease) the impact of the link degradation significantly.

Traditional approaches to the evaluation of these performance reliability indicators require a substantial amount of historical data to determine the exogenous variability of both demand and capacity in the network. An example for this is the sensitivity approach proposed by Bell *et al* (1999). In this approach statistical distributions for link performance (typically link travel time, link delay or link capacity) have to be obtained first. The case study undertaken in New Zealand by Nicholson and Dalziell (2001) presented data about costs for different events, like snowfall or accidents, that can if not successfully managed cause road closure. The data collection for this study was, however, a very tedious and time consuming exercise and has to include several simplifications. Evaluating the impact of an event is possible with modern simulation tools to gain information on the probabilities that these events happen is, however, far more difficult.

In this paper an alternative approach is proposed that does not require data on event probabilities. The idea is to use a normative approach to test the network artificially regarding its robustness. A game is envisaged between network users trying to minimise their expected travel time by using the best possible route, and a network tester trying to penalise the users most by *failing* one link. "Failing" a link can mean link degradation, i.e. capacity reduction, as well as closure. Therefore, this method could be described as looking for an upper (worst case location) lower (users know the best route) bound of impact on the network user travel times in case of a link failure. We analyse how the outcome of this game can help identify the weak links in the network and provide alternative network reliability measures based on what could fail rather than on what would fail. With a case study of the Leicester, England, network it is shown that the approach can identify possible weak links for total traffic, traffic between specific OD-pairs, and traffic coming from specific origins.

METHODOLOGY

Game Theoretic Approach

The basic idea follows from Cassir and Bell (2000), which is an extension of Bell (2000). Imagine a network with n heterogenous users who want to reach their destination during some typical period of the day (the morning rush hour, for example). The question is then, which part of the network (link) will cause the most disruption to the users when failed due to some capacity reducing incident? And knowing that, how would the users adapt their route choice?, etc. We have here the conditions for an artificial game between the users and an imaginary network tester.

This game can be formulated mathematically and solved for the Nash equilibrium (see Nash, 1951) of mixed strategies for the users and the tester. The Nash equilibrium gives the route choice probabilities for the users and the link failure probabilities for the tester. It can be shown that such an equilibrium can be found as a mutually consistent solution to a bi-level problem, where the upper level problem is the maximisation of total expected cost and the lower level problem is a stochastic user equilibrium assignment based on expected link costs (see Cassir and Bell, 2000).

It should be pointed out that the "game" between a hypothetical network spoiler and the network users is only an illustration of the basic minimax problem: The users try to find the best routes (minimising the costs) but are extremely pessimistic in considering failure probabilities at the worst possible location on their journeys (maximising the costs). Therefore, the game should be considered as one way to find the critical links and to estimate an upper bound of failure costs for users with route choice. Depending on the "definition of the spoiler" it is hereby possible to analyse the reliability and find the critical links for the network as a whole (Case A), for each OD pair (Case B) or for traffic coming from a certain origin or travelling to a certain destination (Case C). All studies are interesting in themselves. The first Case A is concerned with the weak points in the network and assumes failure probabilities summing up to one link failure for the whole network. Case B reveals which OD pair suffers most and which link capacity reduction causes the most damage in case there is a chance for a link degradation on exactly one link for each OD pair. Case C analyses which are the critical links for a certain origin, assuming that one link for traffic from each origin might suffer a capacity reduction. This case will give useful reliability information for all traffic coming from one origin (or going to one destination).

Mathematical Formulation and Algorithm

Following Cassir and Bell (2000) a mathematical formulation of the bi-level optimisation problem is presented, whose solution yields a Nash equilibrium to the game described above, for a network with one OD pair.

Notation:

D is the total demand for the network in veh/hr
$\mathbf{h}$ is a vector of route flows (veh/hr)
$\mathbf{q}$ is a vector of link failure probabilities
$g_{jk}(\mathbf{h})$ is the (flow dependant) travel time of route j when link k is failed
$v_u(\mathbf{h})$ is the flow (veh/hr) on link u
t_{uk} is the (flow dependant) travel time of link u when link k is failed
p_j is the route choice proportion for route j
α is the logit dispersion parameter

Bilevel problem : Solve simultaneously

Upper level: $\text{Max}_{\mathbf{q}}\ \Sigma_j\Sigma_k\, q_k g_{jk}(\mathbf{h})h_j$ subject to $\Sigma_k\, q_k=1,\ \mathbf{q}\geq 0$

Lower level: $\text{Min}_{\mathbf{h}}\ \Sigma_j h_j(\ln h_j-1) + \alpha\left(\Sigma_u\Sigma_k q_k \int_0^{v_u(\mathbf{h})} t_{uk}(x)dx\right)$ subject to $\Sigma_j\, h_j=D$

At a joint optimum the following equilibrium conditions apply:

Upper level: $\forall k$

$$q_k = 0 \Leftarrow \Sigma_j g_{jk}(\mathbf{h})h_j < \max_r \Sigma_j g_{jr}(\mathbf{h})h_j$$
$$q_k > 0 \Rightarrow \Sigma_j g_{jk}(\mathbf{h})h_j = \max_r \Sigma_j g_{jr}(\mathbf{h})h_j \qquad (1)$$

Lower level: $\forall j$

$$h_j = D\, p_j(\mathbf{h}) = D \exp(-\alpha\Sigma_k g_{jk}(\mathbf{h})q_k) \,/\, \Sigma_r \exp(-\alpha\Sigma_k g_{rk}(\mathbf{h})q_k) \qquad (2)$$

The conditions (1) are equivalent to the network tester being unable to increase his pay-off by unilaterally changing his strategy. Conditions (2) describe a logit stochastic user equilibrium (SUE) for the flow vector **h**, with expected route costs depending on the network tester strategy. The result is a set of equilibrium route choice probabilities and the corresponding expected average costs, allowing for some random perturbations (link failures) in the network aimed at penalising the users as much as possible. The joint optimum found with this artificial game is therefore a local optimum for both conditions. This means it is not necessarily a global optimum of the upper problem as found with the leader-follower structure in a Stackelberg game.

Both conditions are derived by taking the necessary conditions for an optimum of the corresponding problems. The algorithm used to solve the bi-level problem is based on the Method of Successive Average MSA (see Sheffi, 1985). It is known that the MSA works better for the lower level (stochastic) problem than for the upper one (deterministic). However, experience showed that the MSA did work for the upper problem, because the results where never far from the exact solution.

Step 0: Initialise m = 0 and $\mathbf{q}^m = 1$ / number of links. Set all path flows **h** and link flows to the SUE equilibrium flows (obtained in normal conditions). Set m = m+1.

Step 1: Calculate flow dependent link failure and non-failure costs (travel times).

Step 2: Calculate expected link and route costs.

Step 3: Obtain auxiliary route flows $\mathbf{y}^m$ from the PFE (logit assignment model):
$y_j^m = D \exp(-\alpha\Sigma_k g_{jk}(\mathbf{h}^{m-1})q_k^{m-1}) \,/\, \Sigma_r \exp(-\alpha\Sigma_k g_{rk}(\mathbf{h}^{m-1})q_k^{m-1}) \quad \forall j$

Step 4: Update route and link flows using the MSA:
$\mathbf{h}^m = \mathbf{h}^{m-1} + (\mathbf{y}^m - \mathbf{h}^{m-1})/m$

Step 5: Obtain auxiliary link failure probabilities $\mathbf{z}^m$, by assigning failure probability 1 to the link that has the highest total failure cost, and 0 to other links.

Step 6: Update link failure probabilities by the MSA:
$\mathbf{q}^m = \mathbf{q}^{m-1} + (\mathbf{z}^m - \mathbf{q}^{m-1})/m$

Step 7: If convergence criteria are satisfied stop, otherwise set m := m+1 and return to Step 1.

Reliability Assessment

The outcome of the artificial game described above provides normative measures of network reliability, as opposed to predictive measures of actual reliability. It is important to stress the difference. Such normative measures can be useful in pointing out the critical links or critical OD pairs in the network, in terms of connectivity and performance (travel time). The results of the game will be a) failure probabilities and b) (maybe) different flows for all links. High failure probabilities do not indicate that this link is likely to fail but suggest that if users are uncertain and wary about the prospect of failure, it is likely to cause much damage when failing. The results of this artificial game will therefore indicate the importance of a link (whether it potentially causes much damage) but not the weakness (whether it is actually likely to fail). With the Nash equilibrium solution of the game, we further define 2 measures of reliability for an OD pair:

$R^{con}_{od} \equiv$ Probability of not encountering a link failure between OD pair *od* in the hypothetical testing game.

$R^{perf}_{od}(\tau) \equiv$ Probability that travel time between OD pair *od* is less than some satisfaction criteria τ (τ is defined as the average OD travel time in normal conditions plus a safety margin) in the hypothetical game.

These 2 measures give an indication of the robustness of the network for particular OD pairs. In the same way R^{con}_{o} and R^{perf}_{o} for Case C are defined, i.e. R^{con}_{o} is hence *Probability of not encountering a link failure for journeys from Origin o.*

Example on Single Grid Network

The MSA procedure was run for various levels of demand on a small test network (Figure 1), first for a totally symmetrical network (all links have same capacity) and then for an assymmetrical network (Link 1 leading out of the origin has capacity halved compared to the other links). The results for R^{con} and R^{perf} in both cases are shown in Figures 2 and 3 respectively, while Figure 4 shows the link failure probabilities for Links 1 and 3 in the asymmetrical network. In both examples we looked at the effect of a link degradation instead of total link closure, i.e. the network spoiler can half the capacity of one link at a time.

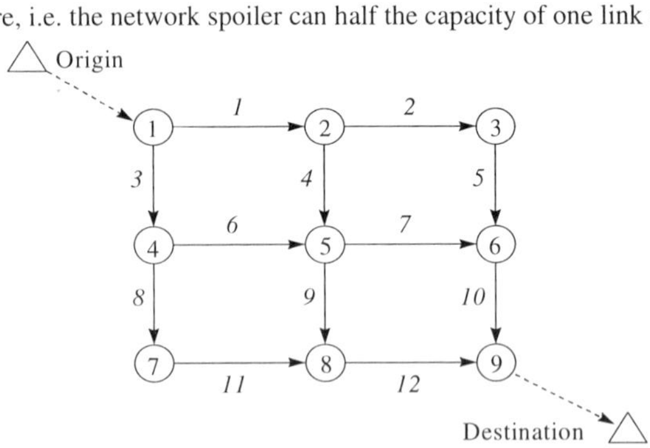

Figure 1 Hypothetical network

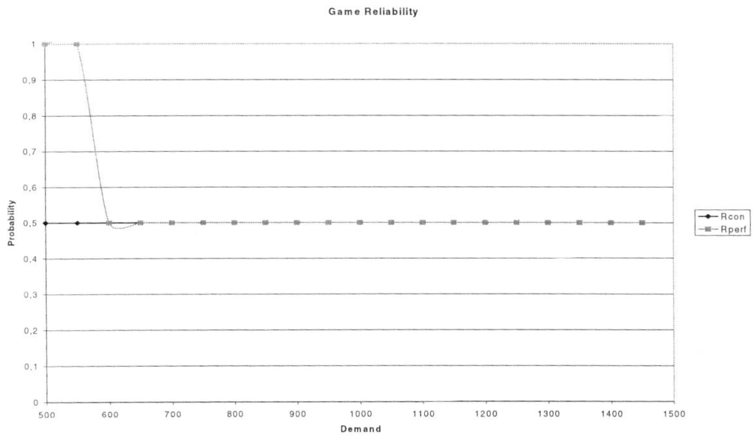

Figure 2 Symmetrical network

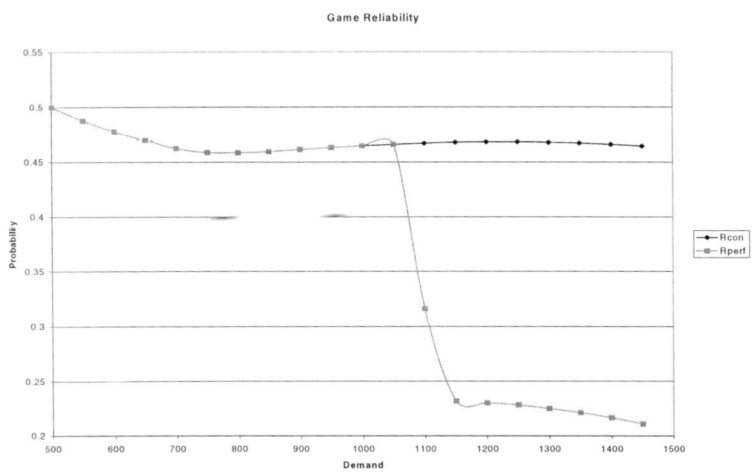

Figure 3 Assymmetrical network

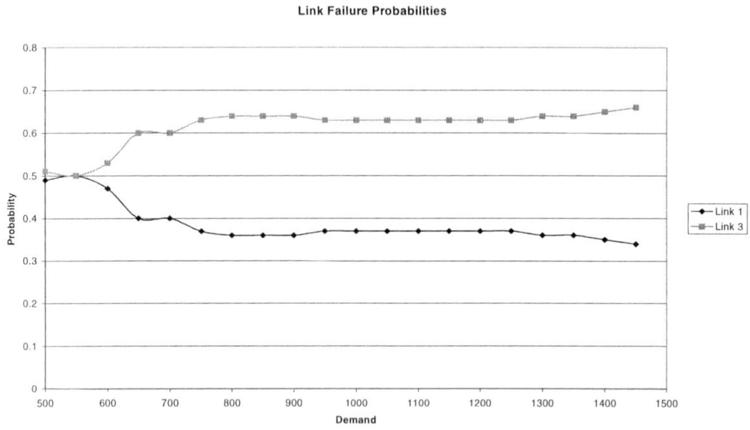

Figure 4 Link failure probabilities

From this simple example, we can draw some interesting first conclusions about the outcome of the network testing game. In the symmetrical case, we see that the probability of a failure (to be interpreted as the *relative fear* of a failure) is constant at 50%. This is to be expected, as the failures are concentrated on the 2 links leading out of the origin and the 2 links leading into the destination, and combine because of the symmetry to give 50% chances of failure on all paths. This symmetrical network and the corresponding connectivity value of 50% can serve as a good benchmark for OD reliability. When such a value is approached for an OD pair, this means that the OD pair has good redundancy, because it can be compared to this symmetrical case where there are 2 equal and independent alternative routes (in the example there are actually 6 routes, but out of the origin there are only 2 equal independent alternatives; routes through Link 1 or routes through Link 3). The performance reliability indicator R^{con} follows R^{perf} except for very low demand where failure does not result in sufficiently high travel times to exceed the performance criteria.

In the asymmetrical case, the former picture is slightly modified. R^{con} drops below 50%. With increasing demand, the asymmetry in the network grows as capacity is reached earlier on Link 1 than on Link 3. Consequently, Link 3 tends to attract more demand, but becomes then a likelier target for failure. This results in an imbalance in failure probabilities and link usage between Links 1 and 3, which can explain the lower value for R^{con} (more people using a link more likely to be failed means higher probability of encountering failure altogether). R^{perf} follows R^{con} until a certain level of demand, where it drops sharply below 25%. This is a bad sign of performance reliability, since it indicates that travel times can exceed the performance threshold, even when no failure is encountered (because of congestion).

LEICESTER NETWORK

Network Description and Assignment Parameters

We now consider a more realistic example with several OD pairs on a small network in Leicester, England. The problem formulation and algorithm for a multi-commodity (several OD pairs) network is very similar to the 1 OD pair network case. However, as mentioned in Section 2.1, there are now 3 possibilities for envisaging the artificial testing game, depending on which level we are interested in:

- Case A: Network level. Game between all network users and one spoiler who can fail at most one link at a time on the whole network.
- Case B: OD level. Several independent OD specific games between some OD users and an OD spoiler who "plays" only against the corresponding OD users. For each OD specific game, the flows on the network coming from other OD pairs are set to the equilibrium flows in normal conditions (given by some equilibrium assignment model)
- Case C: Tree level. Here we have also several independent games between some tree users (coming from one origin) and a tree specific spoiler who "plays" only against the corresponding tree users. Again for each tree specific game, the flows on the network coming from other trees are set to the equilibrium flows in normal conditions.

The Leicester network considered is shown in Figure 5. It consists of 104 links, 9 origins and 9 destinations. It should be noted that the figure does not reveal that some links have 2 lanes, which are then considered as two links, e.g. Link 212S in Figure 5 is modelled as Links 212S1 and 212S2. The network demand (OD matrix) represents the morning peak hour with a total inflow of 4484 veh/h. First some estimates of average normal traffic conditions (travel times and traffic flows) for this network were obtained by applying a Stochastic User Equilibrium Assignment based on the logit model, as implemented in the Path Flow Estimator or PFE (see Bell *et al.*, 1996), with route dispersion factor alpha = 2.0. The resulting equilibrium route travel times (without considering link failures) could then be used to define the performance criteria τ in the hypothetical game to follow.

τ_{od} = average OD equilibrium travel time plus a safety margin (taken arbitrarily as 20% of this average time).

We then used the game procedure for the three cases mentioned above, defining link failure as halving the link capacity. For the sake of simplicity we considered independent link failures. The results are presented below.

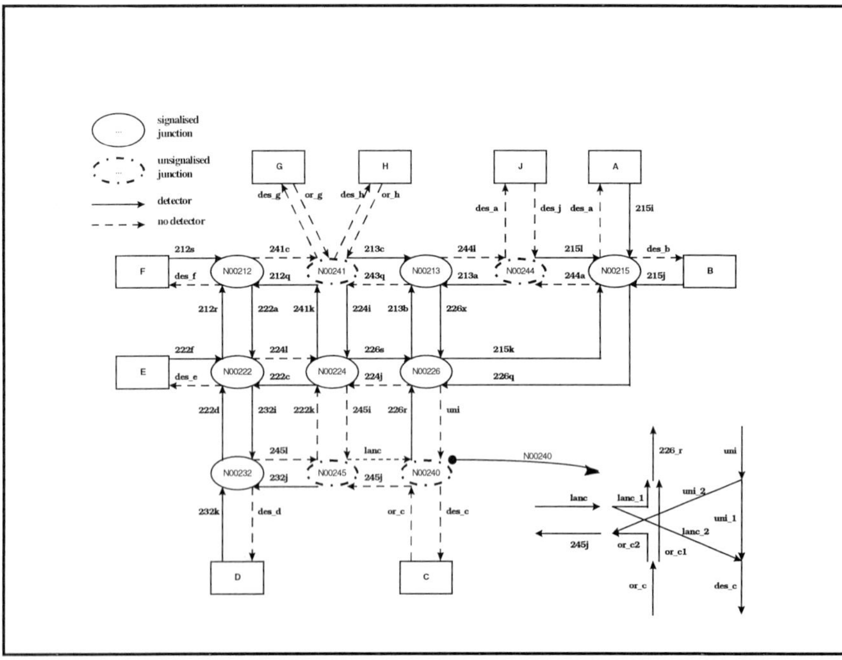

Figure 5 Leicester network

Case A: Single Network Spoiler

First the link failure probabilities for the whole network game are presented in Table 1, along with the travel times and flows (obtained from the game equilibrium), and for comparison the travel times and flows in normal conditions obtained from the Stochastic User Equilibrium (SUE). Then the reliability indicators, calculated on the basis of these link failure probabilities and the game equilibrium flows (and travel times), for the 17 most unreliable OD pairs are listed in Table 2. The table further shows the differences in mean OD travel time for the SUE solution, referred to as "Performance criteria" and the increased "expected travel time" E[Time] for the game equilibrium.

Table 1 Failed links in network spoiler case (Case A)

Link	Failure Proba.	Travel time (s)	Fail. Travel Time (s)	SUE Travel Time (s)	Flow (veh/hr)	SUE Flow (veh/hr)	Capacity (veh/hr)
212S2	0.43	19.80	1461.99	60.31	902.6	999.7	1000.0
215L	0.33	60.99	1739.49	159.25	772.7	842.0	791.7
222A1	0.20	222.20	2196.86	48.45	635.4	547.5	575.0
215K2	0.03	569.64	2901.09	204.54	323.8	267.1	250.0

Table 2 Unreliable OD pairs in network spoiler case (Case A)

Origin	Destination	Nbr of paths	Flow (veh/hr)	R^{con}	R^{perf}	E[Time] (mn)	Perf. Criteria (mn)
OR_C	**DES_B**	2	32.0	0.94	0.06	11.19	6.23
OR_F	DES_D	4	471.0	0.73	0.07	13.48	3.33
OR_F	DES_B	5	678.0	0.33	0.17	23.16	7.16
OR_F	DES_C	5	99.0	0.65	0.25	14.60	1.48
OR_D	**DES_B**	5	111.0	0.88	0.30	12.13	7.20
OR_F	DES_A	4	115.0	0.62	0.30	14.77	4.62
OR_E	**DES_B**	5	177.0	0.87	0.34	11.74	6.71
OR_F	DES_G	2	133.0	0.54	0.47	13.18	2.65
OR_F	DES_J	2	21.0	0.54	0.47	14.00	3.67
OR_F	DES_H	2	21.0	0.54	0.47	13.47	3.00
OR_G	**DES_B**	2	18.0	0.74	0.67	10.38	5.00
OR_H	**DES_B**	2	62.0	0.74	0.67	10.10	4.65
OR_J	**DES_B**	2	29.0	0.73	0.70	9.52	3.95
OR_B	DES_G	2	12.0	1.00	0.74	1.75	2.10
OR_B	DES_H	2	29.0	1.00	0.74	1.50	1.80
OR_D	DES_F	3	187.0	1.00	0.76	1.82	2.18
OR_E	DES_A	4	117.0	1.00	0.77	2.52	2.91

We can see in Table 1 that the network spoiler concentrates only on 4 links, which have thus to be understood as the most critical links in the network. It is not so surprising to find Link 212S2 (micro-link between Links 212S and 241C in Figure 5) as the most failed in this hypothetical game. As can be seen from the table, it is a link with fairly high flow (999 veh/h) under SUE conditions, and is moreover very close to capacity (1000 veh/h). Therefore it is a very attractive link to fail for the spoiler. The other links that are failed are also close to or above long-term capacity in normal conditions and thus provoke very high travel times when failed. It is interesting to note that as a result of the artificial game, flow is reshuffled and is reduced for the two most critical Links (212S2 and 215L) while being increased for the other two (222A1 and 215K2). The travel times increase for these two links and they also have extremely high failed travel times, but this is compensated by the lower risks of failure. This seems in agreement with the logic of the game.

Table 2 complements this analysis. We can see that the only OD pairs for which there is a non-zero probability of encountering failure are those concerned with the four links mentioned above. These are namely the OD pairs with either Origin F or Destination B (highlighted in bold in the table). For trips coming from Origin F, it is indeed rather difficult to avoid either Link 212S2 or Link 222A1 (micro-link leading into Link 224l). The same applies for trips going to Destination B with Links 215L and 215K2 (micro-link leading into Link Des_B). The performance of the worst OD pair OR_C → DES_B can be explained as follows. There are 2 paths for this OD pair: one that goes through Link 215L and another one through Link 215K2. About 90% use the path through Link 215K2 (with 3% failure probability) and about 10% use the path through Link 215L (33% failure probability). This explains why the probability of encountering failure is so low (R^{con} = 0.94). However since the (non-failed) travel time on Link 215K2 is significantly increased compared to the normal conditions (from 204 to 569 seconds), it means that even when there is no failure on Link 215K2, the travel time on the path using that link does not meet the satisfaction criteria. As it happens, the only possibility for the satisfaction criteria to be met is to go through Link 215L when it is not failed. But since this only occurs 67% of the time, a very large majority prefer the much safer (but not so satisfactory) option of going through Link 215K2. This example illustrates that there is a problem with Link 215K2. This link does not need to be failed very often to create problems (long travel times) in the network, due essentially to its very small capacity (250 veh/h) and the addition of traffic flow provoked by failures elsewhere.

In conclusion, the network spoiler case appears to be useful for highlighting the most critical links in the network and also the OD pairs that are most affected by those links. Attention could then be devoted to whether to increase capacity on these critical links for the OD pairs most concerned, in order to make the network as a whole more "reliable".

Case B: OD Spoiler

Next we show the results obtained when playing the game separately for each OD pair against one OD specific spoiler. Table 3 ranks the most "unreliable" OD pairs according to reliability indicator R^{perf} and Table 4 indicates the critical links for each of these OD pairs respectively.

Table 3 OD reliability indicators for OD spoiler case (Case B)

Origin	Destination	Nbr of paths	Flow (veh/hr)	R^{con}	R^{perf}	E[Time] (mn)	Perf. Criteria (mn)
OR_A	DES_B	2	141.0	0.00	0.00	2.80	0.76
OR_F	DES_B	5	678.0	0.33	0.05	25.59	7.16
OR_B	DES_F	3	444.0	0.36	0.18	9.12	2.78
OR_E	DES_A	4	117.0	0.62	0.18	10.88	2.91
OR_F	DES_E	2	122.0	0.41	0.18	11.52	1.91
OR_D	DES_F	3	187.0	0.35	0.20	4.68	2.18
OR_F	DES_D	4	471.0	0.45	0.20	15.19	3.33
OR_F	DES_G	2	133.0	0.49	0.22	17.38	2.65
OR_A	DES_E	4	115.0	0.50	0.23	10.60	3.19
OR_D	DES_A	4	137.0	0.63	0.43	10.92	3.37
OR_C	DES_F	3	138.0	0.44	0.44	10.39	3.37
OR_E	DES_B	5	177.0	0.50	0.44	21.59	6.71
OR_F	DES A	4	115.0	0.50	0.49	18.08	4.62
OR_A	DES_E	4	156.0	0.50	0.50	10.31	3.11
OR_D	DES_B	5	111.0	0.50	0.50	21.68	7.20

Table 4 Link failure and usage probabilities for OD spoiler case (Case B)

Origin	Dest.	Link	Fail	Use	....			....		
OR_A	DES_B	215I1	1.00	1.00						
OR_F	DES_B	212S2	0.21	0.74	215L	0.70	0.70	215K2	0.09	0.30
OR_B	DES_F	226Q	0.03	0.30	212Q	0.77	0.77	222C	0.20	0.23
OR_E	DES_A	215L	0.15	0.20	226R	0.32	0.32	226S	0.53	0.48
OR_F	DES_E	212S1	0.74	0.72	212S2	0.26	0.28			
OR_D	DES_F	212R	0.75	0.76	212Q	0.25	0.24			
OR_F	DES_D	212S2	0.31	0.35	222A1	0.69	0.65			
OR_F	DES_G	212S2	0.59	0.56	222A1	0.41	0.44			
OR_A	DES_E	212Q	0.49	0.50	222C	0.51	0.50			
OR_D	DES_A	215L	0.12	0.21	226R	0.38	0.34	226S	0.50	0.45
OR_C	DES_F	212Q	0.68	0.69	222C	0.33	0.31			
OR_E	DES_B	215L	0.51	0.53	215K2	0.49	0.47			
OR_F	DES_A	212S2	0.51	0.53	222A1	0.46	0.47	226S	0.03	0.65
OR_A	DES_E	212Q	0.46	0.52	222A2	0.02	0.52	222C	0.52	0.48
OR_D	DES_B	215L	0.49	0.51	215K2	0.51	0.49			

The different nature of this hypothetical game played at the OD level makes it not relevant to compare these results directly with the previous case. In this case, it is however interesting to look at the performance of some OD pairs separately. The worst OD pair is between Or_A → DES_B, with no reliability at all. This is actually not very surprising since Origin A and Destination B are so close together. Therefore for this OD pair, the spoiler just concentrates on Link 215I1 (the left-turn off the origin link), because the alternative detour path is so long that it does not attract any demand even though it is "failure free". The second most unreliable OD pair is more interesting. It is the OD pair between Origin F and Destination B. If we recall from the previous case, all worst affected OD pairs either came from Origin F or ended at Destination B. A look at Table 4 also shows that for this OD pair the critical links (215L, 212S2 and 215K2) are among the only 4 links that are failed when the hypothetical game is played at the network level in Case A (see Table 1). Moreover one can see that the value for R^{con} (33%) is exactly the same as in Case A (see Table 2). While such exactness may be a coincidence, the similarity of these results can be explained by the fact that Or_F →Des_B is by far the most important OD pair in terms of trips (672 veh/hr) in the whole network. Therefore it must have an overriding influence on the outcome of the game played at the network level, so it is not surprising to find that the outcome is close to the results when the game is being played for this OD pair only. There is however quite a significant difference for R^{perf} between the 2 cases. It is reduced from 17% in Case A down to 6% in this case. This is to be expected since the spoiler intends to maximise the expected travel time only for the OD users, whereas in Case A the objective is to maximise it for all network users. As a result it seems normal that in terms of travel times, and the probability of reaching the destination within a satisfactory time, the OD users should be worse off in this present case than in case A.

As a conclusion, one can say that this option is useful when one is interested in a particular OD pair, with many alternatives paths. Otherwise, for general information on a whole network, Case A is preferable and provides more interesting insights. This is because for many OD pairs, it is likely that the OD spoiler will only concentrate on some links near the origins or destinations, gathering the entire OD flow and therefore not revealing much about reliability and robustness in the network.

Case C: Tree Spoiler

Finally we look at the case with a spoiler who is determined to cause the most damage to users from a selected origin. Table 5 lists the reliability indicators for each origin and the average increase in expected travel time compared to the "normal" travel time. Compared to Table 3 it can be seen that the increases are not so severe, because in this case the spoiler does not focus on a specific OD pair, but tries to impair all users from one origin, who might have different destinations. Table 6 shows the links with the highest failure probability for each origin.

Table 5 Origin reliability indicators for "tree spoiler" case (Case C)

Origin	Flow (veh/hr)	R^{con}	R^{perf}	E[Time] (mn)	Perf.Criteria (mn)
OR_A	483	0.65	0.62	8.09	1.98
OR_B	800	0.5	0.45	8.26	2.31
OR_C	334	0.7	0.66	6.88	2.69
OR_D	516	0.82	0.76	7	2.94
OR_E	451	0.62	0.56	12.22	3.36
OR_F	1660	0.49	0.47	17.83	4.1
OR_G	54	0.81	0.78	8.46	2.73
OR_H	148	0.71	0.68	11.34	2.97
OR_J	38	0.57	0.55	16.3	3.08

Table 6 Link failure and usage probabilities for tree spoiler case (Case C)

Origin	Link	Fail	Use	....			....		
OR_A	215I2	0.5	0.36	226Q	0.5	0.34			
OR_B	215J1	0.53	0.65	226Q	0.33	0.35	212Q	0.1	0.53
OR_C	212Q	0.55	0.33	222C	0.38	0.22	226R	0.08	0.39
OR_D	226S	0.43	0.19	226R	0.34	0.17	215L	0.23	0.12
OR_E	226S	0.65	0.41	215L	0.35	0.24			
OR_F	212S2	0.65	0.55	222A1	0.35	0.38			
OR_G	215L	0.54	0.19	215K2	0.42	0.15	226S	0.04	0.21
OR_H	215L	0.56	0.42	215K2	0.44	0.34			
OR_J	215L	0.57	0.26	226S	0.38	0.27	215K2	0.05	0.16

Table 5 reveals that Origins D and G seem to be most robust. One reason for this is certainly the relatively low demand, especially for Origin G. If there is a high demand from one origin or if the links close to the origin are all congested, the network spoiler is likely to spoil a link close to the origin where he can "catch" all users. This, however, is not the case for Origin D. Link 232K has a capacity of 1700 veh/h, and therefore halving the capacity still does not cause a severe travel time increase to an inflow of 516 veh/h. Therefore the spoiler will focus on other links as shown in Table 6. However, no destination is very close so that the traffic has many routeing options. This is reflected in a fairly equal failure distribution between 3 links and this results in a high value for R_D^{con}.

Origin J only generates very little traffic (38 veh/h), but is nonetheless not very reliable, neither in terms of connectivity nor travel time. It might be concluded that the origin is not well placed in the network, because the traffic has to cross many links that could suffer much delay in case of unreliability. Most of the vehicles coming from Origin J are travelling to Destinations A or B and will use Link 215L. The link is further used by a lot of traffic coming from Origins F, G and H travelling to the same destinations. This high traffic volume on the link explains that it becomes attractive for the spoiler, because a failure will result in a significant cost increase and a low value for R^{perf}.

Origin F and OD pair OR_F → Des_B in particular have already been mentioned in both of the previous sections, because it is the origin with the highest demand as can be seen from Table 5. Origin F has therefore a big influence on the network results and it cannot surprise that links 212S2 and 222A1 (Table 6, most unreliable links for OR_F) are also listed in Table 1 as the most and 3rd most failed links. Link 212S2 is taken by all users going East from Origin F, traffic going South has to take Link 222A1. Therefore all traffic encounters one link with a considerable failure probability. In case of failure both links do not have sufficient capacity for the Origin F demand and the traffic will be severely delayed. This explains why Origin F has the highest demand and the lowest reliability.

It can be concluded, that this "tree-spoiler" case provides some additional insight into the network reliability. With this analysis it is possible to provide the network analyst with some information about which locations can provide more reliable transport than others. In the example shown vehicles situated at Origin D (e.g. lifeline services) are more likely to reach all destinations without encountering great delays than a vehicle from Origin F, always given uncertainty about a link failure in the network. This case considered the egress from an origin, in the same we could look at the access to a destination.

CONCLUSIONS

The paper presented a normative approach to transportation network reliability based on game theory to analyse the weak points of the network and its performance in case of a link failure. For this purpose two reliability measures are introduced to estimate a) the probability to encounter a failure, R^{con}, and b) the probability to arrive at the destination within an acceptable travel time, R^{perf}. The "acceptable travel time" was defined as the average travel time when users are not worried about link failures, plus a safety margin. In this paper "link failure" was defined as a capacity reduction on a link but the impact of a total link closure could be analysed as well with the same technique, as done for a multi-modal network in Schmöcker and Bell (2002).

The approach was first presented with the help of some tests on a small grid-network. These tests illustrated the game approach and showed the usefulness of the measures R^{con} and R^{perf}. The game finds an upper bound of failure probabilities for users who are aware of all routes in the network. If a link has a failure probability this means its failure has a big impact on the network performance. The link can therefore be considered as a "critical link".

If the reliability measure R^{con} is high, this shows that the network offers alternative routes that are not too expensive and therefore are taken by network users. A high value of R^{perf} indicates that the probability of a link failure does not have a big impact on the travel time of users because the existing route or the alternative routes have enough spare capacity to compensate for the link degradation.

The paper further showed an application of the approach on the Leicester network. Three cases are compared. The first Case A looked at the overall network. It was found which links have the most damaging impact on the network performance in case of failure. Not surprisingly there are a few inner city links which are crucial for the traffic within and through the network. Case B identified the critical links for each OD pair. It can be seen that these links are not necessarily inner city links, but the critical links can also be more towards the beginning or end of the journey, where the users might not have many route alternatives. Finally Case C looked at all users coming from the same origin and their critical links. This analysis could be of special interest if emergency services are located near these origins. The game analysis can then provide information about which links are very important to access the destinations in the network.

It can therefore be concluded that this reliability assessment can be an important tool for network planning. The approach analyses which links are most important for the network performance and gives information about how many users would experience the network disruption and what the consequences in terms of travel time are. As presented in the paper, the network planner can gain these information for the users of every single OD pair as well as for users from a certain origin or for all the network users. This will enable a comparison of different networks and identify which links, OD pairs and locations in a network could suffer most, when users have to be worried that some links not operating with their usual capacity. Further research should find methods to consider several dependent link failures. Assuming one independent link failure as done in this paper is clearly not true for earthquake situations, where some areas of the network are more affected than others, which provokes simultaneous and interdependent link failures. However, for scenarios with several link failures it will be difficult to decide how many links at most are allowed to fail simultaneously.

REFERENCES

Asakura, Y. (1996). Reliability Measures of an Origin and Destination pair in a Deteriorated Road Network with Variable Flow. *In "Transportation Networks: Recent Methodological Advances (Ed M G H Bell)"*, Pergamon Press: Oxford.

Bell, M. G. H., C. Cassir, Y. Iida and W. H. K. Lam (1999). A sensitivity based approach to network reliability assessment. *Proc 14th Int. Symp. on Transportation and Traffic Theory* (ed. A Ceder), Pergamon: Oxford, 283-300.

Bell, M. G. H. (2000). A game theory approach to measuring the performance reliability of transport networks. *Transportation Research*, **34B**, 533-546.

Bell, M. G. H., W. Lam and Y. Iida (1996). A Time-Dependent Multiclass Path Flow Estimator. *Proceedings of the 13th International Symposium on Transportation and Traffic Theory,* Lyon, France, July.

Cassir, C. and M.G.H. Bell (2000). The N+M person game approach to network reliability. *In "Reliability of Transport Networks" (Eds. M.G.H. Bell and C. Cassir)*, Research Studies Press, Baldock (Herts.), United Kingdom, 91 -102

Lam, W.H.K. and N. Zhang, (2000). A New Concept of Travel Demand Satisfaction Reliability for Assessing Road Network Performance. *Paper presented at the Matsuyama Workshop on Transport Network Analysis*, August 2000, Matsuyama, Japan.

Nash, J. (1951). Non-cooperative games. *Annals of Mathematics.* **54**, 286-295.

Nicholson, A. J. and E. Dalziell (2001). Risk Evaluation and Management: A Road Network Reliability Study, *Procs of 1st International Symposium on Transportation Network Reliability (INSTR)*, July/August 2001, Kyoto University.

Nicholson, A. J. and Z. P. Du (1997). Degradable Transportation Systems: An Integrated Equilibrium Model. Transportation Research, **31B**(3), 209-223.

Nicholson, A. J., Schmöcker, J-D., Iida, Y. and Lam, W.H.K. (2002) Assessing Transport Reliability: Malevolence and User Knowledge, *Procs of 1st International Symposium on Transportation Network Reliability (INSTR)*, July/August 2001, Kyoto University.

Schmöcker, J.-D. and M. G. H. Bell (2002). The PFE as a Tool for Robust Multi-Modal Network Planning. *Traffic Engineering & Control*, **44(3)**, 108-115.

Sheffi, Y. (1985). Urban transportation networks: equilibrium analysis with mathematical programming methods, *Prentice-Hall*, Englewoods Cliffs, New-Jersey.

Wakabayashi, H. and Y. Iida (1992). Upper and lower bounds of terminal reliability of road networks: an efficient method with Boolean algebra. *Journal of Natural Disaster Science,* **14**, 29-44.

14

Stochastic Network Design Problem: An Optimal Link Investment Model for Reliable Network

Yasuo Asakura, Graduate School of Science and Technology, Kobe University, Kobe, Japan
Eiji Hato, Civil and Environmental Eng., Ehime University, Matsuyama, Japan
Masuo Kashiwadani, Civil and Environmental Eng., Ehime University, Matsuyama, Japan

Abstract

An optimal network design model is formulated providing a set of link investment pattern for the most reliable network with highest network performance under uncertain conditions. The connectivity probability of a link is assumed to be improved by the investment to the link. The object function is represented as the expected performance measure. The formulated model is categorized in a group of stochastic network design problem in which the existence of a link is probabilistic. The characteristics of the gradient vector of the objective function are analyzed. The derivatives of the objective function can be approximately evaluated without enumerating all possible network state vectors. Numerical examples are calculated for analyzing the sensitivity of optimal investment policies.

Background and Research Objectives

Various approaches can be found in transport network reliability studies such as route choice behaviour under uncertain conditions, flow modelling in a degraded network, evaluation of network reliability and network design and management.

There were not so many studies on an individual's route choice behaviour in a damaged network. As well as theoretical studies, empirical studies were also very few since collecting behavioural data in a degraded network is difficult. However, travel behaviour itself is more or less interpreted as one of the decision making process under uncertainty. Therefore, a stock of probabilistic choice models and risk models in behavioural science is available for describing travel behaviour of an individual when he/she travels in a degraded network. Within a framework of probabilistic choice, Khattak and Palma (1997) studied driver's travel behaviour in abnormal weather conditions. Bonsall (2000) suggested that traveller's attitude for risk should be further studied.

Flow models are essential for evaluating network reliability with traffic congestion. These models intend to describe interactions between traveller's choice behaviour in a degraded network and the performance of the network. Asakura et al. (1999) formulated a Stochastic User Equilibrium model with variable demand for describing traveller's trip making and route choice behaviour. Lee et al. (2000) applied the interference theory and proposed "reliability assignment model" which could be used for evaluating travel time reliability as well as describing network flows.

There have been proposed various concepts of transport network reliability; connectivity reliability (Iida and Wakabayashi, 1989), travel time reliability (Asakura et al., 1991) and capacity reliability (Chen et al., 1999). Yang et al. (2000) discussed the relations of travel time reliability and capacity reliability, and proposed to evaluate and synthesize those two reliability concepts. Lo and Tung (2000) proposed the chance constraint approach that was applied to evaluate capacity reliability of a network. These studies were developed as evaluation tools for network reliability. There are a variety of assumptions on network states and flows. These models should be carefully examined when they are applied in actual networks.

As shown in a general framework proposed by Du and Nicholson (1997), the final goal of network reliability studies must be to provide useful information on network improvement programme to transport policy makers. As well as descriptive based approaches, normative approaches should be investigated to discuss the network improvement programme. When there is no uncertain factor in a network condition, conventional network design problems can be applied to obtain the optimal network with the best performance. This approach is, however, not always appropriate for network reliability analysis. A link in a network may be deteriorated and the existence of the link is uncertain. Thus, the probabilistic factors should be involved in the network design and management problem.

The main objective of this paper is to show a model framework of an optimal link investment problem. This is an optimal network design model that provides a set of link investment pattern for the most reliable network with highest network performance under uncertain conditions. A normative model is formulated so that the uncertainty of the occurrence of a network state is taken into consideration. The formulated model is categorized in a group of stochastic network design problem in which the existence of a link is probabilistic. This paper also analyses the characteristics of the gradient vector of the objective function. Instead of enumerating all possible network state, it is shown that the derivatives of the objective function can be approximately evaluated. Numerical examples are calculated for sensitivity analysis.

DEFINITION OF VARIABLES

In this section, variables are defined for formulating the problem. They are the decision variable, link connectivity and network state probability. The decision variable of the problem is the amount of investment to each link in a network. The investment variable on a link is written as x_a, and vector **x** denotes a set of link investment decision variables. We assume that the investment in a link is continuous and used only for improving the physical robustness of the link. For example, the investment money is used for improving structural toughness against natural disaster. This implies that the investment cannot be used for improving the capacity of the link such as lane widening. This assumption could be further relaxed.

The robustness against the risk of functional failure of a link can be represented by the connectivity of the link. The connectivity of a link p_a is the probability of whether the link is physically connected. When the link is connected, the traffic function of the link is kept at the normal state. The link connectivity satisfies the condition of $0 \le p_a \le 1$. It seems natural that the connectivity of a link becomes better when we increase the investment money in the link. We assume that the link connectivity is represented by an increasing function of the investment in the link. This means that $p_a(x_a)$ is an increasing function of the decision variable of a link x_a. This function is also assumed differentiable for decision variable. For example, we can apply $p_a(x_a) = 1 - \exp(-\lambda x_a)$ as a function describing the relation between link connectivity and link investment. Here λ is a parameter. Figure 1 shows an image of the link connectivity function.

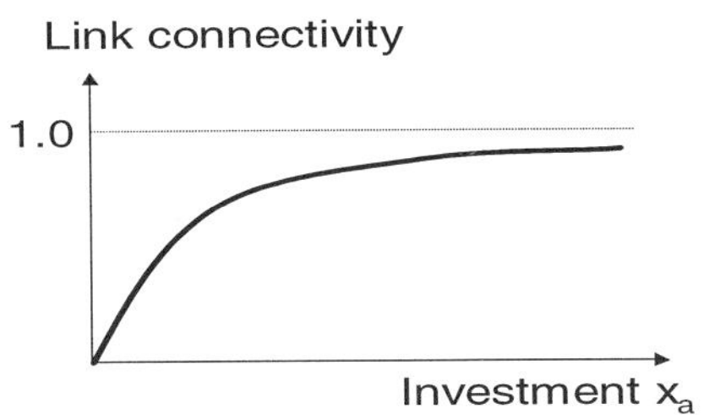

Figure 1 Link Connectivity Function

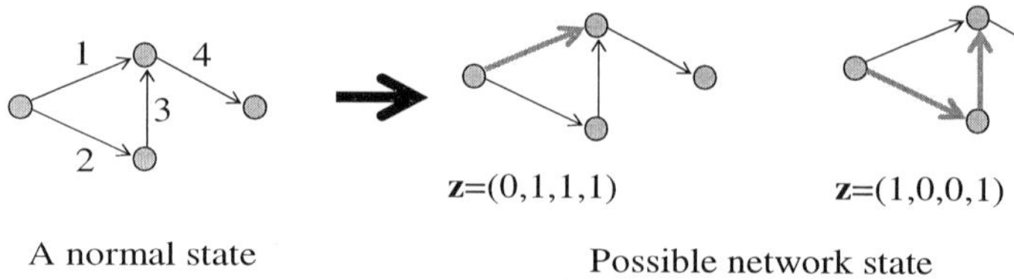

Figure 2 Network State Vector

It is impossible to achieve the perfect condition without link failure even if sufficient amount of money is used for improving the link connectivity. This means that various states of a network can be probable. A state vector $\mathbf{z} = \{z_a\}$ is introduced showing a state of a network with some disconnected links. The element of the state vector z_a is equal to 1 when a link is connected. It is equal to 0 when the link is deteriorated and disconnected. When a network has L links, the number of all possible network states becomes 2^L. Figure 2 shows examples of network state vectors.

The network state probability $\Pr(\mathbf{z}|\mathbf{x})$ represents the conditional probability of the occurrence of a state **z** for a given set of link connectivity probability. When the link connectivity is independent each other, the network state probability is represented by the following equation.

$$\Pr(\mathbf{z}|\mathbf{x}) = \prod_a p_a(x_a)^{z_a} (1 - p_a(x_a))^{1-z_a} \tag{1}$$

The partial derivative of the state probability for a link investment variable is then analytically represented as,

$$\frac{\partial \Pr(\mathbf{z}|\mathbf{x})}{\partial x_a} = \left(\frac{\partial p_a(x_a)}{\partial x_a}\right)^{z_a} \left(1 - \frac{\partial p_a(x_a)}{\partial x_a}\right)^{1-z_a} \prod_{b \neq a} p_b(x_b)^{z_b} (1 - p_b(x_b))^{1-z_b} \tag{2}$$

Eq. (2) is involved in the procedure of sensitivity analysis of the network design problem in the following sections.

FLOWS IN A NETWORK

Various measures can be used for evaluating the performance $PM(\mathbf{z})$ of a network state **z**. One of simple measures is a 0-1 flag identifying network connectivity. The value of $PM(\mathbf{z})$ is equal to 1 when the network is connected, and it is equal to 0 when it is disconnected. Since this measure is flow-independent, network flows are not necessary for evaluating network performance. However, it is necessary to describe flows in a degraded network when we evaluate the network performance under flow dependent conditions. For example, the total travel time of a network can be frequently used as the performance measure in a congested network. Thus, describing flows in a degraded network becomes necessary.

We have already proposed a network flow model, NL-SUE (Nested Logit type Stochastic User Equilibrium) model with variable demand (Asakura et. al, 1999). When a degraded state of a network continues for a while, the User Equilibrium principle for a static condition can be applied. Traffic congestion due to partially closed links in a network can be considered in a model through a link const function. Route choice behaviour of a driver in a degraded network is probabilistic since he may not be familiar with the network. A probabilistic choice model is appropriate to describe a driver's route choice. A driver may cancel his trip when the level of service such as travel cost of a degraded network becomes extremely worse. Then, trip generation process is also probabilistic. We assume a hierarchical structure of a driver's travel choice behaviour. As well as trip generation and route choice, there are various dimensions of a driver's travel behaviour. They can be considered in a multiple hierarchical structure.

When we apply Nested Logit model for describing probabilistic choice behaviour of a driver, we obtain a Stochastic User Equilibrium model with variable travel demand. The equivalent optimisation formulation of the SUE conditions assuming a hierarchical choice structure can be written in the following problem.

<NL-SUE formulation>

$$\min . Z(\mathbf{f},\mathbf{q},\bar{\mathbf{q}}) = \sum_{a\in A}\int_0^{v_a} t_a(w)dw + \sum_{rs\in\Omega}(q_{rs}C_{rs} + \bar{q}_{rs}R_{rs}) + \frac{1}{\theta_1}\sum_{rs\in\Omega}\sum_{k\in K_{rs}} f_k^{rs}\ln\left(\frac{f_k^{rs}}{q_{rs}}\right) + \frac{1}{\theta_2}\sum_{rs\in\Omega}\left[q_{rs}\ln\left(\frac{q_{rs}}{T_{rs}}\right) + \bar{q}_{rs}\ln\left(\frac{\bar{q}_{rs}}{T_{rs}}\right)\right] \tag{3}$$

subject to

$$\begin{aligned} & q_{rs} + \bar{q}_{rs} = T_{rs} && \forall rs \in \Omega \\ & \sum_{rs\in\Omega} f_k^{rs} = q_{rs} && \forall k \in K_{rs}, rs \in \Omega \\ & f_k^{rs} \geq 0 && \forall k \in K_{rs}, rs \in \Omega \\ & q_{rs} \geq 0, \bar{q}_{rs} \geq 0 && \forall rs \in \Omega \end{aligned} \tag{4}$$

Notations of the variables are as follows. $t_a(w)$: travel time function of link a, v_a: traffic volume on link a, f_k^{rs}: path flow on a k-th path of OD pair rs, q_{rs}: revealed OD flow of OD pair rs, $\bar{q}_{rs}$: cancelled OD flow, T_{rs}: potential OD flow (given), C_{rs}: fixed cost of travel, R_{rs}:

fixed cost for cancelling the trip, θ_1 : route choice parameter, θ_2 :travel choice parameter, K_{rs} : path set of OD pair rs, Ω : set of OD pairs, A : set of links.

NL-SUE can be calculated with various methods. MSA (moving successive average) method is one of the simplest algorithms while it may not always converge to the exact solution. Applying Simplicial Decomposition method is also possible when the scale of a network is not so large and the number of path is limited. Entropy decomposition algorithm using link flow variable with origin (or destination) flag can be also applicable.

OPTIMUM INVESTMENT PROBLEM FORMULATION

The performance measure $PM(\mathbf{z})$ of a network state $\mathbf{z}$ is evaluated using the output of the network flow analysis. The performance measure is, for example, the total travel cost or the consumer's surplus. When it is not necessary to consider the flow in a network, the total travel distance can be used as the performance measure. The connectivity of a network as a whole is another performance measure.

Let us consider the formulation of the objective function. When the value of the decision variable of a link x_a is determined, the link connectivity probability $p_a(x_a)$ is also determined. There may occur various network states $\mathbf{z} \in Z$ for this investment decision $\mathbf{x}$. The state probability $\Pr(\mathbf{z}|\mathbf{x})$ denotes the probability of the occurrence of a network state $\mathbf{z}$ for given investment $\mathbf{x}$. Each state of a network has its own performance measure $PM(\mathbf{z})$. We have assumed that the investment money to a link would be used only for improving link connectivity and not for improving link capacity. Thus, the performance measure $PM(\mathbf{z})$ is not necessarily dependent on the decision variable $\mathbf{x}$. In order to evaluate network performance considering all states of a network, the objective function should be the expected value of the performance measure. This is written as;

$$f(\mathbf{x}) = E[PM(\mathbf{z})] = \sum_{\mathbf{z} \in Z} \Pr(\mathbf{z}|\mathbf{x}) PM(\mathbf{z}) \tag{5}$$

Here Z denotes the possible vector space.

The Stochastic Network Design Problem (S-NDP) is formulated as the optimisation problem to find the link investment vector $\mathbf{x} \geq \mathbf{0}$ that optimises the expected performance measure $f(\mathbf{x})$ under the total budget constraint of investment $c(\mathbf{x}) \leq C$. Here, C denotes the upper limit of total investment. This constraint can be simply written as $\sum_a x_a \leq C$. The upper and the lower bounds for the investment to a link, $x_a^{lower} \leq x_a \leq x_a^{upper}$, can be involved as well. As a whole, the S-NDP is the problem to find a link investment pattern that can generate a higher probability of the network state with higher performance.

$$f(\mathbf{x}) = E[PM(\mathbf{z})] = \sum_{\mathbf{z} \in Z} \Pr(\mathbf{z}|\mathbf{x}) PM(\mathbf{z}) \rightarrow \textit{optimise}$$

subject to,

$$\sum_a x_a \leq C$$

$$x_a^{lower} \leq x_a \leq x_a^{upper}$$

For example, when there are three possible states $\mathbf{z}_1, \mathbf{z}_2, \mathbf{z}_3$. The performance measure is evaluated in some cost term, and the order of the performance measure for each state is assumed as $PM(\mathbf{z}_1) < PM(\mathbf{z}_2) < PM(\mathbf{z}_3)$. The optimum strategy to minimise the expected cost is to invest to links so that it will increase the value of $\Pr(\mathbf{z}_1|\mathbf{x})$ at most, then $\Pr(\mathbf{z}_2|\mathbf{x})$ and $\Pr(\mathbf{z}_3|\mathbf{x})$.

The S-NDP can be compared with a typical Deterministic Network Design Problem (D-NDP) as follows. Suppose that the decision variable of the D-NDP is the link investment money $\mathbf{x} \geq \mathbf{0}$ as same as S-NDP. The investment money can be used for improving the link capacity. This means the performance of a state could be improved when there is some investment in a network On the other hand, the state of a network is deterministic in D-NDP. It is not necessary to consider various states of a network. This means that the objective function yields to optimising the value of the performance measure such as total travel time minimization. Thus, the D-NDP is to find $\mathbf{x}$ which optimises $f(\mathbf{x}) = PM(\mathbf{x})$ under the budget constraint $c(\mathbf{x}) \leq C$. It is well known that the D-NDP can be formulated using the bi-level optimisation framework, when User Equilibrium network flow is required to evaluate the performance function (Huang and Bell, 1998). It is the same for the S-NDP if UE flow is assumed. Furthermore, the S-NDP requires the evaluation of the expected value. In this sense, the S-NDP would have the triple-level optimisation structure; namely the evaluation process by taking the expected value is added with the bi-level structure of D-NDP with User Equilibrium flow constraints.

The S-NDP can be extended to discuss the balanced investment between capacity and connectivity. In the previous section, we have assumed that the link investment money could be used only for improving link connectivity $p_a(x_a)$. However, it is more realistic to assume that the link investment money can be used for improving link capacity as well. This means that the performance measure $PM(\mathbf{z}, \mathbf{x})$ should be rewritten as the function of decision variable $\mathbf{x}$ as well as network state $\mathbf{z}$. Therefore, the objective function of the optimisation problem becomes $f(\mathbf{x}) = E[PM(\mathbf{z},\mathbf{x})] = \sum_{\mathbf{z}\in Z} \Pr(\mathbf{z}|\mathbf{x}) PM(\mathbf{z},\mathbf{x})$. The constraints are the same as the S-NDP mentioned above. When this formulation is used, it becomes possible to discuss the balance of the investment between link connectivity and link capacity. In some cases, the strategy of improving link connectivity is more effective to achieve higher value of network reliability. In the other cases, the strategy of improving link capacity is more effective. Although this extension brings mathematical complexity, it is an important and interesting topic for network reliability studies.

SENSITIVITY ANALYSIS

In this section, the possibility of the sensitivity analysis of the S-NDP is discussed. The objective function written in Eq. (5) is found separable. Using the derivative of the state probability shown in Eq. (2), the derivative of the objective function for the decision variable x_a can be written as,

$$\frac{\partial f(\mathbf{x})}{\partial x_a} = \sum_{\mathbf{z}\in Z} \frac{\partial \Pr(\mathbf{z}|\mathbf{x})}{\partial x_a} PM(\mathbf{z}) \tag{6}$$

Enumerating all possible state vectors $\mathbf{z}\in Z$, it is possible to evaluate numerically the above derivative for decision variable x_a. Thus, any gradient methods could be applied to obtain the optimal solution for the formulated S-NDP. However, it is impractical for an actual scale network to calculating derivatives though enumerating all possible state vectors. The number of the possible state vectors for a network with L links becomes 2^L, and calculating $PM(\mathbf{z})$ for all possible state vectors requires huge computational costs.

It is known that the upper and the lower bounds of the expected performance measure can be evaluated using a sub set of state vectors $Z_n = \{\mathbf{z}_1, \mathbf{z}_2, \ldots\ldots, \mathbf{z}_n\}$, where $Z_n \subseteq Z$. The upper and the lower bounds of the expected performance measure are shown as follows.

$$UB(\mathbf{x}) = \sum_{\mathbf{z}_i\in Z_n} \Pr(\mathbf{z}_i|\mathbf{x}) PM(\mathbf{z}_i) + PM(\mathbf{z}^*)(1 - \sum_{\mathbf{z}_i\in Z_n} \Pr(\mathbf{z}_i|\mathbf{x})), \tag{7.a}$$

$$LB(\mathbf{x}) = \sum_{\mathbf{z}_i\in Z_n} \Pr(\mathbf{z}_i|\mathbf{x}) PM(\mathbf{z}_i) + PM(\mathbf{z}^{**})(1 - \sum_{\mathbf{z}_i\in Z_n} \Pr(\mathbf{z}_i|\mathbf{x})), \tag{7.b}$$

$PM(\mathbf{z}^*)$ and $PM(\mathbf{z}^{**})$ denote the best and the worst values of the performance measure, respectively. The best value generally corresponds to the performance of the network state in which all of the links are kept connected. If all possible state vectors are involved in the sub-set, the upper and the lower bounds yield to the expected performance measure.

The derivatives of the upper and the lower bounds can be calculated as;

$$\frac{\partial UB(\mathbf{x})}{\partial x_a} = \sum_{\mathbf{z}_i \in Z_n} \frac{\partial \Pr(\mathbf{z}_i|\mathbf{x})}{\partial x_a} PM(\mathbf{z}_i) + PM(\mathbf{z}^*)(1 - \sum_{\mathbf{z}_i \in Z_n} \frac{\partial \Pr(\mathbf{z}_i|\mathbf{x})}{\partial x_a}), \qquad (8.a)$$

$$\frac{\partial LB(\mathbf{x})}{\partial x_a} = \sum_{\mathbf{z}_i \in Z_n} \frac{\partial \Pr(\mathbf{z}_i|\mathbf{x})}{\partial x_a} PM(\mathbf{z}_i) + PM(\mathbf{z}^{**})(1 - \sum_{\mathbf{z}_i \in Z_n} \frac{\partial \Pr(\mathbf{z}_i|\mathbf{x})}{\partial x_a}). \qquad (8.b)$$

Here, the derivative of the state probability $\partial \Pr(\mathbf{z}_i|\mathbf{x})/\partial x_a$ is also represented by the same equation of Eq. (2). When the difference of the upper and the lower bounds is small enough, the derivative of the objective function can be approximated using Eqs.(8.a) and (8.b).

$$\frac{\partial f(\mathbf{x})}{\partial x_a} \cong \left(\frac{\partial UB(\mathbf{x})}{\partial x_a} + \frac{\partial LB(\mathbf{x})}{\partial x_a} \right) \Big/ 2 \qquad (9)$$

Eq.(9) gives a good approximation when both upper and lower bounds are smooth and close each other sufficiently. Instead of using Eq. (6), Eq. (9) can be used as the gradient of the objective function of S-NDP. When this equation is used in any gradient methods, it is possible to find the approximated optimum solution.

Yang (1997) and Bell et al. (1999) have investigated the sensitivity analysis of the User Equilibrium and its applications. The approximated gradient function of the objective function represented in Eq. (9) is also useful in the sensitivity analysis of the S-NDP. When there is a present link investment pattern $\mathbf{x}$ and the gradient vector of the objective function $\nabla f(\mathbf{x})$ is analytically obtained, it becomes possible to compare the effects of one unit of additional investment in a link. This information can be used to find a set of candidate links for improvement.

NUMERICAL EXAMPLES

In this section, the above-formulated S-NDP is numerically analysed using a toy network. Figure 3 depicts a toy network with 4 nodes, 5 directed links. A pair of origin and destination is assumed from node 1 to node 4. Each link has its own capacity shown in the figure.

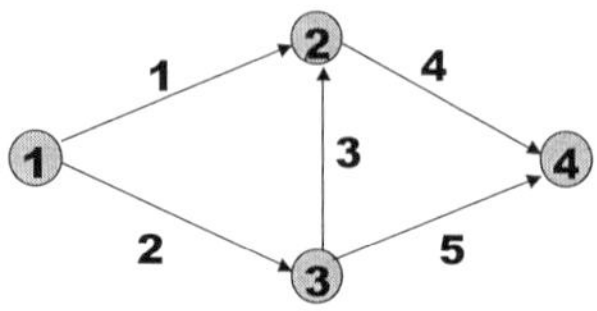

Link#	Capacity
1	100
2	200
3	100
4	50
5	100

Figure 3 Full Network and Link Capacity

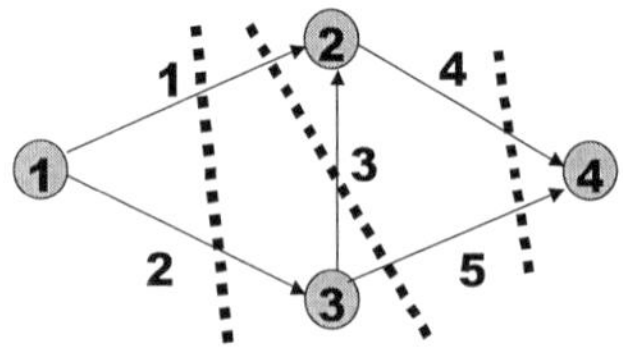

Figure 4 Cut Sets in the Network

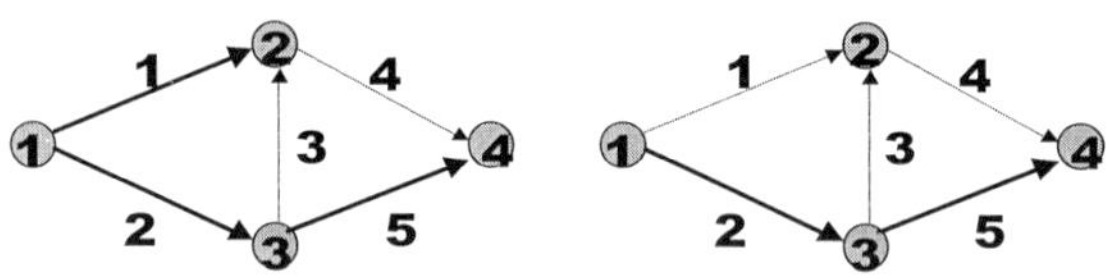

Figure 5 Identical Network States

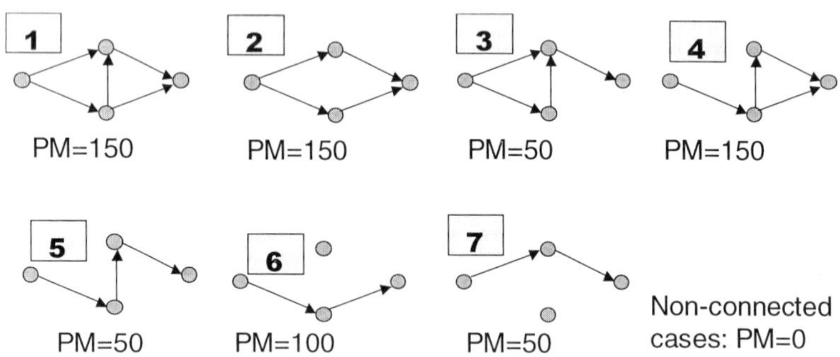

Figure 6 Possible Network States and the Performance Measure

The performance measure of the S-NDP is the maximum capacity of a network. This is defined as the maximum number of trips that could be loaded onto the network. The minimum value of the capacities of cut sets corresponds to the capacity of the network. For the full network, there exist three cut sets as shown in Figure 4, which are {1,2}, {1,3,5} and {4,5}. The numbers in the parenthesis are the link number of the cut set. Since the capacities of the cut sets are C{1,2}=300, C{1,3,5}=300 and C{4,5}=150, the network capacity is equal to 150 that is the minimum of the cut set capacities.

As the full network has 5 links, the number of the possible states of the network becomes $2^5 = 32$. Among these network states, there are disconnected networks. Furthermore, some states are identical in terms of connectivity. For example, the network state **z**={1,1,0,0,1} is identical with the network state **z**={0,1,0,0,1} as shown in Figure 5, since link 1 cannot be used even if it exists.

Thus, there are seven network states in which the OD pair from node 1 to node 4 is connected. They are shown in Figure 6. The capacities of those networks are calculated, and the values are shown in the same figure. It is needless to mention that the capacity of a disconnected network is zero.

The following link connectivity function is used, which defines the relations between the link connectivity probability $p_a(x_a)$ and the investment in the link x_a.

$$p_a(x_a) = 1 - \exp(-0.1x_a)$$

We assume that the present investment pattern to the network is given as **x**=(23, 16, 16, 12, 9). This means that the link connectivity at present is calculated as **p**=(0.9, 0.8, 0.8, 0.7, 0.6). Under this condition, the state probability $\Pr(\mathbf{z}|\mathbf{x})$ is calculated as shown in Table 1. Note that the values of the state probability for network state 6 and 7 contain the values of the state probabilities of the identical states, respectively.

Table 1 State Probability and Maximum Capacity

Network State	State Probability	Network Capacity
1	0.24192	150
2	0.06048	150
3	0.16128	50
4	0.02688	150
5	0.01792	50
6	0.04704	100
7	0.09828	50
non-connected	0.3462	0

The expected value of the performance measure $E[PM(\mathbf{z})]$ is then calculated. It is the expected value of the network capacity weighted by the state probability, and is equal to 67.97. This value is less than the half of the performance measure of the full network (state 1) that corresponds to the normal network state. It is found that the existence of uncertainty would further reduce the expected network performance.

The values of the partial derivatives of the objective function are then calculated. In order to do so, the value of the partial derivative of the link connectivity function is calculated. As the number of the possible states is small, the approximation using the upper and the lower bounds is not necessary. The exact values of the derivatives can be calculated. The values of the derivatives are shown in Table 2. The order of the values of the derivatives is link 1 > link 3> link 5 >> link 4 > link 2. This suggests that the investment strategy for the present condition would be to invest money into links 1 and 3 in order to improve the expected network capacity under uncertainty

Table 2 Results of the Sensitivity Analysis

Link	x_a	p_a	$\frac{\partial p_a(x_a)}{\partial x_a}$	$\sum_{z \in Z} \frac{\partial \Pr[\mathbf{z}/\mathbf{x}]}{\partial x_a} PM[\mathbf{z}]$
1	23	0.9	0.01	52.75
2	16	0.8	0.02	3.75
3	16	0.8	0.02	49.57
4	12	0.7	0.03	6.25
5	9	0.6	0.04	26.03

CONCLUSION

In this paper, we have presented a model framework of Stochastic Network Design Problem (S-NDP). Assuming that the investment money into the network can improve link connectivity, the S-NDP determines the optimum link investment pattern for improving the expected network performance. A variety of the network performance measure can be applied to the objective function that is represented as the expected value of the performance measure weighted by the state probability of a network. They are, for example, the connectivity reliability of a network, the expected network capacity, the expected total travel time in a network and so on. The User Equilibrium conditions can be considered when UE flows are required for evaluating the performance measure such as the total travel time.

The objective function is represented as the products of the performance measure of a state and the probability of the occurrence of the state. The performance measure of a state is assumed independent with the decision variable since the link investment is only used for improving link connectivity. On the other hand, the partial derivative of the state probability for the decision variable can be analytically represented when link connectivity is independent each other. Thus, the partial derivative of the objective function can be obtained as the products of the derivative of the state probability and the performance measure of the network state. This can be used to solve the S-NDP.

One of the difficulties in numerical calculation is to enumerate all possible network state vectors for evaluating the expected performance measure. A practical method is to approximate the expected value with the upper and the lower bounds calculated using a sub set of possible state vectors. This is also applied to evaluate the derivatives of the objective function with the decision variable. The resulting approximation equations were shown in this paper.

The sensitivity analysis using the partial derivative of the expected performance measure seems one of the interesting topics for discussing investment strategies under the present investment pattern. The comparison of the partial derivatives of the objective function for the present link investment would give the important information for the investment in the future. The numerical examples using a toy network suggested the effectiveness of the sensitivity analysis. However, we have not yet investigated numerically the effectiveness of the approximation. This should be examined using larger scale networks. The selection of optimal link investment patterns were not yet studied as well. Using the approximated gradient vector of the objective function, the formulated S-NDP can be solved and the optimal link investment patterns can be determined. The whole algorithm for solving S-NDP and corresponding numerical examples are now under development and will be presented elsewhere.

The independence assumption of link connectivity is not always the necessity condition for the S-NDP formulation. It is possible to assume any interdependence of link connectivity probabilities. When the network state probability is represented by a closed form, it would be possible to approximate the derivatives of the network state probability as shown in Eq.(2). The statistical analysis on link connectivity probabilities may help to introduce the interdependence of them. This should be discussed when we consider the external impacts that might occur at the same time.

More general formulation of S-NDP is found important. It is a combined investment problem of improving both connectivity and capacity of a link. When a network has a sufficient capacity, the investment policy of improving link connectivity is more effective for achieving a more reliable network. In contrast with the case, the investment strategy of improving link capacity may be more effective. It depends on the types of uncertainty concerned and the goals of the transport network management. Although the mathematical handling of the extended problem becomes complicated, this would be a challenging topic in network reliability design studies.

REFERENCES

Asakura, Y. and M. Kashiwadani(1991) Road Network Reliability Caused by Daily Fluctuation of Traffic Flow. *Proceedings of the 19th PTRC Summer Annual Meeting, Seminar G*, Brighton, pp.73-84.

Asakura, Y., M. Kashiwadani and E. Hato (1999) Flow Model and Performability of a Road Network under Degraded Conditions. *Proceedings of the 14th ISTTT* (Ceder, A. ed.), pp.257-281, Pergamon.

Asakura, Y. and E. Hato (2000) A Route Recognition Model for Transport Network Reliability Analysis. *Reliability of Transport Networks* (Bell and Cassir eds.), pp.203-216, Research Studies Press

Bell, M.G.H., C. Cassir, Y. Iida and W.H.K Lam (1999) A Sensitivity Based Approach to Network Reliability Assessment. *Proceedings of the 14th ISTTT* (Ceder, A. ed.), pp.282-300, Pergamon.

Bonsall, P. (2000) Traveller's Response to Uncertainty. *Reliability of Transport Networks* (Bell and Cassir eds.), pp.1-10, Research Studies Press.

Chen, A.H., H.Yang, H. Lo and W.H. Tang (1999) A Capacity Related Reliability for Transportation Networks. *Journal of Advanced Transportation*, **Vol.33, No.2**, pp.183-200.

Du, Z.P. and Nicholson, A. (1997) Degradable Transportation Systems: Sensitivity and Reliability Analysis, *Transpn. Res.*, **Vol.31B, No.3**, pp.225-237.

Huang, H.J. and Bell, M.G.H.(1998) Continuous Equilibrium Network Design Problem with Elastic Demand: Derivative-Free Solution Methods. *Transportation Networks: Recent Methodological Advances* (Bell ed.), pp.175-193, Pergamon.

Iida, Y. and H. Wakabayashi (1989) An Approximation Method of Terminal Reliability of a Road Network Using Partial Minimum Path and Cut Set. *Proceedings of the 5th WCTR*, Yokohama, pp.367-380.

Khattak A.J. and Palma, A.D.(1997) The Impact of Adverse Weather Conditions on the Propensity to Change Travel Decisions: A Study of Brussels Commuters. *Transpn. Res.*, **Vol.31A, No.3**, pp.181-203.

Lee, S., Moon, B. and Asakura, Y. (2000) Reliability Analysis and Calculation on Large Scale Transport Networks. *Reliability of Transport Networks* (Bell and Cassir eds.), pp.173-189, Research Studies Press.

Lo, H. and Tung, Y.K. (2000) A Chance Constrained Network Capacity Model. *Reliability of Transport Networks* (Bell and Cassir eds.), pp.159-172, Research Studies Press.

Yang, H. (1997) Sensitivity Analysis for the Elastic Demand Network Equilibrium Problem with Applications. *Transpn. Res.*, **Vol.31.B**, pp.55-70.

Yang, H., K.K. Lo and W.H.Tang (2000) Travel Time Versus Capacity Reliability of a Road Network. *Reliability of Transport Networks* (Bell and Cassir eds.), pp.119-138, Research Studies Press.

15

Improving the Reliability of Street Networks in Highly Densely Populated Urban Areas

Yan Li, Ritsumeikan Asia Pacific University, Oita, Japan
Hiroshi Tsukaguchi, Ritsumeikan University, Shiga, Japan

Abstract

Emergency vehicles should be planned to have smooth access to people's residences after a large scale disaster. This study proposes ANR, the Accessible Node Rate, which represents the ratio of the number of accessible nodes to total nodes in the network, as an indicator for evaluating the reliability of a local street network. Using the ANR, this study evaluates the urban districts of Nada Ward in Kobe City where the earthquake caused extensive damage, and proposes planning changes to provide the local street networks with higher reliability. Furthermore, in order to see if the ANR can be used for analyzing another city's reliability, a case study of Beppu City which is supposed to have the possibility of the same magnitude of earthquake, is examined. A system that supports traffic network editing and analysis for reducing the workload in a wider area is developed.

Introduction

After the Great Hanshin-Awaji Earthquake occurred on 17th, January, 1995, in the Kobe area of Japan, not only roads but also basic facilities of transportation such as the railways, harbors and airports were widely damaged. Especially, in the highly densely populated urban area, emergency vehicles could not access smoothly because of the disruption to the roads, obstructing the lifesaving activities immediately after the earthquake disaster. The urgency for improving the reliability of the traffic networks was reconfirmed and many investigations as well as research papers have reported on the reliability of the local street network at the district level, which was a field of less interest in the previous research.

Most of the reports concerning the reliability of the local street networks are mainly concentrated on the disruption to the streets after the Great Hanshin-Awaji Earthquake. The report of Ieda et al (1995) pointed out that the disruption of local streets wider than 8 meters was quite different from area to area. Tsukaguchi et al (1995) and Li & Tsukaguchi (1996) reported that local streets wider than 12 meters are merely disrupted and the disruption of other local streets mainly depends on factors such as the magnitude of the earthquake, the width of the streets, and the proportion of the street length with wooden housing alongside. Odani et al. (1996) reported similar results through the investigation of a different part of the stricken area.

Instead of concentrating on analyzing the disruption of single streets, some reports studied the function of the stricken *network*, such as Uno et al. (1997), Wakabayashi (1996) and the previous research by the authors (Li & Tsukaguchi, 1997A, 1997B, Tsukaguchi et al, 1999).

So far, regarding the connection ability and the service level of the road networks in an emergency case including a disaster, the concept of reliability has been used (Iida & Wakabayashi, 1989, Asakura, Y., 1996, Bell & Iida, 1997). The reliability indicates the probability of accessibility higher than a certain service level between any pair of origin and destination nodes of a road network, in a given period of time. This index emphasizes the importance of the evaluation between both nodes, and in addition has the problem of calculation time. Many studies were carried out by estimating the reliability for important sections within large areas such as the level of intercity highways.

In this study, we assume that emergency vehicles should be planned to have smooth access to people's residences even when a large scale disaster arises, and this study focuses on the densely populated local street network surrounded by the city's district distributors. When we consider the reliability improvement at the local street network level, the accessibility (connection ability) of the emergency vehicles from a non-stricken area to *all nodes* in the stricken road network is becoming more important. In this study, from the view of reliability, we propose the Accessible Node Rate (ANR) as an indicator of the reliability of the local street network at the district level. Using the ANR, we will examine the urban districts of Nada Ward in Kobe City where the earthquake caused extensive damages, and propose planning changes to provide the local street networks with higher reliability. Furthermore, in order to see if the ANR can be used for the reliability analysis of another city where an earthquake might occur in the future, we will make a case study of Beppu City, Oita Prefecture, Japan. As well, we will develop a system that supports traffic network editing and analyzing for reducing the workload involved in an analysis of a large area.

THE DEFINITION OF ANR

Accessible and Inaccessible Nodes

Even if some of the links that connect a node are blocked, the node can still be accessible; see (a), (b), (c) in Figure 1. But if all the links connected to the node are blocked, it will become inaccessible; see (d) of the Figure. Inaccessibility could happen to multiple nodes if all of the links connected to an area are blocked; see (f). In this case, if only one of these blocked links is open, all the nodes become accessible; see (e). It can also be said from Figure 1 that the bundles of inaccessible nodes could be found according to the locations of blocked links.

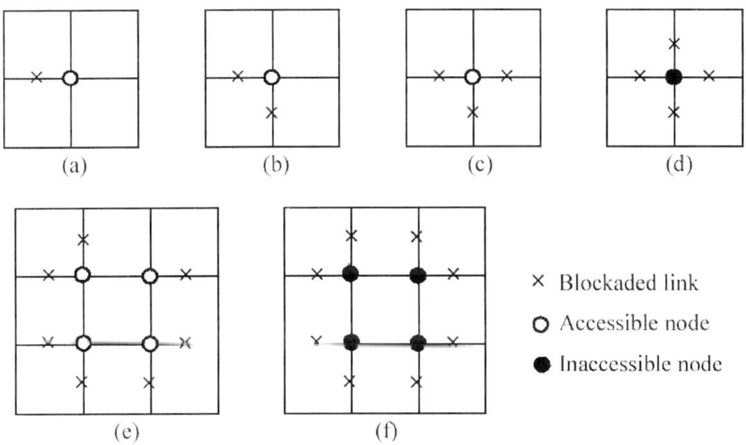

Figure 1 The accessible node and inaccessible node

The ANR and its Calculation Method

From the view point of smooth access for emergency vehicles from a non-stricken area, we define ANR, the Accessible Node Rate, to be the ratio of the number of accessible nodes to the total nodes in the network. That is,

$$ANR = n / N$$

where N means the total number of nodes, and n means the number of nodes that can be accessed from a non-stricken area.

We suppose every link has a probability of being blocked according to the strength of the disaster and the situation along the street. So, the given conditions are the link blockage probability l_a, where $0 \le l_a \le 1.0$ and $a \in A$. Here, A is the set of links that composes the local street network. Then the ANR can be estimated as the average accessible node rate when the disaster hits the study area, based upon a number of Monte Carlo simulations.

We recommend the Monte Carlo direct sampling method for the calculation of the ANR. The ANR is calculated using the following algorithm.

Step 1. We establish m as the simulation time of the direct sampling. The initial number is 0.

Step 2. We establish $m = m + 1$.

Step 3. For each link, we produce a random number r_a, which is a uniformly distributed deviate indicating the strength of the disaster, where $0 \le r_a \le 1.0$ and $a \in A$. By comparing with the given blockage probability l_a, we decide whether the link a is a blockage or not at this time of direct sampling. In the case of $r_a < l_a \le 1.0$, link a is blocked and in the case of $0 \le l_a \le r_a$, link a is not blocked.

Step 4. We check every node's accessibility from the optional nodes of the network of a non-stricken area, and then we calculate the total number of accessible nodes n_m. For this simulation time m, the accessible node rate is $anr_m = n_m / N$.

Step 5. We calculate the ANR up to m times, that is $ANR_m = 1/m\sum_{i=1}^{m} anr_i$.

Step 6. If it satisfies the convergence condition $\frac{1}{ANR_{m-1}}\left|ANR_m - ANR_{m-1}\right| \le \varepsilon$, we stop calculating. If not we go back to step 2. In this study, we establish the convergence criterion ε to be 0.001.

Applying ANR to the Stricken Area of Kobe City

For the study area, we selected the urban area surrounded by four principal district distributors in the Nada Ward in Kobe City, which has suffered damage from the magnitude 7 earthquake. The size of area is approximately 1300 meters in the East-West direction and 600 meters in the North-South direction. A study network consists of all the local distributors that are wider than 4 meters. By using aerial photos taken on 18th, January, 1995, the day after the great earthquake, we deciphered the situation (blocked or otherwise) of each link on the network. The link blockage situation can be divided into two levels, one in which cars have traffic access and another where the cars do not have traffic access because of the collapse of the houses along the street. 20 out of 116 total nodes in the street network were inaccessible nodes by cars, and the ANR was 82.7% (Figure 2).

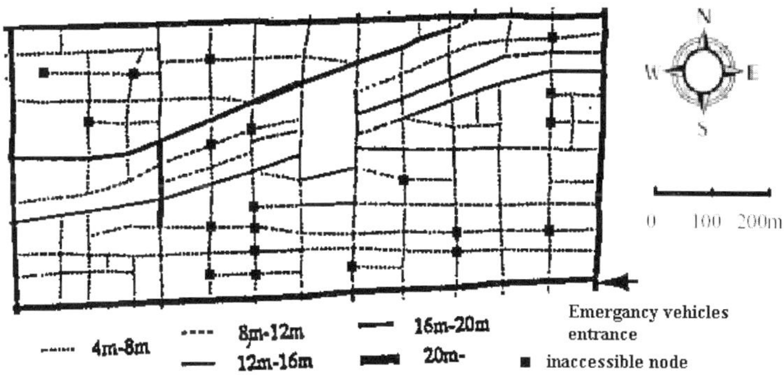

Figure 2 The local street network and the blockage situation of the node

The Link Blockage Probability and the Emergency Vehicle Entrance

According to research by the authors (Li and Tsukaguchi, 1997A), the link blockage situation in the area is mainly affected by two factors, that is, the street width and the proportion of the street length with wooden housing alongside (Table 1). Below, we use the figures in Table 1 as the link blockage probability l_a, which is necessary for the ANR calculation.

Table 1 Link blockage probability l_a

<table>
<tr><td rowspan="2"></td><td colspan="5">Street Width</td></tr>
<tr><td>4m - 6m</td><td>6m - 8m</td><td>8m - 10m</td><td>10m - 12m</td><td>12m -</td></tr>
<tr><td>Wooden housing proportion from 0 to 40%</td><td>0.481</td><td>0.379</td><td rowspan="2">0.381</td><td rowspan="2">0.190</td><td rowspan="2">0.000</td></tr>
<tr><td>Wooden housing proportion from 40% to 100%</td><td colspan="2">0.701</td></tr>
</table>

Source: Li and Tsukaguchi, 1997A

As we mentioned above, accessible nodes are those from which there is smooth access to the nodes in the non-stricken area. So here we have to set the entry point of the emergency vehicles. From Table 1, we can see the links wider than 12 meters have never been blocked. It means that the network composed of links wider than 12 meters was physically accessible after the Great Hanshin-Awaji Earthquake. Since the district distributors surrounding the study network are wider than 12 meters, and they are in a network that is physically accessible from the non-stricken area, we suppose the entrance of the emergency vehicles is at an optional node on the district distributors (Figure 2). Although heavy traffic jams on the network after the earthquake occurred meant poor accessibility even on the physically connected network (Odani et al., 1996), there is a question about how to manage the traffic after a disaster, which will not be discussed in this study.

The Real ANR and the Estimated ANR

Using the link blockage probability shown in Table 1, we calculated the ANR of the present street network for verification. After 245 simulations, the estimated ANR converged to 83.1%, which is close to the figure deciphered from the photos mentioned above (82.7%).

Reliability Improvement Planning of the Study Area

As shown in Table 1, increasing the local streets' width and decreasing the wooden housing proportion along the street will improve the strength of the links, thus improving the reliability of the local street network. In this section, we will prepare alternative plans using these two factors and maximize the ANR.

As a result of a simulation on the local street network proposed by the authors (Li and Tsukaguchi, 1997b), sub-district distributors at an interval of 300 meters offer higher reliability. Thus we set sub-principal distributors as in Figure 3.

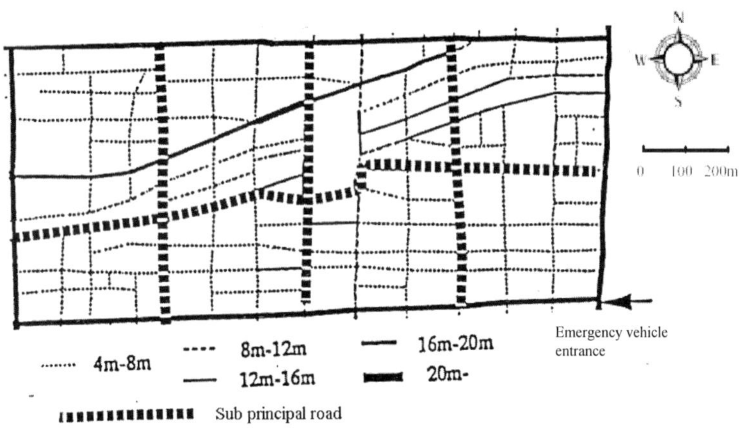

Figure 3 Sub principal distributors

We will establish three scenarios considering the widening of the sub-district distributors from their present width. The three scenarios are 8 meters, 10 meters and 12 meters. From Table 1, as far as the link blockage probability is concerned, there is no difference between 12 meter-wide streets and those wider than 12 meters. This is the reason we do not consider sub-district distributors wider than 12 meters.

Since we can also tell from Table 1 that the blockage probability of a local street with a width larger than 8 meters seems to be unrelated to the proportion of roadside wooden housing, we will only consider 2 scenarios involving local streets with a width less than 8 meters:

- The wooden housing proportion is maintained at the present level.
- The wooden housing proportion is improved to less than 40%.

Cosequently, we have seven alternative cases as is shown in Table 2. We also indicate the calculation results of the ANR for each of these 7 cases in this table.

According to the calculation results, establishing the width of the sub-principal district distributors at less than 10 meters width decreases the value of ANR (case A). On the other hand, when the width of the sub-principal district distributors becomes more than 10 meters (case B and C), the ANR reaches around 95%. The four cases (case D, E, F and G) shows that decreasing the proportion of wooden housing on the local streets to less than 40%, leads to high ANRs; all reaching over 98%. In conclusion, when the wooden housing proportion of the local streets is below 40%, it seems that widening of the street is unnecessary.

Table 2 The alternative plans and their ANR

Case	Width of the sub district distributors	Wooden housing proportion of streets with width less than 8 meters	ANR
A	8m - 10m	The present	85.65%
B	10m - 12m	The present	94.35%
C	12m -	The present	95.09%
D	The present	Less than 40%	98.23%
E	8m - 10m	Less than 40%	98.69%
F	10m - 12m	Less than 40%	98.73%
G	12m -	Less than 40%	98.78%

CASE STUDY OF RELIABILITY ANALYSIS TO BEPPU CITY

In order to see if ANR can be used for another city's reliability improvement analysis where an earthquake might occur in the future, a case study of Beppu City, Oita Prefecture, Japan was made.

In the case study of Kobe City, we only examined local streets wider than 4 meters. That is because we used air photos to decipher the blockage situation after the earthquake, in which the narrow passages were difficult to distinguish. But actually, emergency vehicles can enter two-meter-wide narrow passages if necessary. This creates two problems. One is how to get the very detailed network data for ANR calculations, which usually requires a tremendous amount of time. The other problem is how to decide the link blockage probability of the streets narrower than 4 meters. For the first problem, we developed an interactive, effective network editing system for the Windows operating system. For the second problem, we expect more researches in the future and here we just simply extend the data in Table 1 to the streets with width less than 4 meters.

In the case study of Kobe City, we tried to improve the reliability by building sub-district distributors. Here, we will try another method, to find the nodes that are likely to be closed, which might also be helpful for planning.

Network Data Editing System

Usually, creating traffic network data for simulation is done manually and requires a lot of work, depending on the size of the network. We developed a system that could create these data on a screen interactively with a mouse and dialog boxes (Li and Dalhan, 2001). This system was developed as an application for the Windows operating system by using Microsoft Visual C++. It can import a Bitmap image or draw the GIS (Geographical Information System) data as the background. Users can draw nodes and links of the network with a mouse, and each node and link is automatically given a number. Users can also delete links or nodes at any time, and all of the existing nodes or links will be re-numbered simultaneously.

Figure 4 shows the network of the case study area of Beppu City using a Bitmap image as background. The size of the area is approximately 700 meters in an East-West direction and 650 meters in a North-South direction. All the streets wider than 2 meters, which is considered to be the minimum width for an emergency vehicle to enter, were drawn. There are 202 nodes and 698 links in the network. Since the wider, black streets shown in Figure 4 are on the over-12-meter-network, we choose any node on the circumferential street as the entrance for emergency vehicles.

In this system, if a user wants to add some information to a certain link, he just needs to click the link, and then the dialog box will be ready to accept the input. Here we use the dialog box for inputting link width and the proportion of wooden housing, which are obtained by field investigations (see Figure 5).

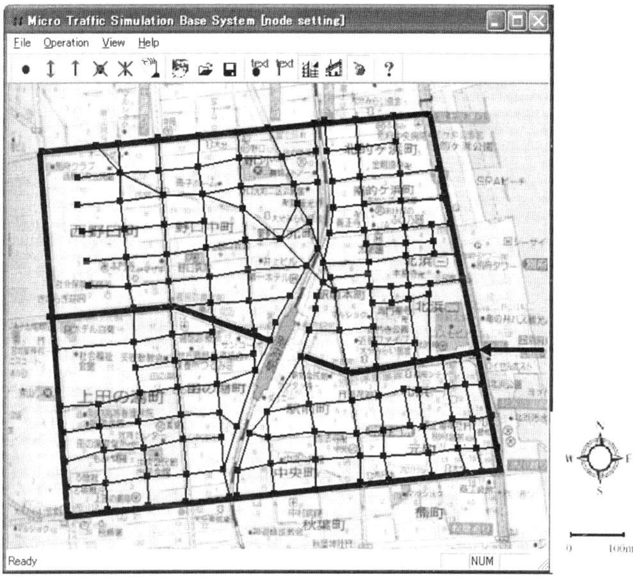

Figure 4 Street network built by the network editing system

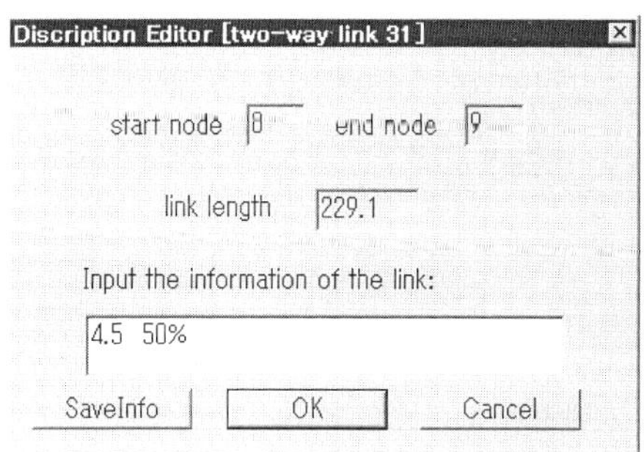

Figure 5 Dialog box for setting link attributes

The Link Blockage Probability

For a stricken area, it is easy to calculate the link blockage probability because every link's blockage situation is obvious. But for the reliability improvement planning of other areas, a disaster may possibly occur in the future and we don't have the blockage situation data now.

There are many factors that could affect the link blockage probability; the strength of the disaster, the strength of the houses along the street, and the width and length of the streets, etc. For Beppu City, since it is also in the earthquake zone, we suppose that they suffer an earthquake of the same strength as Kobe City. The link blockage probabilities are assumed to be transferable between cities, and we will try to analyze the network's reliability if an earthquake occurs.

We will use the results of Table 1 for determining the link blockage probability in Beppu City. Since we don't have the data for links with width of 2m to 4m, we will extend the data of Kobe as shown in Figure 6. From the Figure, for the links with a width of 2m to 4m, we suppose the blockage probability is 60% when its wooden housing proportion is from 0 to 40%, and 90% when its wooden housing proportion is from 40% to 100%.

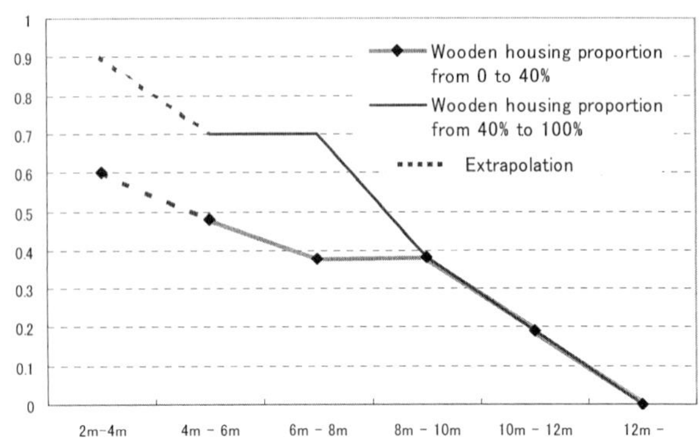

Figure 6 Link blockage probability for Beppu City

The ANR and Inaccessible-likely Nodes in Beppu City

We calculated the ANR of the network. The ANR of the network is 76.0%, excluding the nodes on the over-12-meter network that are supposed to be accessible all the time. There are 19 nodes that were inaccessible during more than half of the simulations (see Figure 7). Therefore, if an earthquake similar to the Great Hanshin-Awaji Earthquake happened in Beppu City, nearly one-fourth of the area will be inaccessible to emergency vehicles.

CONCLUSION

In this study, on the basis of the reliability perspective, we assumed that the emergency vehicles should be planned to have smooth access to residences after a large scale disaster, and we propose the ANR, Accessible Node Rate, which represents the ratio of the number of accessible nodes to total nodes in the network, to be the indicator for evaluating the reliability of a local street network.

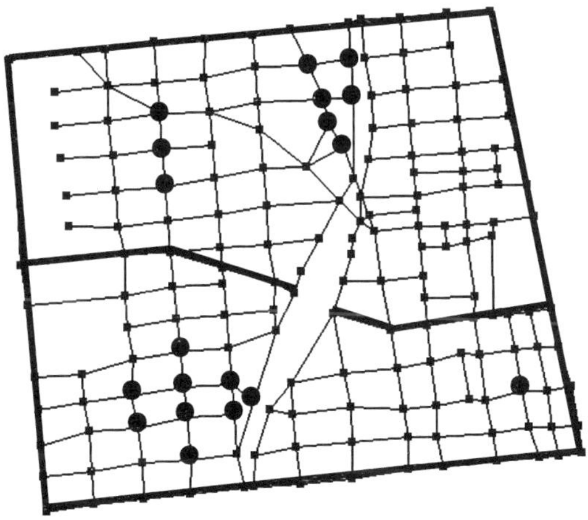

Figure 7 Inaccessible nodes in Beppu City (estimated)

Using the ANR, we examined the urban districts of Nada Ward in Kobe City where the magnitude of the earthquake was very strong, and proposed the planning changes to provide the local street networks with higher reliability.

Furthermore, in order to see if the ANR can be used for analyzing another city's reliability, we made a case study on Beppu City, which is supposed to have the possibility of the same magnitude of earthquake. We also developed a system that supports traffic network editing and analysis for reducing the workload in a wider area.

This paper showed that the ANR is a very useful indicator for reliability analysis in highly densely populated areas.

REFERENCES

Asakura, Y. (1996). Reliability measures of an origin and destination pair in a deteriorated road network with variable flows. Proceeding of the 4th Meeting of the EURO working Group in Transportation.

Bell, M.G.H., and Iida, Y. (1997). Transportation network analysis, Chapter 8: Network reliability (pp.179-192). John Wiley & Sons.

Ieda, H., Kaminishi, S., Itoyama, T. and Suzuki, T. (1995). Street blockades in the Hanshin earthquake, Proceedings of The 18th Conferences of Infrastructure Planning.

Iida, Y. and Wakabayashi, H. (1989). An approximation method of terminal reliability of a road network using partial minimal path and cut set. Proceedings of the 5th WCTR, Yokohama, pp.367-380.

Li, Y and Tsukaguchi, H. (1996). A study on the capability of the local road network against the great Hanshin-Awaji Earthquake Disaster, International Conference on Urban Engineering in Asian Cities in the 21 Century, Bangkok.

Li, Y., and Tsukaguchi, H. (1997A). Evaluating disaster prevention ability of local road network. In Hanshin-Awaji Great Earthquake Researches, pp.407-414.

Li, Y., and Tsukaguchi, H. (1997B). A basic study on the indicator and structure of the road network concerning disaster prevention. Infrastructure Planning, Vol.14, pp.361-370.

Li, Y., and Dalhan, N. (2001). A base system for the micro traffic simulation using the GIS database. Proceedings of The 24th Conference of infrastructure Planning.

Odani, M., Matsumoto, M., Sako, T., Imai, H.(1996). Analysis on actual situation of use of private cars by residents after the Great Hanshin-Awaji Earthquake, Proceedings of Traffic Engineering Conference.

Tsukaguchi, H., Totani, T. And Nakatsuji, K. (1995). Study of Impassable roads during the Great Hanshin Earthquake, Proceedings of The 18th Conference of Infrastructure Planning.

Tsukaguchi, H., Vandebona, U., Li, Y. (1999). Planning of residential street network for disaster prone urban areas, World Transport Research, Proceedings of 8th world conference on transport research, Pergamon, pp.411-424.

Uno,N.,Iida,Y., Taniguchi,E., Ushiba,T.(1997). Deterioration of traffic function of road network and influence traffic regulation after the Great Hanshin-Awaji Earthquake, Proceedings of the Great Hanshin-Awaji Earthquake Research, 201-208.

Wakabayashi, H. (1997). Reliability of highway network and probability importance in the 1995 Great Hanshin Earthquake, Proceedings of the Great Hanshin-Awaji Earthquake Research, pp.237-246.

16

Street Network Reliability Evaluation following the Chi-Chi Earthquake

Yung - Lung Lee, Department of Land Management and Development, Chang Jung Christian University, Taiwan, R. O. C.
Kuang - Yih Yeh, Department of Urban Planning, National Cheng Kung University, Taiwan, R. O. C.

Abstract

Street closure following an earthquake makes life-saving and rescue work more difficult, especially in congested urban areas. After the great Chi-chi earthquake, recovery and reconstruction work became necessary. In particular, we have to investigate the street network damages and other effects. Considering the road improvement plan, not only malfunction recovery but also a comprehensive evaluation should be considered in future urban planning. The reliable street network against earthquakes is the central concern of this survey.

This study attempts to implement a case study survey of damaged areas, and analyze the effecting factors on street closure. We also try to evaluate the road function against earthquake integrated with the street-damaged characteristics. Using a discriminant model of street closure, some design guidelines for future transportation planning decisions are proposed.

Nan-tou City network has 365 links and 66.6% of them are less than 6 meters in width. Ton-shi Town network has 248 links and 35% of them are less than 6 meters in width. The same definition of street closure (impassable by vehicles) of 'street width less than 4 meters' was applied to these two case studies. This research considered several street network alternatives, the inaccessible nodes for all alternatives and analysis in order to determine a more reliable street network for increased safety. Results of both surveyed areas suggest that when we decrease the interval distance of the 8-M street, the percentage of the differences of inaccessible nodes will also decrease.

The Chi-chi earthquake did great harm in the disaster area. Fortunately we find the street damages of our surveying area were not too severe to maintain their functions in times of earthquake. However, it is still meaningful for a transport planner to evaluate the reliability of residential streets because the main target of planning a residential street network is to restrain through-traffic in this area to ensure safe and comfortable conditions for pedestrians while meeting residential access requirements.

INTRODUCTION

The great Chi-chi earthquake, magnitude Richter 7.3, occurred Sept. 21, 1999 and caused serious damages in the central Taiwan area. This study aims to investigate and construct a anti-hazardous street network system against earthquake disasters. Based on the experiences of the Chi-chi great earthquake in Nan-tou City and Ton-shi Town, we make some comparisons between urban sizes to contribute to a more reliable and safe city transportation planning in the future. Using aerial photos and on-site interviewing, with a historic review of the earthquake, main factors determining street closure effects were analyzed by developing a discriminant model of the street closure. We also applied a street network simulation to evaluate the increased safety against earthquakes.

The street closure after an earthquake made life relief and rescue works more difficult, especially in congested urban areas. The reliable street network against earthquake is the central point of concern in this survey. The purpose of this study is to determine the following:

1. Proposing an effective method of an analysis on street network closures and damages after an earthquake.
2. The important factors to explain street closures and damages are discussed and analyzed by a discriminant model.
3. A simulation model is developed to evaluate the alternatives of street network improvement by the inaccessible nodes reduction.
4. We take Nan-tou City and Ton-shi Town as the comparison area and try to make some suggestion for the future planning.

METHODOLOGY

Survey

This study proposes survey contents to measure the effect of Chi-chi earthquake on local and district transportation. Survey concepts for data collection from the damaged area are summarized as Table 1. This study utilized the map explaining the damages of the building made by the government, and the aerial photo taken after the earthquake (1999/9/22) from National Cheng-kung University Satellite Information Center. We set up the surveying map with the street network and the possible closure location in order to have a clearer picture of the street network. Figure 1 and Figure 2 demonstrate the survey area.

The surveyed items are shown in Table 1 and candidates of the factors explaining street closures were road links, street width, the number of floors of alongside buildings, structures of them, street wall length, closure width, street trees, facility poles and street lights.

Table 1 Survey table of street network closure in Chi-chi earthquake

Items	Contents	Descriptions
1	Surveying subjects	Street network, street width, street closure factors (construction structure, floor height, street trees, street wall), closure width
2	Surveying damages category	We categorized the road services damaged in the earthquake according to 4 ranks: Rank 1 ： Almost no damage Rank 2 ： Little damage but accessible by vehicles Rank 3 ： Impassable by vehicles but accessible by pedestrians Rank 4 ： Impassable by neither vehicles nor pedestrians
3	Surveying objects	The construction structures related to the street closure or the persons.
4	Surveying contents	1. Street closure 2. Road facility damages 3. Roadside damages 4. Construction damages
5	Surveying methods	1. Determine the most seriously affected area from the data published by the Construction and Planning Administration Bureau. 2. Verify the possible closure area in the air photo. 3. Map the street network from the air photo. 4. Establish the street map and planned street map for the on site surveying.

As well, we had some problems with the accuracy of the data extracted from air photos. Even when we used every possible means to identify the damaged area for the on site survey, we still encounter some troubles like finding nothing in field surveying. However, the residents' warm assistance was proved to be really helpful for us.

Figure 1 The aerial photo of surveying area in Nan-tou City includes the existing street network and damaged area

Figure 2 The aerial photo of surveying area in Ton-shi town includes the existing street network and damaged area

Street Closure Modeling

A discriminant model is developed with dummy variables in which the external criteria are the 4 mentioned ranks of street closure and explanatory variables are street width, the numbers of floors of the alongside buildings, structures of the buildings, street wall length, closure width, street trees, utility poles and street lights. Because the street trees did not affect on the closure in the research samples, we excluded it from the candidates of the explanatory variables. As for the building constructions, we excluded some houses as having space more than 3 meters in front of their adjacent streets, for the reason that it contributes less to the street closure.

Although the model developed here can be used to clarify the factors affecting the street closures, it is not suitable for estimating the condition of a given link. For practical use, we re-categorize the road service ranks into two, which represent whether the vehicles can pass or not. By putting ranks 1 and 2, and rank 3 and 4 together, the model is estimated. The conformity level is used to compare the fitness of estimated results.

Simulation Approach to Alternative Street Network

Tsukaguchi (1995) proposed a desirable density of distribution streets in sprawled area as 200-250 meter interval from a viewpoint of accessibility of emergency vehicles in an ordinary condition. Recently Tsukaguchi (1999) also mentioned that it is necessary to arrange 12-meter-width streets by 200-meter interval if we wanted to keep the percentage of inaccessible nodes less than a few. Also almost the same effect can be expected if the street network consisted of 8-meter width streets at 200-meter intervals and also with solidly built frontage buildings.

Yeh et al. (2000) took Chi-chi town as the surveying area of street damages and found that 8-meter newly improved roads contributed a great deal to the rescue and relief works. Lee et al. (2000) proposed that a 6-meter-width road with 200-meter interval would be more efficient than the planned street with 60-meter interval. That explains some other critical viewpoints regarding cultural asset preservation.

Based on these findings, this research considers several street network alternatives whose details are explained in the following. A simulation model is developed to investigate whether each node in the study area remains connected for vehicles from surrounding arterial streets. The detailed procedure of calculating the percentage of inaccessible nodes is shown in Figure 3. Inaccessible nodes for all alternatives are summarized and analyzed to determine a more reliable street network for safety.

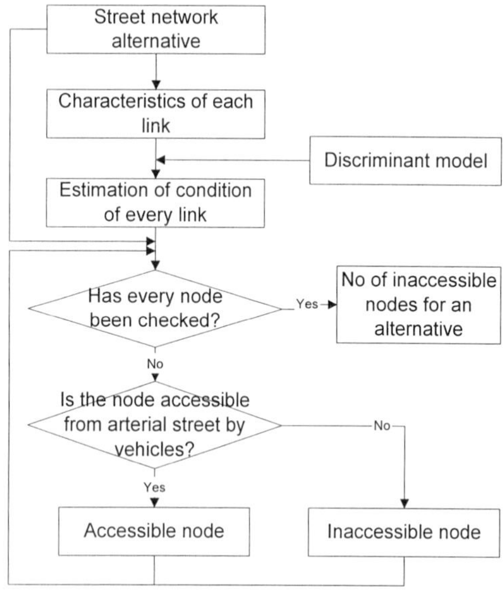

Figure 3 Flow chart of calculation of percentage of inaccessible nodes (Tsukaguchi, 1999)

Five alternatives are considered for final evaluation. The features of alternatives are as follows:

Case1: The present street network and possible inaccessible nodes simulated by the procedure.
Case2: District and local distributors with 8 meters width are managed at 300 meters intervals.
Case3: District and local distributors with 8 meters width are managed at 200 meters intervals.
Case4: District and local distributors with 8 meters width are managed at 100 meters intervals.
Case5: District and local distributors with 8 meters width are managed at 50 meters intervals.

MODELING THE STREET CLOSURE

Nan-tou City Case Study

We also undertook a field survey and interviews. There are 196 street closure samples collected and analyzed (See Figure. 4). In the preliminary surveying results in Nan-tou City , factors effecting the street closure under 4 ranks of road service analysis were street width of 1~4 M, the alongside buildings that has four floors, and the buildings built with strengthened brick.

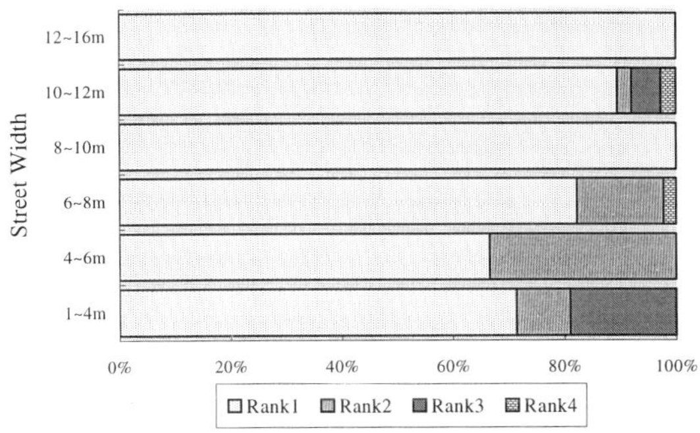

Figure 4 Street closures by street width of Nan-tou City

However in order to clarify some complexity, we use this model to estimate the two ranks, that is, when putting ranks 1 and 2, rank 3 and 4 together. The conformity level improved to about 88.3%, meaning the samples are well classified into 2 ranks and the significant variable is "street width 1-4M" (Table 2).

There are 365 links taken into consideration in this surveying area. Most of streets (85%) in this district are 4 to 8 meters in width except for surrounding streets. Figure 5 shows 34.3% of streets were below 4 meters in width, which should be taken into serious consideration from a planning viewpoint.

Table 2 Discriminant model of Nan-tou (Fisher's linear discrimiant equation)

Explanation variables	Rank1 and 2 Coefficient of discrimant Vehicles passable	Rank 3 and 4 Coefficient of discrimant Vehicles impassable
Street width 1-4 M	1.027	5.107
Constants	-0.745	-1.970

Remark: 88.3% of the samples are classified in 2 ranks

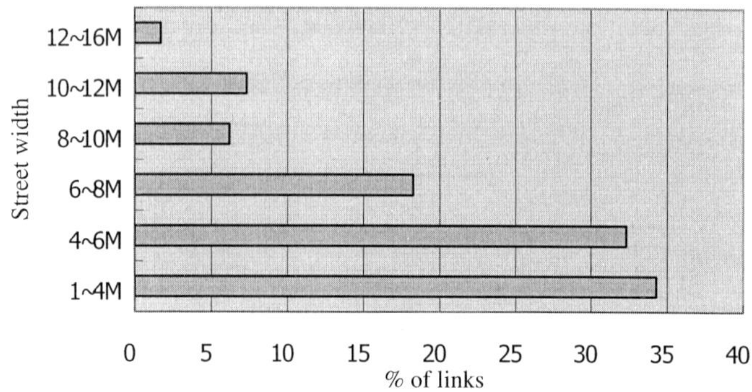

Figure 5 Composition of street width in Nan-tou City

Ton-shi Town Case Study

We also took a field survey and interviews in Ton-shi Town. There are 248 street closure samples collected and analyzed (Figure 6). It was found that the major factor effecting on street closures under 4 ranks road service analysis was street width of 1~4 M.

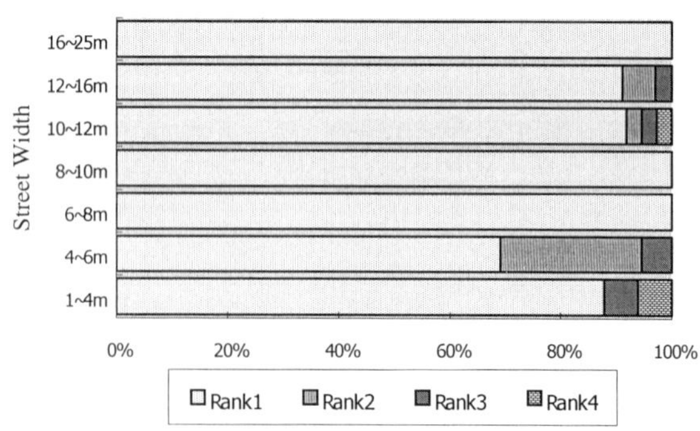

Figure 6 Composition of street width in Ton-shi Town

From the results of two-rank analysis, the conformity level became about 85.9%, meaning that the samples are well classified into 2 ranks and the significant variable is "street width 1-4M" (Table 3), which is the same result as Nan-tou city case study.

Table 3 Discriminant model of Ton-shi（Fisher's linear discrimiant equation）

Explanation variables	Rank1 and 2 Coefficient of discrimant Vehicles passable	Rank 3 and 4 Coefficient of discrimant Vehicles impassable
Street width 1-4 M	1.076	3.531
Constants	-0.759	-1.399

Remark: 85.9% of the samples are classified in 2 ranks

There are 248 links taken into consideration in this surveying area. Half of the streets in this district are 4 to 8 meters in width (51%). Figure 7 shows some of street below 4 meters in width (13%), which should also be taken into serious consideration from a planning viewpoint.

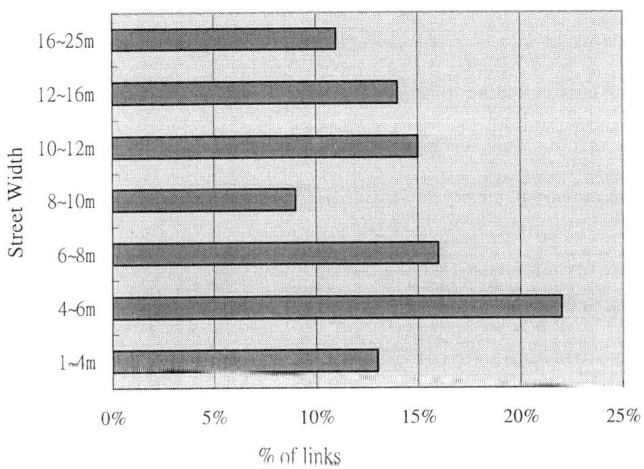

Figure 7 Composition of street width in Ton-shi town

SIMULATION ANALYSIS TO EVALUATE ALTERNATIVES

Inaccessible Nodes in Nan-tou City

Following the flow shown in Figure 3, we find the inaccessible nodes that are demonstrated in Figure 8. Then 5 alternatives of district and local distributors with 8 meters width managed at 50, 100, 200 and 300 meters intervals are also evaluated by the same method. Figure 9 demonstrates the result with 100 meters interval as an example.

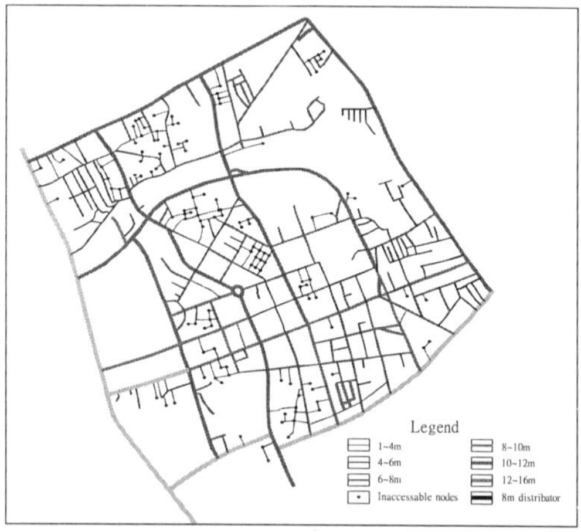

Figure 8 The location of inaccessible nodes in Nan-tou

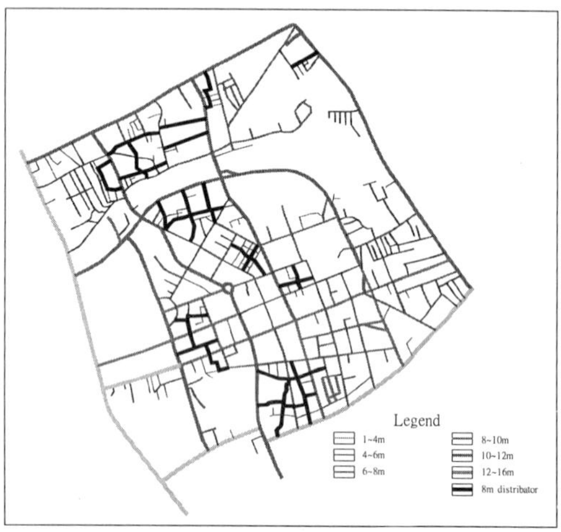

Figure 9 The street network alternative (8 meter width 100 meters interval)

The inaccessible nodes for all alternatives are summarized as Table 4. We would like to analyze the following to determine:

1. The development of 8-meter width road improvements decreases the number of inaccessible nodes.
2. When we introduce different kinds of street network alternatives, maximal 64.75% nodes decrease compared with the present condition.
3. The street network alternative 8-meter in width with 50 meters interval is the most

obvious improvement. But it also means government has to invest big money. Taking cost into consideration, we would like to propose that the 200-meter interval might be more efficient.

4. Considering whether other factors such as traffic management could be regarded as the more suitable network.

Table 4 The inaccessible nodes comparison with street networks alternative in Nan-tou

Items	Descriptions	Inaccessible nodes	Differences of inaccessible nodes compared with present	Percentage of the differences of inaccessible nodes (compared with present)
Present	Present condition	122	-	-
Case 1	8 meter width 300 meters interval	86	36(122-86)	Minus 29.51%
Case 2	8 meter width 200 meters interval	74	48(122-74)	Minus 39.34%
Case3	8 meter width 100 meters interval	59	63(122-59)	Minus 51.64%
Csae4	8 meter width 50 meters interval	43	79(122-43)	Minus 64.75%

Inaccessible Nodes in Ton-shi Town

In the same way, we find the inaccessible nodes in Ton-shi town network and were demonstrated in Figure 10. Figure 11 demonstrates the result with 100 meters interval. The other cases are estimated the same way. The inaccessible nodes for all alternatives are summarized in Table 5, and analyzes the following:

1. The present inaccessible nodes are not so many as in Nan-tou City. The development of 8-meter width road improvements is also useful.
2. The street network alternative shows maximal 54.41% nodes decrease compared with the present condition.
3. The street network alternative 8-meter width 50 meters interval is the most obvious improvement.

Figure 10 The location of inaccessible nodes in Ton-shi

Figure 11 The street network alternative (8 meter width 100 meters interval)

Table 5 The inaccessible nodes comparison with street networks alternative in Ton-shi

Items	Descriptions	Inaccessible nodes	Differences of inaccessible nodes compared with present	Percentage of the differences of inaccessible nodes (compared with present)
Present	Present condition	68	—	—
Case 1	8 meter width 300 meters interval	51	17(68-51)	Minus 25.00%
Case 2	8 meter width 200 meters interval	45	23(68-45)	Minus 33.82%
Case3	8 meter width 100 meters interval	39	29(68-39)	Minus 42.64%
Csae4	8 meter width 50 meters interval	31	37(68-31)	Minus 54.41%

The Comparison of the Result of Nan-tou City and Ton-shi Town

From the figure analysis and on-site survey, we can conclude the characteristics of the urban scale and street network in Nan-tou City and Ton-shi Town as Table 6.

Table 6 The comparable differences of surveying area

Items \ Places	Nan-tou City	Ton-shi Town
Urban scale	City	Town
Population (year 2000)	104,160	57,858
Household	29,554	16,130
Area (KM^2)	71.6	117.4
Population density (persons/ KM^2)	1458.9	492.8
Arterial road	Pass through the center area	Pass through the center area
Grid type street	Locate in center area	Not significant
Sprawled area	Yes	No
Links numbers under 6-meter width	66%	35%
Isolation of simulation	122	68

Buildings in Ton-shi Town were heavily damaged causing many deaths. In the area where the damages were serious, most of the buildings were falling and sinking to the ground vertically. Therefore, collapsed buildings did not block the streets in many places. The population density of Ton-shi is lower. And the street network depicted the branch from the arterial road. So the simulated isolation nodes will also be less than Nan-tou. We compare the street closure discriminant model in terms of 4 ranks. Nan-tou City shows some affecting factors, but only one variable 'street width 1-4 M' was significant in Ton-shi. However the variable 'street width 1-4 M' is the major factor of the street closure. The improvement of the street width less than 4-M is the most important job.

Next we would like to compare the improvement rate of the inaccessible nodes when 8M- wide streets are allocated. We take the data from Table 4 and Table 5, and compare the improvement of inaccessible nodes in %. Figure 12 reveals that the two surveying areas have the same tendency of improvement. That means when we decrease the interval distance of the 8-M street from 300 M to 50 M, the percentage of the improved inaccessible nodes (%) will also increase. The 8-M planned road functioned well in times of Chi-chi Earthquake. But if we take the 50 M interval, the local government requires a very large budget for this improvement. Even with the best situation attained we still could not avoid the isolations in the simulation results. We have to use some other tools for safety reasons. The open spaces, neighborhood parks and strict construction codes all require further considerations.

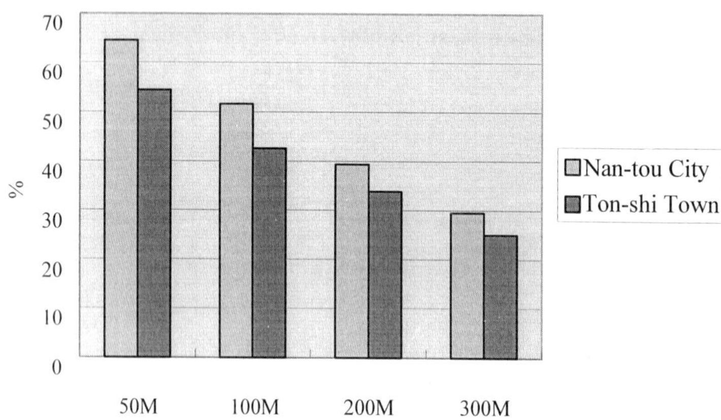

Figure 12 The improved inaccessible nodes (%) comparison with the existing situation indifferent areas

CONCLUSION

The Chi-chi earthquake did great damages in the disaster area. We find the street damages were not too severe to keep functioning in times of earthquake. However, it is still meaningful for transport planners to evaluate an accessibility of residential streets because the main target of planning residential street network is to restrain through traffic from this area, ensure safe and comfortable conditions for pedestrians while satisfying residential access requirements.

Nan-tou City was composed of 365 links and it illustrated 66.6% of all less than 6 meters. Ton-shi Town was composed of 248 links and it illustrated 35% of all less than 6 meter. And the same characteristic of the street closure (impassable by vehicles) of the two places discriminated from these samples is 'street width fewer than 4 meters'. We must seriously consider that such closure of streets and the resulting inaccessibility of the network greatly endanger lives and increase property damages, as disaster relief activities and search and rescue activities are delayed. Learning from this great disaster about insufficient data and possible new information will provide us with a very good lesson to avoid more dangers in the future.

The street network evaluations offer the planner very good decision procedures of urban planning road improvements under strict budgetary concerns. We would like to do the right thing at the right time according to the evaluation.

ACKNOWLWDGEMENT

We would like to thank the support of National Science Foundation (NSC 89-2218–E–309-004-) in Taiwan and the International Association of Traffic and Safety Science (IATSS, H266 Research Project) in Japan, as well as the useful comments from Professor Tsukaguchi at the University of Ritsumeikan Japan which are appreciated very much.

REFERENCES

Lee Y. L., Huang K. C., Yeh K. Y., (2000), The Street Network Reliability Against Disaster in Historic Area - An-pin District Survey in Tainan City, International Symposium on City Planning 2000, July 17-18, p57-65.

Tsukaguchi H. and Li Y., (1999), District and Local Distributor Network to Ensure Disaster-resilient Urban Planning, Shanghi International Symposium on Transportation Proceeding.

Tsukaguchi H., Vandebona U. and Li Y., (1999), Planning of Residential Street Network for Disaster Prone Urban Area, Proceedings of WCTR.

Yeh K. Y., Lee Y. L., Wu Y. L., (2000), The Street Closure Condition of Local Roads – The Case Study in Chi-chi Town, The Survey and Analysis of Road Damage and Traffic Condition for Transportation Networks – Case Study of Chi-chi Great Earthquake, Sponsored by IATSS, pp3-1 to 3-14.

Yeh K. Y., Lee Y. L., (2001), The street network closure survey in times of Chi-chi earthquake – Nan-tou city case study, part 3 in Traffic management systems in time of earthquake disaster based on the experience of Chi-chi and Hanshin-Awaji Earthquake, Sponsored by IATSS, pp3-1 to 3-31.

17

EVALUATION OF ROAD NETWORK RELIABILITY CONSIDERING TRAFFIC REGULATION AFTER A DISASTER

Fumitaka Kurauchi, Department of Civil Engineering, Kyoto University, Kyoto, Japan
Yasunori Iida, Department of Civil Engineering, Kyoto University, Kyoto, Japan
Hirofumi Shimada, Ministry of Land, Infrastructure and Transport, Tokyo, Japan

ABSTRACT

Road network performance when a large disaster happens depends on how the road traffic is regulated. To evaluate the performance of the road network, one should consider the traffic regulation in an emergent condition. Authors have proposed the idea of area traffic regulation when a large disaster occurs (Iida, et al., 2000). In the former paper, a bi-level optimisation model to calculate the optimal regulation ratio is proposed. This paper proposes an efficient and fast method to calculate regulation ratios for two-stage road traffic regulation. The simplified model adopts linear programming method instead of bi-level optimisation method, which is quicker and simpler. The simpler algorithm contributes to conduct various kinds of case studies, and helps finding a reliability of the road network considering traffic regulations after a disaster.

INTRODUCTION

The road network capacity deteriorates enormously when a large disaster occurs, as observed when the Great Hanshin-Awaji Earthquake occurred in 1995. The road network suffered from the destruction of expressways and bridges, as well as the blockade of major arteries, resulting in the closure of roads and traffic jams. This caused serious difficulty in travelling, not only for ordinary cars, but also for emergency vehicles. Analysis of changes in the OD traffic volume for automobiles within road networks of the Hanshin region after the earthquake showed that the OD ratio of short-distance trips increased greatly (Kurauchi and Iida, 1998). This increase was directly related to activities such as verifying the safety of relatives or acquaintances, purchasing goods and materials, and evacuating victims, and was generated in large part by the need to rely on automobiles due to a lack of availability of other modes of transportation normally used for everyday activities (IATSS, 2000). This indicates the desirability of traffic regulation that affects short-distance trips as little as possible.

The traffic congestion observed after the Great Hanshin-Awaji Earthquake indicated that the linear regulation of traffic within specific segments or routes of road is not sufficient for improving traffic conditions. Based on this fact, the Disaster Prevention Basic Act was revised to enable the overall regulation of traffic within a particular area. In response to this revision, the authors have proposed a 2-stage area regulation method to relieve traffic jams, and a model to calculate the regulation rates (Shimada, Iida and Kurauchi, 2001). It is possible to evaluate the effect of link blockade on the total network capacity considering traffic regulation after an earthquake using this model. However, as the proposed model is structured as a bi-level optimisation problem, it is a time consuming model. A simpler model is needed for application to various case studies to evaluate the reliability of network after a disaster.

Upon these backgrounds, this study proposes another model to calculate network capacity easily and quickly. Firstly, we will indicate the idea of two-stage area traffic regulation after a disaster. Then, model formulations proposed by the previous study (Shimada, Iida and Kurauchi, 2001), BLM/UE and BLM/SO, and the remained problems in these models are shown. Thirdly, the model formulation by linear programming method (LPM) is proposed as an alternative model. Then, the proposed model is compared with previous model. Also using LPM, the importance of a link is evaluated by calculating the reduction in the road network capacity as a case study.

TRAFFIC MANAGEMENT WITH TWO-STAGE AREA REGULATION

Revisions to the Disaster Prevention Basic Act

In Japan, under normal conditions, road traffic is managed based on the Road Traffic Act. Once a large disaster occurs, the Disaster Prevention Basic Act becomes effective regulation for managing traffic. Emergency roads for the transport of rescue and emergency vehicles and emergency supplies are designated under the latter act during period following the disaster. After the Great Hanshin-Awaji earthquake, the emergency roads that were exclusively used by rescue team seemed to be free of congestion. Nevertheless, the access/egress roads to/from the emergency routes were heavily congested with passenger cars so the emergency vehicles needed a long time to reach the destination. Learning from this experience, revisions to the Disaster Prevention Basic Act that would manage traffic within an entire affected area, in addition to a linear regulation of traffic of the specific routes or roads were appended. It is anticipated that by regulating larger region more efficient and effective traffic conditions would be realised.

Concept of the Two-stage Area Regulation

After the Great Hanshin-Awaji Earthquake, a large number of vehicles from the surrounding areas poured into the damaged region. The incoming flow of traffic was a mix of emergency vehicles (e.g. ambulance or tracks loaded with food) considered as urgent trips and private vehicles (e.g. onlookers or reporters), which were non-urgent trips. The latter group of vehicles represented a massive portion of the overall demand and was responsible for the severe traffic congestion succeeding the disaster. Therefore, a regulation that would limit traffic demand of private vehicles to the level that can be serviced by the network needs to be defined.

Considering the latter requirement and the prescription of revised Disaster Prevention Basic Act, this paper proposes a two-stage area regulation to effectively relieve traffic congestion after the disaster. The regulation aims in facilitating emergency rescue and relieving activities as well as in mitigating the spread of damage in response to on-site conditions. Furthermore, it accommodates the enactment of traffic regulation not only within the affected area but also in conjunction with unified regulation of traffic in surrounding areas.

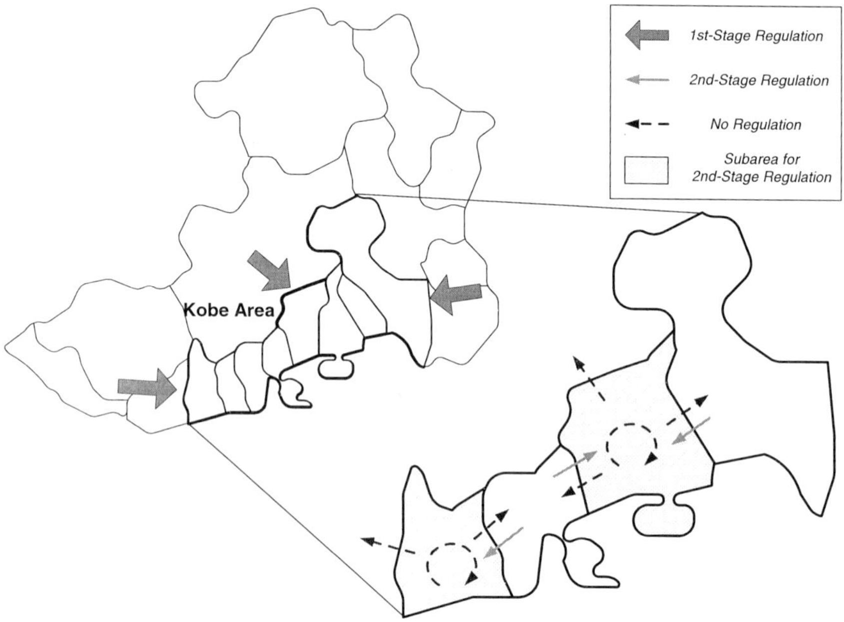

Figure 1 Image of the Two-stage Area Regulation

The proposed scheme for the two-stage area regulation for Hanshin region is illustrated in Figure1. The depicted areas were the most severely damaged by the Hanshin-Awaji earthquake. The first-stage in the proposed regulation involves a strict entry prohibition for vehicular traffic from regions outside the damaged area except for emergency vehicles. The second-stage regulation executes specification of the cities stated above as regulation sub-areas, and prohibits the entry of traffic into these sub-areas. Under the second-stage area regulation, traffic could move freely both within and out of the sub-areas to unregulated areas. The methods for implementing area regulation requires a thorough planning of dividing the regulated region into zones or area-blocks before a disaster occurs as it would be difficult to designate zones during the state of confusion directly after the outbreak of a disaster.

Mathematical Modelling for Selecting the Two-stage Area Regulation Rate

A model for computing regulation rate can be formulated in accordance with the concept of a network capacity model. What we need to decide is the regulation ratio (or entry permission ratio) for first- and second-stage regulations. One option is to maximise inflowing vehicles from the outside (namely, maximising an entry permission ratio for first stage regulation), and then to maximise the vehicles moving between sub-areas using remained capacities. Alternatively, it is also possible to calculate first- and second-stage regulation ratio together by weighing them with appropriate ratio. In this paper, the latter concept is adopted because it is a more flexible model and the former policy can also be installed by giving a large weight to vehicles of first-stage regulation. The proposed two-stage area regulation algorithm calculates both the first- and the second- stage regulation rates to maximise generation and concentration of traffic volume within the regulated area, subject to acceptable level of congestion selected by a planner. The success of implementing the two-stage area regulation varies depending on factors such as the driver's route choice behaviour or the extent to which dynamic route guidance information is provided to the drivers.

It is quite difficult how drivers choose their route in case of emergency, when their prior knowledge about road network is not valuable. Therefore, in this study, to evaluate the potential ability of the network, we assume that drivers always select the shortest route in terms of travel time. It is expected that ITS technology could obtain the traffic information in terms of the travel time, update it regularly and present it to the drivers. Assuming that the drivers select the shortest route based on the information provided by the traffic management in both the first- and the second-stage of area regulation, the user equilibrium condition would be attained. Then, the proposed model can be regarded, as a Stackelberg problem in a game theory in which player 1 is traffic manager, and player 2 are drivers. An objective function of the upper problem, i.e. the strategy of player 1, is to maximise a total traffic volume under a condition that each link traffic volume is less than a specified ratio of the volume and the road capacity. An objective function of the lower problem, i.e. the strategy of player 2, is to minimise a sum of an integral of link performance function of each link that calculates the traffic flow patterns of user equilibrium condition. In other words, the drivers attempt to minimise the cost related to travel time. Let us call this model as BLM/UE (Bi-Level optimisation Model under User Equilibrium condition).

Output variables in the proposed model are $\boldsymbol{\theta}$, which represent an entry permission rate for first and second area regulation, and link flow pattern $\mathbf{v}$. Path flow pattern $\mathbf{h}$ is not uniquely determined because user equilibrium condition is assumed. Regulation rates for first and second regulation are calculated by subtracting θ_1 and θ_2 from 1. BLM/UE can be formulated as follows:

{Formulation of BLM/UE}

=Upper Problem=

Maximise $Z_1(\boldsymbol{\theta}) = \Xi \cdot \boldsymbol{\theta}' \cdot \mathbf{R} \cdot \mathbf{w}$ (1)

Subject to

$\mathbf{v} \le \mu \mathbf{c}$ (2)

$\boldsymbol{\theta} \le \mathbf{1}_W$ (3)

$\boldsymbol{\theta} \ge \mathbf{0}_W$ (4)

v is obtained by solving following lower problem.

=Lower Problem=

$$\text{Minimise } Z_2(\mathbf{v} \mid \boldsymbol{\theta}) = \sum_{a=1}^{l} \int_0^{v_a} t_a(x)dx \tag{5}$$

Subject to

$$\mathbf{B} \cdot \mathbf{h} = \boldsymbol{\theta}^t \cdot \mathbf{R} \cdot \mathbf{W} \tag{6}$$

$$\mathbf{v} = \mathbf{A} \cdot \mathbf{h} \tag{7}$$

$$\mathbf{h} \geq \mathbf{0} \tag{8}$$

where

i	:	number of OD pairs,
l	:	number of links,
$\boldsymbol{\theta}$	:	entry permission rate vector $=(\theta_1, \theta_2, 1)^t$,
θ_k	:	entry permission rate for kth regulation,
Ξ	:	3 x 3 weight matrix for θ_k $\begin{pmatrix} \xi_1 & 0 & 0 \\ 0 & \xi_2 & 0 \\ 0 & 0 & 1 \end{pmatrix}$
$\mathbf{w}$	:	OD traffic volume vector,
$\mathbf{W}$	:	the matrix whose diagonal components are OD traffic volume w_k,
$\mathbf{h}$	:	path traffic volume vector,
$\mathbf{c}$	:	link capacity vector,
$\mathbf{A}$	:	link-path incidence matrix,
$\mathbf{B}$	:	OD pair-path incidence matrix,
$\mathbf{R}$	:	regulation policy-OD pair incidence matrix (size of 3 x i),
μ	:	allowable congestion level,
$\mathbf{1}_W$	:	vector whose size is the number of OD pairs and all components are 1,
$\mathbf{0}_W$	:	vector whose size is the number of OD pairs and all components are 0,
v_a	:	link traffic volume for link a,
$t_a(.)$	:	cost performance function for link a.

Another model that assumes system optimal condition as a network loading principle can be formulated similarly. By substituting eq.(5) into system optimal principle, the following BLM/SO can be formulated. Apparently, BLM/SO assumes that traffic manager can force the drivers to use designated routes. Since this model is applied for emergency conditions after an earthquake and it is technically possible using the recent progress in ITS, we set these assumptions.

{Formulation of BLM/SO}

=Upper Problem=

Maximise $Z_3(\boldsymbol{\theta}) = \boldsymbol{\Xi} \cdot \boldsymbol{\theta}^t \cdot \mathbf{R} \cdot \mathbf{w}$ (9)

Subject to

$\mathbf{v} \leq \mu \mathbf{c}$ (10)

$\boldsymbol{\theta} \leq \mathbf{1}_W$ (11)

$\boldsymbol{\theta} \geq \mathbf{0}_W$ (12)

$\mathbf{v}$ is obtained by solving below lower problem.

=Lower Problem=

Minimise $Z_4(\mathbf{v} \mid \boldsymbol{\theta}) = \mathbf{v}^t \cdot \mathbf{t}(\mathbf{v})$ (13)

Subject to

$\mathbf{B} \cdot \mathbf{h} = \boldsymbol{\theta}^t \cdot \mathbf{R} \cdot \mathbf{W}$ (14)

$\mathbf{v} = \mathbf{A} \cdot \mathbf{h}$ (15)

$\mathbf{h} \geq \mathbf{0}$ (16)

Although the BLMs are rather strict models in the sense of traffic engineering because it assumes the user equilibrium/system optimal condition to assign traffic, the heuristic solution method must be applied because the feasible region of the upper problem cannot be defined explicitly. Therefore, the solution may fall into a local optimum and the computation is time consuming. To utilize this model for various cases to evaluate network reliability, a faster algorithm is needed.

MODEL FORMULATION OF LPM

To simplify the problem, it is assumed that the travel time for each link is independent of the traffic volume, although the traffic volume cannot exceed the capacity. If the path sets for all OD pairs are known, the model to maximise the total traffic volume that can be produced is formulated as a linear programming problem (LPM) as follows. In this formulation, unknown variables are the entry permission rate and path traffic volume.

Maximize $Z_5(\boldsymbol{\theta}, \mathbf{h}) = \boldsymbol{\Xi} \cdot \boldsymbol{\theta}^t \cdot \mathbf{R} \cdot \mathbf{w}$ (17)

Subject to

$\mathbf{A} \cdot \mathbf{h} \leq \mathbf{c}$ (18)

$\mathbf{B} \cdot \mathbf{h} = \boldsymbol{\theta}^t \cdot \mathbf{R} \cdot \mathbf{W}$ (19)

$\mathbf{h} \geq \mathbf{0}$ (20)

$\boldsymbol{\theta} \leq \mathbf{1}_W$ (21)

$\boldsymbol{\theta} \geq \mathbf{0}_W$ (22)

To use this model, a path enumeration method is needed. If all paths are not enumerated, then the network capacity will be smaller than its maximum. To avoid this problem, Eqns. (17) to (22) are rewritten as follows by using traditional summation expressions, referring to LeBlanc et al. (1975).

$$\text{Maximize } Z_6(\boldsymbol{\theta}, \mathbf{Y}) = \sum_k \xi_k \, \theta_k \cdot \sum_{ij} r_{ijk} \cdot x_{ij} \tag{23}$$

Subject to

$$\theta_{k_{js}} \cdot x_{js} + \sum_a \delta_{aj} y_{as} = 0 \quad \text{(for all nodes } j \text{ except } j{=}s\text{, and destinations, } s.) \tag{24}$$

$$\sum_s y_{as} \le c_a \quad \text{(for all links } a) \tag{25}$$

$$y_{as} \ge 0 \quad \text{(for all links } a\text{, and destinations, } s.) \tag{26}$$

$$0 \le \theta_k \le 1 \quad (k=1,2) \tag{27}$$

where,

x_{ij}	:	OD traffic volumes from i to j
ξ_k	:	weight for kth regulation (k=1, 2)
r_{ijk}	:	regulation-OD incidence matrix (=1 if x_{ij} is for kth regulation),
k_{js}	:	is 1 if x_{js} is for first regulation, 2 for second, and 3 for no regulation.
δ_{aj}	:	link-node incidence matrix (=1 if link a is inflowing to node j,-1 if link a is outflowing from node j; otherwise 0.)
y_{as}	:	link traffic volume for link a whose destination is s.

We do not have to enumerate paths in solving eqns. (23) to (27), although we have lots of unknown variables, y_{as}.

INTERPRETATION OF THE MODELS

Before comparing the calculation results, relationships between three models, BLM/UE, BLM/SO and LPM are discussed. Let us discuss with quite simple network that has only one OD pair and two routes. In all three models, upper problem tries to maximise the total traffic volume, i.e., to widen the width of Figure 2(a). The lower problem of BLM/UE tries to find the equilibrium point where the cost performance functions of two routes meet (A in Figure 2(a)), and the lower problem of BLM/SO tries to find the equilibrium point where the cost performance function plus the marginal cost of two routes meet (B in Figure 2(a)). As is found in Figure 2(b), both BLM/UE and BLM/SO will stop calculation when traffic volume of either route reaches to its capacity. On the other hand, LPM tries to maximise the total traffic volume until traffic volumes of both routes reaches to their capacities.

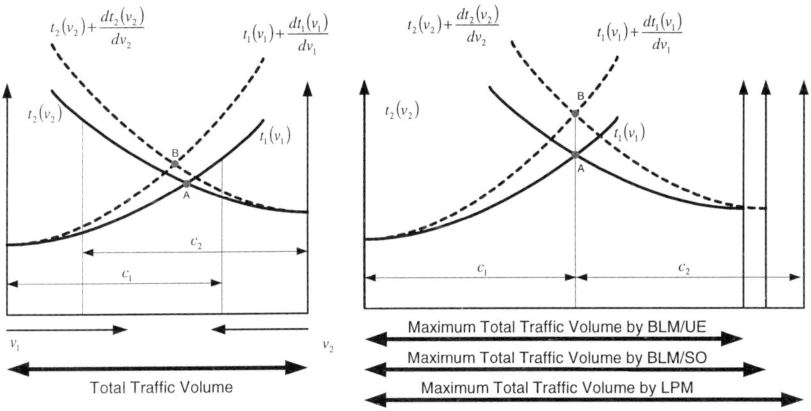

(a) (b)

Figure 2 Interpretation of the Models

As BLMs do not have to memorise the path set and path traffic volume in solving lower problem but to memorise link traffic volume, the memory area needed for calculation is rather small. On the other hand, as LPM needs to memorise the link traffic volume for each destination, in general, large memory area is needed.

EXAMPLE

Comparison of the Calculation Results

To check the validity of the LPM, the total traffic volume estimated by the BLMs and LPM are compared. The three models are applied to the example road network shown in Figure 3. There are 148 OD pairs and three sub-areas for second stage regulation in this example network. In this case study, link blockade is not considered, and allowable link congestion level, μ, is set to be 1.0. Also the weighing factor for regulation (ξ) is also set 1.0 for both first- and second-stage regulations. The total traffic volumes calculated by the three models are shown in Table 1. From the results, the total traffic volume increases when the system optimal condition is assumed. Therefore, it can be said that in an emergent condition after an earthquake, traffic managers should be able to determine the routes of drivers. By comparing BLM/SO and LPM, it is found that LPM gives a larger total traffic volume. As LPM is an easier algorithm and the computation time is shorter, this model is suitable for analysing the reliability of the road network.

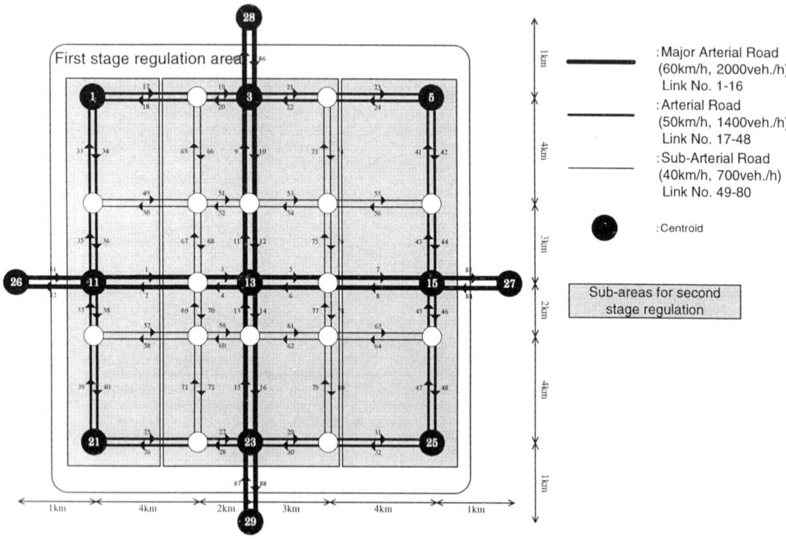

Figure 3 Study Network

Table 1 Calculation Results

Model	Total traffic volume that can be produced	First stage entry permission rate	Second stage entry permission rate
BLM/UE	20663.8	0.775	0.627
BLM/SO	21895.9	0.797	0.700
LPM	26344.5	0.792	0.867

To check how the network is utilised, it is useful to see the congestion levels of links. Figure 4 represents the cumulative distribution of links according to link congestion level. It is found that BLM/UE and BLM/SO stopped calculation when only one link reaches its capacity. On the other hand, in LPM, congestion levels of almost 40% of the links are 1.0. This result implies that if traffic manager can force the drivers to use specific route, the total network capacity increases.

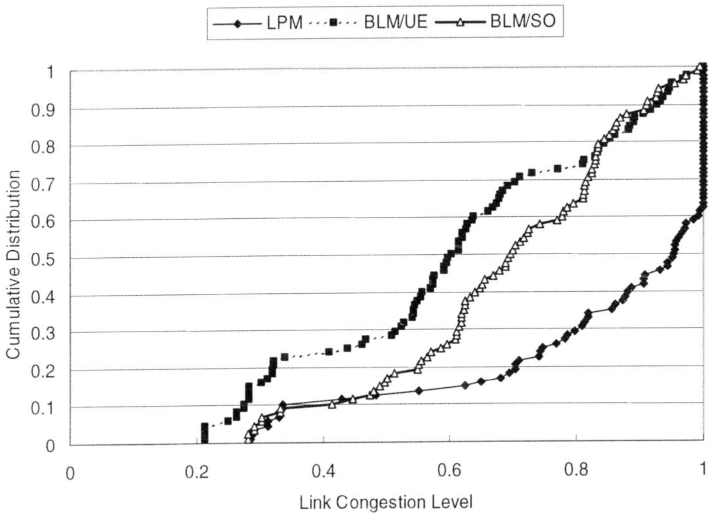

Figure 4 Cumulative Distribution of Link Congestion Level

Link Importance Index from the Perspective of an Anti-Disaster Concept

As a case study for evaluating network reliability by using LPM, total traffic volume is calculated when each link is blockaded, and is compared with the basal case where all links are not blockaded. Figure 5 shows the rate of reduction of the total traffic volume together with the link congestion level (link traffic volume / link capacity) in the basal case. From this figure, it can be said that even though the link congestion level at the basal case is 1.0, the rate of reduction of the total traffic volume differs. For instance, although the congestion levels of links 1 and 2 are 1.0, the respective rates of reduction are 0.08 and 0.28. This may be because of the OD traffic pattern, regulation policy, and network topology. Considering traffic regulation after an earthquake, the rate of reduction in the total traffic volume indicates the importance of each link when a disaster happens.

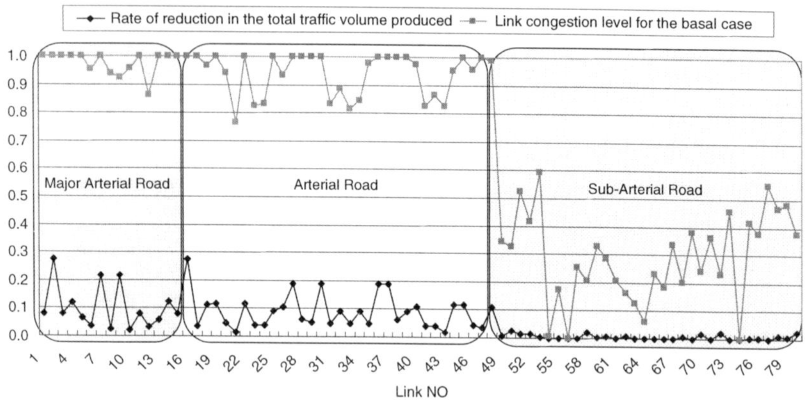

Figure 5 Rate of Reduction in the Total Traffic Volume when a Link Is Blockaded

CONCLUSION AND FURTHER STUDIES

This study proposed a simple and faster method to calculate the capacity of a road network. The model proposed here can calculate the network capacity by considering traffic regulation after a disaster. By analysing the rate of reduction in the total traffic volume, it is possible to evaluate the importance of links. As a further study, the proposed model will be applied to the networks of real world to evaluate the anti-hazardous ability.

REFERENCES

Kurauchi, F. and Iida, Y. (1998) Study on Origin/Destination Flow Fluctuations in the Period following the Great Hansin-Awaji Earthquake, *IATSS Review*, **23(3)**, pp.146-154.(Japanese)

IATSS, Research Report (2000) *Study on Personal Passenger Car Traffic Regulation following the Great Earthquake Disaster* (Japanese).

Iida, Y., Kurauchi, F. and Shimada, H. (2000) Traffic Management System Against Major Earthquakes, *IATSS Research*, **24(2)**, pp. 6-17.

LeBlanc, L. J., Morlok, E. K. and Pierskalla, W. P. (1975) An Efficient Approach to Solving the Road Network Equilibrium Traffic Assignment Problem, *Transportation Research,* **9**, pp. 309-318.

Shimada, H., Iida, Y. and Kurauchi, F. (2001) Traffic Regulation after an Earthquake Using ITS, *Proceedings of the 8th World Congress on Intelligent Transport Systems*, CD-ROM.

18

RELIABLE VEHICLE ROUTING AND SCHEDULING WITH TIME WINDOWS TOWARDS CITY LOGISTICS

Eiichi Taniguchi, Department of Civil Engineering Systems, Kyoto University, Kyoto, Japan
Tadashi Yamada, Department of Social and Environmental Engineering, Hiroshima University, Higashi-Hiroshima, Japan

ABSTRACT

This paper presents integrated models of probabilistic vehicle routing and scheduling model with time windows and dynamic traffic simulation. The probabilistic model incorporates the uncertainty of travel times. After applying the models to test road network, incorporating the uncertainty of travel times using the models not only allows freight carriers to reduce their total costs, but also improves the environment in terms of decreasing CO_2 emissions. As well, the paper investigated the effects and profitability of co-operative freight transport systems using a simulation model based on forecast vehicle routing and scheduling model. Results showed co-operative freight transport systems can lead to a substantial reduction in total delivery costs and total travel times within the whole road network.

INTRODUCTION

Recently shippers and freight carriers in urban areas face many difficulties in trying to provide higher levels of service to consumers such as strict designated pickup/deliver time as well as reducing freight costs. Transporting freight within urban areas using trucks has led to many problems, including; traffic congestion, global and local environment, energy and labour costs. Proposals have been made to drastically reduce these problems using Intelligent Transport Systems (ITS). For example, an advanced information system using satellite technology is being used for fleet management of trucks by a growing number of companies in Japan. It is expected that ITS will enable urban logistics systems to become more efficient and reduce their environmental and social costs.

Some researchers (e.g. Ruske, 1994; Kohler, 1997; Taniguchi and van der Heijden, 2000a; Taniguchi *et al.*, 2001a) have proposed the idea of "City Logistics" to solve these difficult problems. The definition of city logistics can be stated as, "the process for totally optimising the logistics and transport activities by private companies with the support of advanced information systems in urban areas considering the traffic environment, its congestion, safety and energy savings within the framework of a market economy" (Taniguchi *et al.*, 2001b). Although some of the city logistics initiatives listed below have only been proposed, others have already been implemented in several cities.

(a) Advanced information systems
(b) Cooperative freight transport systems
(c) Public logistics terminals
(d) Load factor controls
(e) Underground freight transport systems

This paper focuses on vehicle routing and scheduling procedures using advanced information systems as well as cooperative freight transport systems in urban areas. Freight carriers have depots and their pickup/delivery trucks depart from the depot and visit customers with designated time windows for collecting or delivering goods and return to the depot. This paper presents a vehicle routing and scheduling model with time windows, in particular with variable travel times on roads as well as cooperative freight transport systems. The uncertainty of travel times affects the identification of the optimal routes and schedules of pickup/delivery trucks on very congested urban roads. Recently advanced information systems such as VICS (Vehicle Information Communication Systems) in Japan measures the change of travel times of vehicles on roads for allowing freight carriers to use historical travel time data on urban roads. This paper develops a model to quantify the benefits of considering the uncertainty of travel times in order to rationalise logistics systems and reduce the negative impacts of goods movement on the environment. As well, the cooperative freight transport systems are effective to reduce the number of pickup/delivery trucks that are required to collect and deliver goods in urban areas. Application of the vehicle routing and scheduling model to cooperative freight transport systems will be presented.

A number of researchers (e.g. Solomon, 1987; Russell, 1995; Bramel *et al.*, 1996; Taniguchi *et al.,* 1998) have investigated vehicle routing problems with time windows (VRP-TW). Other researchers have studied stochastic vehicle routing and scheduling problems (e.g. Jaillet and Odoni, 1988; Powell *et al.*, 1995; Gendreau *et al*, 1996). Most research in this area has focused on dynamic routing and scheduling that considers the variation in customer demands. However, there has been limited research on routing and scheduling with variable travel times (Laporte *et al.,* 1992; Taniguchi *et al.,* 2000b).

THE PROBABILISTIC AND FORECAST MODEL

Framework

Figure 1 presents a framework of the model presented in this paper. The model is composed of two sub-models; (a) a model for probabilistic vehicle (pickup/delivery truck) routing and scheduling problem with time windows (VRP-TW-P) for each company and (b) a dynamic traffic simulation model for the fleet of pickup/delivery trucks and passenger cars on the road network within the city.

The model for VRP-TW-P is defined as follows. A depot and a number of customers are defined for each freight carrier. A fleet of identical vehicles collects goods from customers and delivers them to the depot or delivers the goods to customers from the depot. For each customer a designated time window, specifying the desired time period to be visited is also specified. For example, in the case of collecting goods, vehicles depart from the depot and visit a subset of customers to pick up goods in sequence and return to the depot to unload them. A vehicle is allowed to make multiple trips per day. Each customer must be assigned to exactly one route of a vehicle and all the goods from each customer must be loaded on the vehicle at the same time. The total weight of the goods for routes must not exceed the capacity of the vehicle. This problem is used to determine the optimal assignment of vehicles to customers and the departure time as well as the order of visiting customers for a freight carrier. VRP-TW-P explicitly incorporates the distribution of travel times for identifying the optimal routes and departure times of vehicles.

The optimal assignment of vehicles to customers and the departure time as well as the visiting order of customers for each freight carrier, becomes input to the dynamic traffic simulation model. The dynamic traffic simulation model is based on a macroscopic dynamic simulation BOX model. This model estimates the distribution of travel times on each link in 1-hour intervals. The VRP-TW-P model is then re-solved using the updated distribution of travel times on each link obtained from the BOX model. Thus, the distribution of travel times for each link is represented by a normal distribution, in 1-hour time intervals. The model therefore, incorporates time dependent travel times.

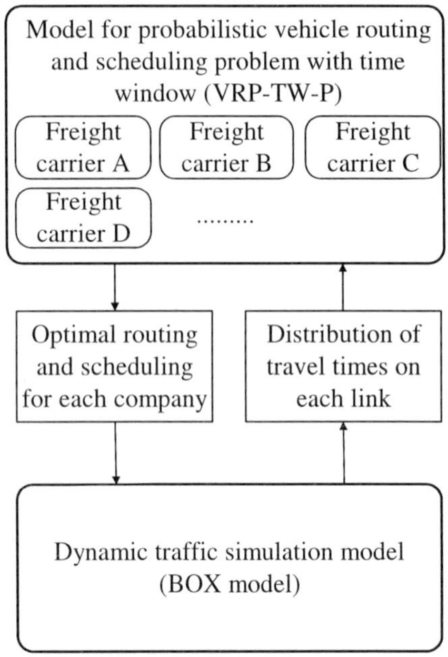

Figure 1 Model framework

VRP-TW-P Model

This section describes a mathematical model of the VRP-TW-P. The model minimises the total cost of distributing goods with truck capacity and designated time constraints. The total cost is composed of three components; (a) the fixed cost of vehicles, (b) vehicle operating costs, that are proportional to the time travelled and spent waiting at customers, and (c) delay penalties for designated pickup/delivery time at customers. The fixed cost of vehicles should be incurred even if they are not used by chance on the day. The model can be formulated as follows:

Minimise

$$C(t_0, \mathbf{X}) = \sum_{l=1}^{m} c_{f,l} \cdot \delta_l(\mathbf{x}_l) + \sum_{l=1}^{m} E\left[C_{t,l}(t_{l,0}, \mathbf{x}_l)\right] + \sum_{l=1}^{m} E\left[C_{p,l}(t_{l,0}, \mathbf{x}_l)\right] \tag{1}$$

where,

$$E\left[C_{t,l}(t_{l,0}, \mathbf{x}_l)\right] = c_{t,l} \sum_{i=0}^{N_l} \left\{ \bar{T}(\bar{t}_{l,n(i)}, n(i), n(i+1)) + t_{c,n(i+1)} \right\} \tag{2}$$

$$E\left[C_{p,l}(t_{l,0},\mathbf{x}_l)\right]=\sum_{i=0}^{N_l}\int_0^{\infty} p_{l,n(i)}(t_{l,0},t,\mathbf{x}_l)\left\{c_{d,n(i)}(t)+c_{e,n(i)}(t)\right\}dt \tag{3}$$

Subject to

$$n_0 \geq 2 \tag{4}$$

$$\sum_{l=1}^{m} N_l = N \tag{5}$$

$$\sum_{n(i)\in\mathbf{x}_l} D(n(i)) = W_l(\mathbf{x}_l) \tag{6}$$

$$W_l(\mathbf{x}_l) \leq W_{c,l} \tag{7}$$

$$t_s \leq t_{l,0} \tag{8}$$

$$t'_{l,0} \leq t_e \tag{9}$$

where

$$t'_{l,0} = t_{l,0} + \sum_{i=0}^{N_l}\left\{\bar{T}(\bar{t}_{l,n(i)},n(i),n(i+1))+t_{c,n(i+1)}\right\} \tag{10}$$

$C(\boldsymbol{t}_0,\mathbf{X})$: total cost (yen)

$\boldsymbol{t}_0$: departure time vector for all vehicles at the depot

$\boldsymbol{t}_0=\left\{t_{l,0}\,\middle|\,l=1,m\right\}$

$\mathbf{X}$: assignment and order of visiting customers for all vehicles

$\mathbf{X}=\{\mathbf{x}_l|\,l=1,m\}$

$\mathbf{x}_l$: assignment and order of visiting customers for vehicle l

$\mathbf{x}_l=\{n(i)\,|\,i=1,N_l\}$

$n(i)$: node number of i th customer visited by a vehicle

$d(j)$: number of depot (= 0)

N_l: total number of customers visited by vehicle l

n_0: total number of $d(j)$ in $\mathbf{x}_l$

m: maximum number of vehicles available

$c_{f,l}$: fixed cost for vehicle l (yen /vehicle)

$\delta_l(\mathbf{x}_l)$: = 1; if vehicle l is used, = 0; otherwise

$C_{t,l}(t_{l,0},\mathbf{x}_l)$: operating cost for vehicle l (yen)

$C_{p,l}(t_{l,0},\mathbf{x}_l)$: penalty cost for vehicle l (yen)

$c_{t,l}$: operating cost per minute for vehicle l (yen /min)

$t_{l,n(i)}$: departure time of vehicle l at customer $n(i)$

$\bar{T}(\bar{t}_{l,n(i)}, n(i), n(i+1))$: average travel time of vehicle l between customer $n(i)$ and $n(i+1)$ at time $\bar{t}_{l,n(i)}$

$t_{c,n(i)}$: loading/unloading time at customer $n(i)$

$p_{l,n(i)}(t_{l,0}, t, \mathbf{x}_l)$: probability in which a vehicle that departs the depots at time $t_{l,0}$ arrives at customer $n(i)$ at time t

$c_{d,n(i)}(t)$: delay penalty cost per minute at customer $n(i)$ (yen/min)

$c_{e,n(i)}(t)$: early arrival penalty cost per minute at customer $n(i)$ (yen/min)

N : total number of customers

$D(n(i))$: demand of customer $n(i)$ (kg)

$t'_{l,0}$: last arrival time of vehicle l at the depot

t_s : starting of possible operation time of trucks

t_e: end of possible operation time of trucks

$W_l(\mathbf{x}_l)$: load of vehicle l (kg)

$W_{c,l}$: capacity of vehicle l (kg).

The problem specified by equations (1) – (10) involves determining the variable $\mathbf{X}$, that is, the assignment of vehicles and the visiting order of customers and the variable $\boldsymbol{t}_0$, the departure time of vehicles from the depot. Note, that $n(0)$ and $n(N_l+1)$ represent the depot in equations (2) and (3).

The distribution of travel times is required in equation (1) for determining the expected value of operating costs and penalty costs. This distinguishes the Probabilistic model (VRP-TW-P) from the Forecast model (VRP-TW-F). The Forecast model, VRP-TW-F adopts only one value to represent travel time instead of a probability distribution. The dynamic traffic simulation calculates the distribution of travel times, which can be approximated by the normal or lognormal distribution for every hour. Then the updated normal or lognormal distribution can be input to the probabilistic model.

Figure 2 shows the penalty for vehicle delay and early arrivals at customers. The time period $(t^e_{n(i)} - t^s_{n(i)})$ of the penalty function defines the width of the soft time window in which vehicles are requested to arrive at customers. If a vehicle arrives at a customer earlier than $t^s_{n(i)}$, it must wait until the start of the designated time window and a cost is incurred during waiting. If a vehicle is delayed, it must pay a penalty proportional to the amount of time it was delayed. This type of penalty is typically observed in goods distribution to shops and supermarkets in urban areas. Multiplying the penalty function and the probability of arrival time as shown in

Figure 3 can identify the penalty of early arrivals and delay at customers for the Probabilistic model. The Forecast model uses a single point estimate of the arrival time of a truck at a customer. Therefore, the penalty for early arrival and delay can be estimated by multiplying the penalty function by the predicted amount of waiting or delay time.

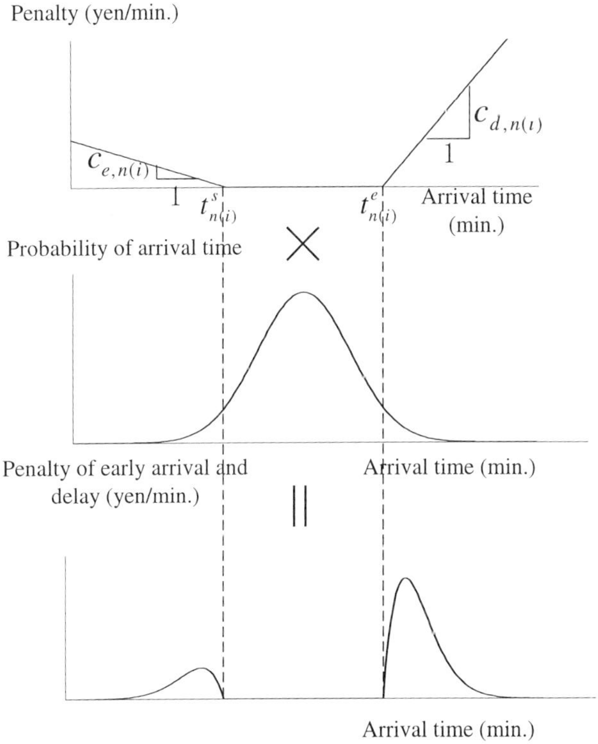

Figure 2 Penalty of early arrival and delay at customers for the probabilistic model

The problem described here is a NP-hard combinatorial optimisation problem. It requires heuristic methods to efficiently obtain a good solution. Recently several researchers have applied heuristic algorithms such as Genetic Algorithms (GA), Simulated Annealing (SA) and Tabu Search (TS) to obtain approximate solutions for the VRP-TW-F. The model described in this paper uses a GA to solve the VRP-TW-P. GA was selected because it is a heuristic procedure that can simultaneously determine the departure time and the assignment of vehicles as well as the visiting order of customers.

Dynamic Traffic Simulation Model

The dynamic traffic simulation model is based on the BOX model that was originally developed by Fujii *et al.* (1994). The modified BOX model explicitly describes the flow of pickup/delivery trucks that depart from a depot and return to the same depot. Pickup/delivery trucks are converted to passenger car units and the first-in-first-out rule is assumed on all links. The model was further modified to identify the arrival of specific vehicles at assigned nodes (customers).

The travel times on each link vary within the day. The output of the BOX model is the updated distribution of travel times on each link. The BOX model calculates the distribution of travel times based on link travel times. The distribution of travel times representing the interval of one hour was formulated by using the four-hour data on travel times. For example, the distribution of travel times representing the time interval, 8:00-9:00 a.m. can be formulated by using data between 6:30 – 10:30 a.m. Then this empirical distribution was used in the VRP-TW-P model.

Application to Test Network

Test Conditions. The model described in the previous section was applied to a test network with 25 nodes and 40 links as shown in Figure 4. This road network is comprised of same type of roads with free running speed of 40 km/h. Any node within the network can generate and attract passenger car traffic. These nodes are referred to as centroids and are also candidate nodes to be visited by the pickup/delivery trucks. Ten freight carriers are assumed to operate a maximum of 12 pickup/delivery trucks in this network. Each freight carrier has one depot at the location indicated in Table 1. Three different types of trucks, having a capacity of 2, 4 and 10 tons respectively can be used. However, up to four trucks of each type can only be operated by each carrier. The passenger car equivalence rates, operating costs and fixed costs for each type of pickup/delivery truck are based on results from recent studies of truck operations in Japan as shown in Table 2. The number of customers for each carrier was generated randomly between 14 and 22. The actual nodes to be visited for each carrier were also determined randomly from all nodes in the network. The freight demand at each customer was determined based on the distribution of freight demand in Kobe City.

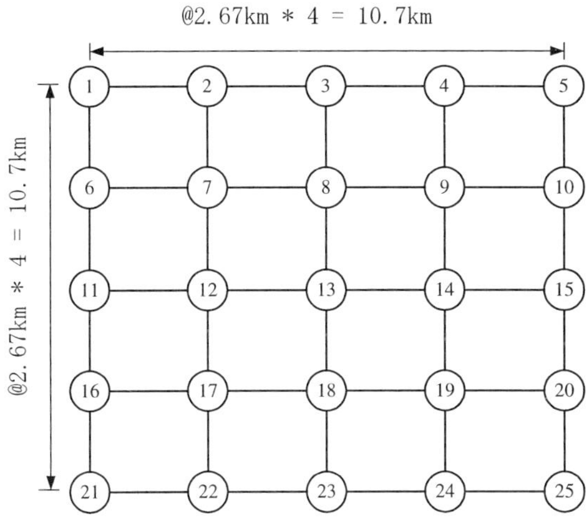

Figure 4 Test road network

Table 1 Location of freight carriers' depots and number of customers

Freight carriers	Node number of depot	Number of customers
A	19	17
B	13	22
C	3	16
D	24	17
E	1	18
F	2	15
G	15	16
H	6	19
I	18	14
J	17	20

Table 2 Characteristics of pickup/delivery trucks

Capacity of truck (ton)	Passenger equivalent rate (pcu/vehicle)	Operation cost (yen/day)	Fixed cost (yen/day)
2	1	14.02	10,417.5
4	1.5	17.54	11,523.1
10	2	23.27	13,789.7

Three types of time windows were permitted in this study, time windows with one hour, time windows for a.m. (9:00-12:00) or p.m. (13:00-17:00), and no time window. The type and starting time of each customers time window was based on a recent survey in Kobe and Osaka area. The dynamic traffic simulation provides the distribution of travel times on each link for the scanning interval. In this study the scanning interval is 1 hour. When initially calculating the optimal routes and schedules, the average travel times on each link were assumed to be equal to the travel times using free running speeds.

Dynamic traffic simulation requires information on passenger car behaviour, as well as optimal routes and schedules of pickup/delivery trucks, produced by the VRP-TW-P model. This includes the departure time and visiting order of customers. Passenger cars in this study include actual passenger cars and trucks other than those that are considered in the optimal routing and scheduling model. Passenger car origin-destination (OD) tables for every hour were estimated using traffic generation rates at each centroid and the probability of O-D choice. The number of passenger cars for each hour was generated using a temporal demand pattern based on the traffic census conducted in Kobe City.

The model described here uses an iterative procedure for day after day. Therefore, the travel time provided by the dynamic simulation fluctuates with the day. Figure 5 shows the calculation procedure. Here, the generation of total passenger cars was varied between 4375 vehicles/day ± 10% (175 vehicles/day) at all nodes. After 10-days operation, the iterative procedure was stopped. The fluctuation of travel times on each link was within 5%. At the end of 10th day's operation, freight carriers produced their optimal planning routes and schedules. They then encountered three different traffic conditions, cases A, B and C. The total generation of passenger cars per day was 6500, 5450 and 4375 (vehicles/day) at all nodes for cases A, B, and C respectively. The generation for case C was the same as in the previous 10-day operation. This generation was uniformly located at all nodes. Case A was more congested than case B that is more congested than case C, with the average speeds of 30, 35 and 38 km/h for cases A, B and C respectively.

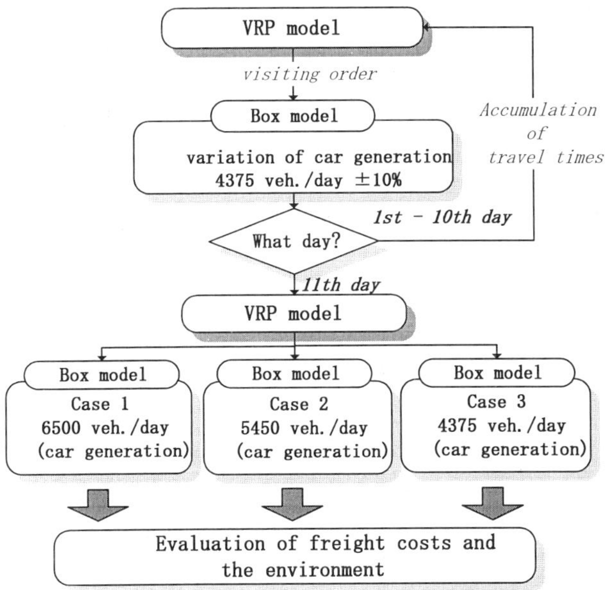

Figure 5 Flowchart for the calculation

Results. Table 3 shows the change in total costs for the 10 freight carriers for the three traffic conditions. This indicates that the probabilistic model reduced the total costs compared with the Forecast model in all cases. The reduction of total costs from Forecast model increases in cases A and B with the higher level of congestion than case C. It means that freight carriers can obtain more benefits from using the Probabilistic model when the traffic congestion becomes worse. The value of stochastic solution (VSS) is defined as the reduction in total costs from using the Probabilistic model instead of the Forecast model. The VSS for the three cases was between 11% and 17%, which is considerable.

Table 3 also indicates that the delay penalty decreased by 24-35% using VRP-TW-P model, which means that it provides better service to customers by reducing the risk of delay. However, the early arrival penalty increased by 8-11%. The fixed cost in all three cases increased by approximately 5% and the operation cost remained at the same level as the VRP-TW-F model. The small increase in the fixed cost is due to the slight increase in the number of trucks used by 10 freight carriers as shown in Table 4. Here, a one 2-ton truck was removed and two 4-ton trucks were added in the optimal routing of the VRP-TW-P model.

Table 3 Change of total costs for 10 freight carriers

Model	Costs	Case 1	Case 2	Case 3
VRP-TW-F model	fixed cost	238,043	238,043	238,043
	operation cost	214,207	197,634	188,051
	early arrival penalty	20,608	22,496	24,356
	delay penalty	789,125	507,169	363,604
	total cost	1,261,983	965,342	814,054
VRP-TW-P model	fixed cost	250,671	250,671	250,671
	change (%)	5.3	5.3	5.3
	operation cost	212,495	195,822	187,991
	change (%)	-0.8	-0.9	0.0
	early arrival penalty	22,764	24,970	26,287
	change (%)	10.5	11.0	7.9
	delay penalty	601,797	331,945	256,786
	change (%)	-23.7	-34.5	-29.4
	total cost	1,087,742	803,423	721,748
	change (%)	-13.8	-16.8	-11.3

unit: yen/day

Table 4 Change in the number of trucks (10 freight carriers, case A, B, and C)

Model	Type of vehicle	Number of vehicles
VRP-TW-F model	2 ton truck	1
	4 ton truck	3
	10 ton truck	14
	total	18
VRP-TW-P model	2 ton truck	0
	4 ton truck	5
	10 ton truck	14
	total	19

Table 5 shows the total travel time. The total travel time using VRP-TW-P model is slightly larger than that using VRP-TW-F model. The actual running time both for passenger cars and trucks does not increase so much, but the waiting time of trucks increased by 12-15%. This is due to the tendency of the VRP-TW-P model to determine optimal routes where trucks arrive at customers early to avoid delay penalties.

Table 5 Total travel time

Model		Type of vehicle	Case 1	Case 2	Case 3
VRP-TW-F model	Running time	Passenger car	66,497	49,001	36,134
		2 ton truck	65	40	39
		4 ton truck	499	447	417
		10 ton truck	3,398	3,111	2,964
		Subtotal of trucks	3,962	3,598	3,419
	Waiting time	Truck	957	1,049	1,139
	Total		71,416	53,648	40,693
VRP-TW-P model	Running time	Passenger car	68,198	49,082	36,111
		Change (%)	2.6	0.2	-0.1
		2 ton truck	0	0	0
		4 ton truck	932	813	761
		10 ton truck	3,125	2,856	2,716
		Subtotal of trucks	4,057	3,670	3,477
		Change (%)	2.4	2.0	1.7
	Waiting time	Truck	1,093	1,208	1,271
		Change (%)	14.2	15.2	11.6
	Total		73,348	53,960	40,859
	Change (%)		2.7	0.6	0.4

unit: minute/day

Table 6 CO_2 emissions

Model	Type of vehicle	Case 1	Case 2	Case 3
VRP-TW-F model	Passenger car	1686	1313	988
	2 ton truck	2	1	1
	4 ton truck	19	18	17
	10 ton truck	789	745	721
	Subtotal of trucks	809	764	739
	Total	2495	2077	1727
VRP-TW-P model	Passenger car	1709	1314	987
	Change (%)	1.4	0.1	0.0
	2 ton truck	0	0	0
	4 ton truck	34	32	31
	10 ton truck	723	683	662
	Subtotal of trucks	756	715	693
	Change (%)	-6.5	-6.4	-6.3
	Total	2466	2029	1680
	Change (%)	-1.2	-2.3	-2.7

unit: kg-C/day

Table 6 shows the predicted change in CO_2 emissions for both the VRP-TW-P and VRP-TW-F models, based on the average running speed of trucks. The figure indicates that the CO_2 emissions by trucks decreased by approximately 6% using VRP-TW-P model compared with

VRP-TW-F model, whereas the CO_2 emissions by passenger cars remained almost the same level for both models. The reduction of CO_2 emissions by trucks was mainly due to 10-ton trucks. As shown in Table 7 the travel time of 10-ton trucks decreased using VRP-TW-P model, which contributed to reduce the total CO_2 emissions by trucks, because the unit emission rate of 10-ton trucks is larger than smaller trucks. This is the reason for the reduction of CO_2 emissions by trucks was achieved in spite of the increase of total travel time of trucks.

The total CO_2 emissions by passenger cars and trucks decreased by 1-3% using VRP-TW-P model compared with VRP-TW-F model (Table 6). Therefore, incorporating the uncertainty of travel times using VRP-TW-P model not only allows freight carriers to reduce their total costs, but also improve the environment in terms of reducing CO_2 emissions.

THE CO-OPERATIVE MODEL

Overview

Co-operative freight transport systems have the potential to satisfy the needs of companies (i.e. shippers and freight carriers) aiming at cost reduction as well as to reduce the total social and environmental costs for the whole community (e.g. Ogden, 1992; Ruske, 1994). However, in spite of these possible benefits to companies and the community as a whole, there have not been many cases where the co-operative freight transport systems have been successfully implemented due to a number of issues relating to the difficulties in introducing and operating them (e.g. Yamada *et al.*, 1999a).

The focuses of this section are on modelling a co-operative vehicle routing and scheduling model with optimal location of logistics terminals and also on investigating the effects and financial viability of co-operative freight transport systems. The model developed in this section incorporates the possibility of using co-operative freight transport systems within the VRP-TW-F. This model considers the behaviour of three stakeholders: freight carriers, co-operative freight organisation and administrators (i.e. public sector).

Modelling Framework

Figure 6 shows the structure of the model. Given a location pattern of logistics terminals and the rates for using the co-operative systems, each freight carrier determines either the amount of goods to be delivered directly to customers or to be transported to a logistics terminal to use the co-operative systems. Freight carrier's decisions are assumed to be on the basis of them each minimising their total costs independently of the other carriers. The behaviour of each freight carrier for vehicle routing and scheduling can be formulated as follows:

Minimise

$$C(t_0, \mathbf{X}, \boldsymbol{a}) = \sum_l \{a_l \times SC_l + (1 - a_l) \times CC_l\} \tag{11}$$

where,

$$SC_l = c_{f,l} \cdot \delta_l(\mathbf{x}_l) + C_{t,l}(t_{l,0}, \mathbf{x}_l) + C_{p,l}(t_{l,0}, \mathbf{x}_l) \tag{12}$$

$$CC_l = c_{f,l} \cdot \delta_l(\mathbf{x}_l) + C_{t,l}(t_{l,0}, \mathbf{x}_l, \mathbf{y}) + C_{u,l}(t_{l,0}, \mathbf{x}_l, \mathbf{y}, \mathbf{p}) \tag{13}$$

$\boldsymbol{a}$: vector representing the possibility of using co-operative freight transport systems; $\boldsymbol{a} = \{a_l \mid l = 1, m\}$

a_l : =1; if vehicle l transports goods directly to customers

=0; if vehicle l transports goods indirectly via a logistics terminal for using co-operative systems

SC_l : cost incurred by vehicle l for transporting goods directly to customers (yen)

CC_l : cost incurred by vehicle l for transporting goods to a logistics terminal for using co-operative systems (yen)

$C_{u,l}$: cost incurred for using co-operative systems by vehicle l (yen)

$\mathbf{y}$: location pattern of logistics terminals

$\mathbf{p}$: vector of the rates for using co-operative systems (yen/ton)

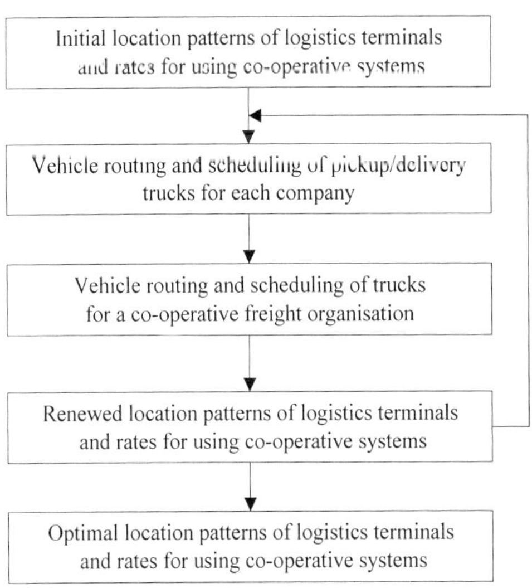

Figure 6 Structure of the model

Equations (4)-(9) can also be used as constraints. In the case of a_l =1, the mathematical problem with equations (11)-(13) and (4)-(9) represents the VRP-TW-F. In fact, AVRSS (Advanced Vehicle Routing and Scheduling Systems) represented using the VRP-TW-F is however not always used by freight carriers in vehicle routing and scheduling planning. Practical vehicle routing and scheduling of freight carriers must therefore be represented for investigating the effects of co-operative freight transport systems. It can be undertaken by incorporating a constraint represented by equation (14) within the model represented by equations (11)-(13) and (4)-(9), since the actual value of loading factors within urban areas in Japan is reported to be approximately 30% (e.g. Giannopoulos and McDonald, 1997).

$$W_l(\mathbf{x}_l)/W_{c,l} \leq W^{MAX} \tag{14}$$

W^{MAX} : upper limit of loading factor

After the decision making of freight carriers, the co-operative freight organisation determines the optimal departure times and the order of visiting customers for all vehicles owned so that the total travel time for them is minimised for alleviating the traffic congestion on the road network and reducing the environmental impacts. These can also be described using the VRP-TW-F. The objective function for the co-operative freight organisation is given below:

Minimise

$$TT(\boldsymbol{t}_{0,l'}, \mathbf{X}_{l'}) = \sum_{l'} T_{l'}(\mathbf{x}_{l'}, \mathbf{y}, \boldsymbol{t}_{0,l'}) \tag{15}$$

TT : total travel time (min)

T_l : travel time for vehicle l (min)

l' : vehicle number for the co-operative freight organisation

The constraints are identical to those of the vehicle routing and scheduling model for each freight carrier (i.e. equations (4)-(9)), but if the profitability of the co-operative freight organisation is taken into account, the following constraint is added:

$$\sum_{l'} \left(c_{f,l'} \cdot \delta_{l}(\mathbf{x}_{l'}) + C_{t,l'}(t_{l',0}, \mathbf{x}_{l'}) + C_{p,l}(t_{l,0}, \mathbf{x}_l) \right) + LC(\mathbf{y}, \boldsymbol{a}) \leq \sum_{u} \sum_{l} C_{u,l}(t_{l,0}, \mathbf{x}_l, \mathbf{y}, \mathbf{p}) \tag{16}$$

LC : logistics terminal cost (yen)

u : freight carrier number

Logistics terminal costs include the construction, operating and management costs and depend on the land price and the amount of goods handled. Equation (14) is not incorporated in this case, since the loading factor would increase after implementing co-operative systems.

Administrators' decision making involves determining the optimal location of logistics terminals and the rates for using the co-operative systems so that the total travel time of all vehicles (i.e. those owned by all the freight carriers and the co-operative freight organisation) is minimised. Logistics terminals are required for implementing co-operative freight transport systems, and their location and the rates for using the systems strongly influence whether freight carriers utilise such systems. The rates for using them also depend on the location of logistics terminals (i.e. usage rate and land price). The following objective function for administrators can therefore be formulated:

Minimise

$$TT(\mathbf{y},\mathbf{p}) = \sum_{l} T_l\ (\mathbf{x}_{l'},\mathbf{y},\mathbf{p},\boldsymbol{a},\boldsymbol{t}_{0,l}) + \sum_{l'} T_{l'}\ (\mathbf{x}_{l'},\mathbf{y},\boldsymbol{t}_{0,l'}) \tag{17}$$

This procedure involves determining an equilibrium solution between administrators (i.e. public sector) and freight transport agents (i.e. freight carriers and co-operative freight organisation) within the framework of Nash equilibrium. The model structure is similar to existing models developed for optimising the location of logistics terminals (Taniguchi *et al.*, 1999; Yamada *et al.*, 1999b). Each of the above mathematical programming problems is a complex and difficult NP-hard combinatorial problem, and the use of heuristic techniques is appropriate for determining optimal solutions. Genetic Algorithms (GA) were applied to obtain good solutions within reasonable computational times.

Applications

Test Conditions. Applications of the model were performed relating to goods delivery in an urban area. Figure 7 shows the test road network with uniform distances between nodes. The area size is 400 km^2 (20 km$\times$20 km). Link travel times on the road network are assumed to vary depending on the time period (i.e. peak hours and off-peak hours) instead of using traffic simulation model. The shadowed portion of the road network represents the centre of city where the traffic conditions are relatively congested. There are three candidate nodes for logistics terminals on the road network. One (candidate node 12) is located around the city centre, with its land price being higher than the others that are in the suburbs. Eight freight carriers are assumed to operate delivery trucks, each with one depot and eleven customers. The customers are mainly distributed within the city centre. Freight demand is larger if customers are located closer to the city centre. Designated time windows are categorised into three types, time windows of two hours length, time windows for a.m. (9:00-12:00) or p.m. (15:00-17:00) and no time window.

Two different types of trucks, having a capacity of 2 and 4 tons respectively, can be used by the co-operative freight organisation, while each freight carrier can only operate 2-ton trucks.

The operating costs and fixed costs for each type of delivery trucks are based on the results from recent studies of truck operations in Japan. Unit penalty costs for early arrivals were assumed to be equivalent to the unit operating cost, while those for late arrivals were set at 5 times the unit operating cost for 4-ton trucks. This type of penalty is typically observed in Just-In-Time transport systems.

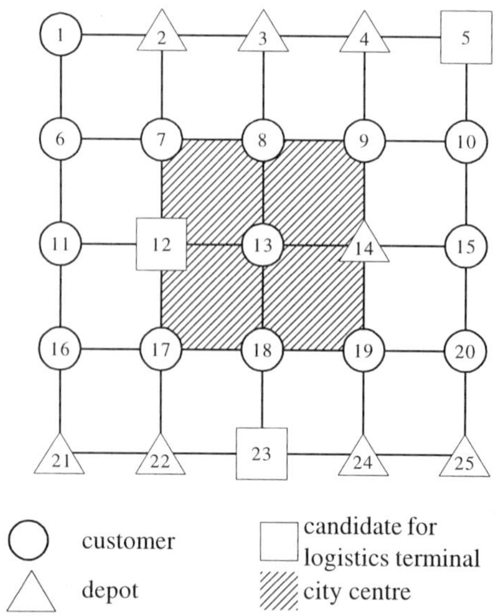

Figure 7 Test road network

Different types of logistics systems should be compared with each other for investigating the effects of the co-operative systems. Case (a) represents the current urban delivery systems, where all goods are delivered directly to customers by an individual freight carrier without the co-operative systems and AVRSS. The results from Case (a), such as the average loading factor and delivery routes of trucks for freight carriers, showed that Case (a) was able to represent an actual situation of urban goods delivery that trucks with low utilisation deliver goods many times within an urban area (Yamada *et al*., 2001).

In Case (b), the co-operative systems are introduced in conjunction with Case (a). Case (b') represents the case where profitability for co-operative freight organisation is not taken into account. Case (c) assumes that freight carriers can use the AVRSS. In this case, the AVRSS is introduced in conjunction with Case (a). In Case (d), both the co-operative systems and AVRSS are introduced in conjunction with Case (a), while Case (d') disregards the profitability of the co-operative freight organisation.

Results. Applications of the model to all cases were undertaken. The results of Case (b) identified that constructing logistics terminals only in candidate node 12 was optimal. This result indicates that it is desirable to construct logistics terminals around the city centre for decreasing total travel time considering the profitability of the co-operative freight organisation. Such logistics terminals can allow the co-operative freight organisation to efficiently deliver the consigned goods due to better accessibility to the city centre.

Figure 8 estimates the total travel times for the whole road network, and the estimated average total costs of freight carriers are shown in Figure 9 for the four cases. The effects of co-operation can be seen from the results of Case (b), but they are smaller than those of AVRSS (i.e. Case (c)). However, in Case (b'), both the total cost and travel time are much smaller than those in Case (c). Case (d') led to the same results as Case (b'). These results indicate the potential benefits of co-operative systems, though it is not realistic that the financial viability of co-operative freight organisation is disregarded. Co-operative freight transport systems could therefore offer the considerable benefits to the city as a whole as well as to freight carriers, if the monetary support for the co-operative freight organisation could be granted. In Case (d), the co-operative systems were not financially viable. This is a normal result, because the delivery costs incurred by freight carriers substantially decrease using the AVRSS and therefore they do not need to use the co-operative systems.

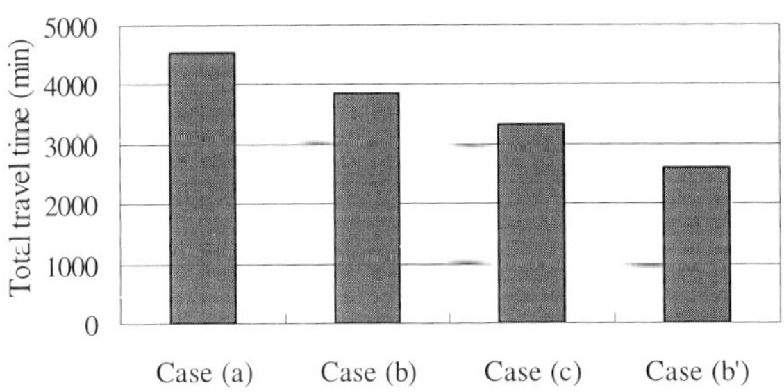

Figure 8 Case comparison: Total travel times

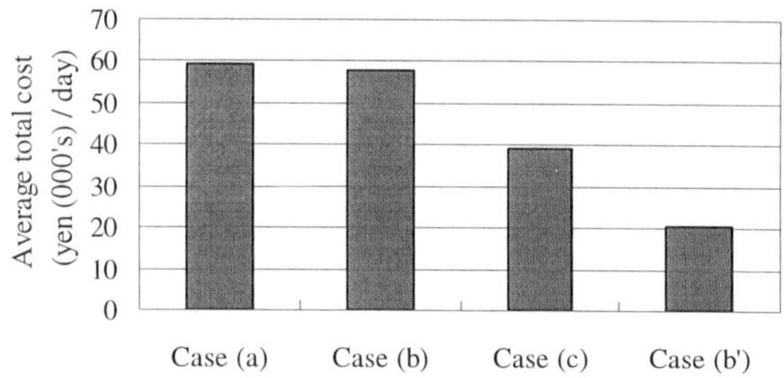

Figure 9 Case comparison: Average total costs

CONCLUSIONS

This paper presented an integrated model of probabilistic vehicle routing and scheduling for urban pickup/delivery trucks and dynamic traffic simulation. The probabilistic vehicle routing and scheduling model incorporates the uncertainty of travel times. After applying the model to test road network the following conclusions were derived.

(a) The total cost for freight carriers decreased by 11-17% using VRP-TW-P model compared with VRP-TW-F model. The main reason for the cost reduction is that trucks tend to arrive at customers earlier to avoid delay penalties using VRP-TW-P model. This leads to the lower risk of delay when the travel time is increased due to unexpected traffic congestion. The lower likelihood of delay provides better service to customers.

(b) Incorporating the uncertainty of travel times using VRP-TW-P model not only allows freight carriers to reduce their total costs, but also improves the environment in terms of reducing CO_2 emissions by about 6% compared with VRP-TW-F model. This means that the probabilistic approach in vehicle routing and scheduling can generate benefits to residents and the community by improving the environment as well as to freight carriers by cost reduction.

This paper also investigates the effects and profitability of co-operative freight transport systems using a simulation model based on the VRP-TW-F. The model was applied to an urban delivery system. Applications of the model led to the following conclusions:

(c) Co-operative freight transport systems could lead to a substantial reduction in total delivery cost and total travel time within the whole road network. This means that the benefits to the whole community could be gained as well as to freight carriers by implementing such systems.

(d) The assistance from the public sector to the co-operative freight organisation for management costs would be appropriate for successfully operating the systems.

REFERENCES

Bramel, J. and Simchi-Levi, D. (1996). Probabilistic analysis and practical algorithms for the vehicle routing problem with time windows. *Operations Research* **44**, 501-509.

Fujii, S., Iida, Y. and Uchida, T. (1994). Dynamic simulation to evaluate vehicle navigation. *Vehicle Navigation & Information Systems Conference Proceedings*, pp.239-244.

Gendreau, M., Laporte, G. and Seguin, R. (1996). Stochastic vehicle routing. *European J. Oper. Res.* **88**, 3-12.

Giannopoulos, G.A. and McDonald, M. (1997) Developments in transport telematics applications in Japan; traffic management, freight and public transport. *Transport Reviews* **17(1)**, 37-59.

Jaillet, P. and Odoni, A.R. (1988). The probabilistic vehicle routing problem. In B.L. Golden, A.A. Assad (eds.), *Vehicle routing; Methods and studies*, North-Holland, Amsterdam, 293-318.

Kohler, U. (1997). An innovating concept for city-logistics. *4th World Congress on Intelligent Transport Systems*, Berlin, Germany, CD-ROM.

Laporte, G., Louveaux, F. V. and Mercure, H. (1992). The vehicle routing problem with stochastic travel times. *Transportation Science* **26**, 161-170.

Ogden, K.W. (1992) *Urban goods movement: A guide to policy and planning*, Ashgate, Aldershot.

Powell, W.B., Jaillet, P. and Odoni, A. R. (1995). Stochastic and dynamic network and routing. In M.O. Ball, T.L. Magnanti, C.L. Monma, G.L. Nemhauser (eds.). *Network Routing*, North-Holland, 141-295.

Ruske, W. (1994). City logistics --- Solutions for urban commercial transport by cooperative operation management. *OECD Seminar on Advanced Road Transport Technologies*, Omiya, Japan.

Russell, R.A. (1995). Hybrid heuristics for the vehicle routing problem with time windows. *Transportation Science* **29**, 156-166.

Solomon, M. M. (1987). Algorithms for the vehicle routing and scheduling problems with time window constraints. *Operations Research* **35**, 254-265.

Taniguchi, E., Noritake, M., Yamada, T. and Izumitani, T. (1999) Optimal size and location planning of public logistics terminals. *Transportation Research E* **35(12)**, 207-222.

Taniguchi, E. and van der Heijden, R.E.C.M. (2000a). An evaluation methodology for city logistics. *Transport Reviews* **20(1)**, 65-90.

Taniguchi, E., Thompson, R.G. and Yamada, T. (2001b). Recent advances in modelling city logistics. In: E. Taniguchi and R.G. Thompson (eds.). *City Logistics II,* Institute of Systems Science Research, Kyoto, 3-33.

Taniguchi, E., Thompson, R.G., Yamada, T. and van Duin, R. (2001a). *City Logistics - Network Modelling and Intelligent Transport Systems*, Pergamon, Oxford.

Taniguchi, E., Yamada, T. and Tamagawa, D. (2000b). Probabilistic routing and scheduling of urban pickup/delivery trucks with variable travel times. In: M.G.H. Bell and C. Cassir (Eds.) Reliability of Transport Networks (Chapter 6), Research Studies Press, Baldock, 73-89.

Taniguchi, E., Yamada, T., Tamaishi, M. and Noritake, M. (1998). Effects of designated time on pickup/delivery truck routing and scheduling. *Urban Transport and the Environment for the 21st Century IV*, WIT Press, 127-136.

Yamada, T., Taniguchi, E., Noritake, M. and Horie, A. (1999a) Attitudes of companies towards introducing co-operative freight transport systems. In: E. Taniguchi and R.G. Thompson (eds.). *City Logistics I,* Institute of Systems Science Research, Kyoto, 219-232.

Yamada, T., Taniguchi, E. and Noritake, M. (1999b) Optimal location planning of logistics terminals based on multiobjective programming method. *Urban Transport and the Environment for the 21st Century V*, WIT Press, 449-458.

Yamada, T., Taniguchi, E., Itoh, Y. (2001b). Co-operative vehicle routing model with optimal location of logistics terminals. In: E. Taniguchi and R.G. Thompson (eds.). *City Logistics II,* Institute of Systems Science Research, Kyoto, 139-153.

19

Generalized Travel Cost Reliability in a Simple Dynamic Network under Advanced Traveler Information System

Tsz Hang Lam, Department of Civil Engineering, The Hong Kong University of Science and Technology, Kowloon, Hong Kong, P. R China
Hai Yang, Department of Civil Engineering, The Hong Kong University of Science and Technology, Kowloon, Hong Kong, P. R China
Wilson H. Tang, Department of Civil Engineering, The Hong Kong University of Science and Technology, Kowloon, Hong Kong, P. R China

Abstract

This paper provides a day-to-day analysis of the reliability of commuting time and trip scheduling under the Advanced Traveler Information System (ATIS). A simple network with parallel routes and bottleneck congestion is used to simulate the departure time and route choice decisions of commuters to minimize total travel time and scheduling delay cost. There are two major factors influencing the decisions of drivers in their departure time and route choices: their accumulated travel experience and information provided by ATIS. A simple experiment is carried for investigating trip-scheduling reliability of this network system.

INTRODUCTION

Reliability of a transportation network can be measured from different perspectives: connectivity reliability (Wakabayashi and Iida, 1992; Bell, 1999), travel time reliability (Asakura, 1996) and capacity reliability (Chen *et al.*, 1999, 2002) have so far been proposed and studied. These three types of reliability analysis focus on different aspects of the performance of congested road networks. A comprehensive review of the existing studies can be found in Iida (1999). Furthermore, a general theoretical framework and a review of reliability analysis of degradable transportation systems can be found in Du and Nicholson (1997).

Consider the morning peak-period commuting problem. Most commuters try to reach their work destinations within a particular range of work start times, or in a preferred arrival time window. Meanwhile, the road network has a limited capacity that is not able to accommodate all traffic within a particular short time period. Thus traffic congestion occurs and it unavoidably causes some commuters to go through the bottlenecks of limited capacity earlier or later than the time window. In this case, a commuter may have a choice between an on time arrival with a long travel or queuing time, and a late or an early arrival with shorter travel time. The commuter has a tradeoff between travel time and schedule delay. Therefore, the journey time and its reliability may be quite different for a different departure time. Also, a commuter's journey time and its reliability are of different importance with a different departure time. Thus, the aforementioned travel time reliability is not sufficient to explain the desire to lower the likelihood of arriving at the destination at an inconvenient time. Indeed, the inability of accurately predicting arrival times may be more costly to travelers than the extra travel time due to congested traffic. Commuters will generally modify their departure time schedules or route choices to minimize both the cost of travel time and the costs associated with schedule delay, especially late arrival. Disutility felt by commuters should be included in costs owing to travel time and schedule delay. In this case it is important to consider the reliability of generalized travel cost. This paper introduces a concept to investigate the performance of a network in terms of travel cost reliability. It is defined as the probability that a trip between an origin-destination (O-D) pair can be completed within a certain generalized travel cost.

One objective of our study is to try to extend the traditional travel time reliability concept by considering both travel time and trip scheduling reliability based on generalized travel cost. The current paper focuses on a day-to-day analysis of the reliability of commuting time and trip scheduling under ATIS. A simple dynamic traffic network with parallel routes and bottleneck congestion is used to simulate drivers' departure time and route choice decision to minimize total travel time and scheduling delay cost.

The organization of this paper is as follows: the description of the simulation experiment settings in our study is presented in Section 2. Section 3 provides the mechanism of driver's adjustments of departure time and route choices with information provided. After that, the results of the experiments are shown in Section 4 and conclusions are in Section 5.

DESCRIPTION OF SIMULATION MODEL

Network in the Model

In our study, we adopt a simulation approach to analyze the driver's departure time and route choice behavior in a dynamic road network. Figure 1 shows the network, which consists of 4 nodes as well as 4 links. For simplicity, all the traffic links are assumed to have zero travel times on running sections so that all those links can let vehicles pass without causing any travel time except at the bottlenecks near the end of the link. On each traffic link, there is a bottleneck of which the capacity is smaller than the in-coming flow rate in the morning peak period, and hence queues may develop. Our simulation tries to capture movements of individual vehicles with a limited number of sample size of 50 vehicles. All vehicles move from the same origin node 1 to destination node 4. At node 2 and node 3, a vector of two side flows $X_j = \{X_1, X_2\}$ is generated. Every day the sample vehicles are loaded into the network according to their departure times, while the side flows are loaded into the network according to headways generated with a given probability distribution. To model the impact of day-to-day traffic flow variation, the total daily side flow is assumed to follow a Normal distribution.

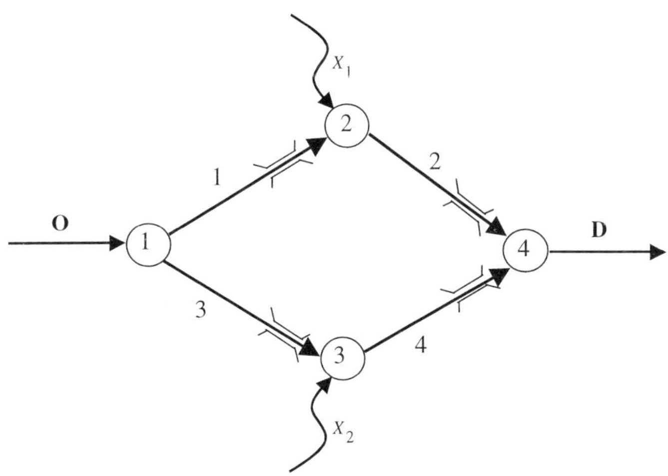

Figure 1 The simple network used for simulation experiment

Basic Assumptions

In conducting the simulation, the following assumptions are made.

- Drivers equipped with an ATIS are provided with both pre-trip (travel time and scheduling delay cost on previous day or the current day) and in-vehicle (instantaneous or predictive route travel time) information and drivers without ATIS decide their route choice and departure times by their own experience only. Each driver does not know others' decisions. That is, user equilibrium is not assumed.
- The preferred arrival times for all drivers are assumed to be within the same time window. They are assumed to have zero schedule cost if they reach their destinations within a given time window, and a loss for both early arrival as well as late arrival.
- All drivers are assumed to choose a departure time and routes aiming at minimizing their individual travel time and schedule delay cost.
- Drivers make their day-to-day departure time and route choice decisions based on their individual accumulated travel experience and travel information provided by ATIS.
- Bayesian updating is used to model the behavior of accumulated experience of commuters in our experiment. It is assumed that all sample commuters can remember all traveling experience and the importance of each day's experience is equal.

Simulation Procedure

The general procedure is shown in Figure 2 and can be described below.

Step 0: Identify the type of traveler information provided by ATIS, input the network and demand data, as well as the characteristic of each driver.
Step 1: Set day number DAY : = 1.
Step 2: Generate a vector of side flows $X_j = \{X_1, X_2\}$ and headway distributions.
Step 3: Generate the departure time pattern of all drivers.
Step 4: For drivers departing in each interval, generate their route choices according to a certain learning mechanism based on past experience and the information provided by ATIS (for drivers equipped with ATIS).
Step 5: Collect statistics to calculate travel time and schedule delay cost for drivers departing in each departure time interval.
Step 6: Update each vehicle's departure time on the next day based on their experience and departure time information provided by ATIS.
Step 7: If the proposed simulation horizon is reached (i.e. DAY: = MAX_DAY), then stop; otherwise DAY: = DAY+1, go to Step 2.

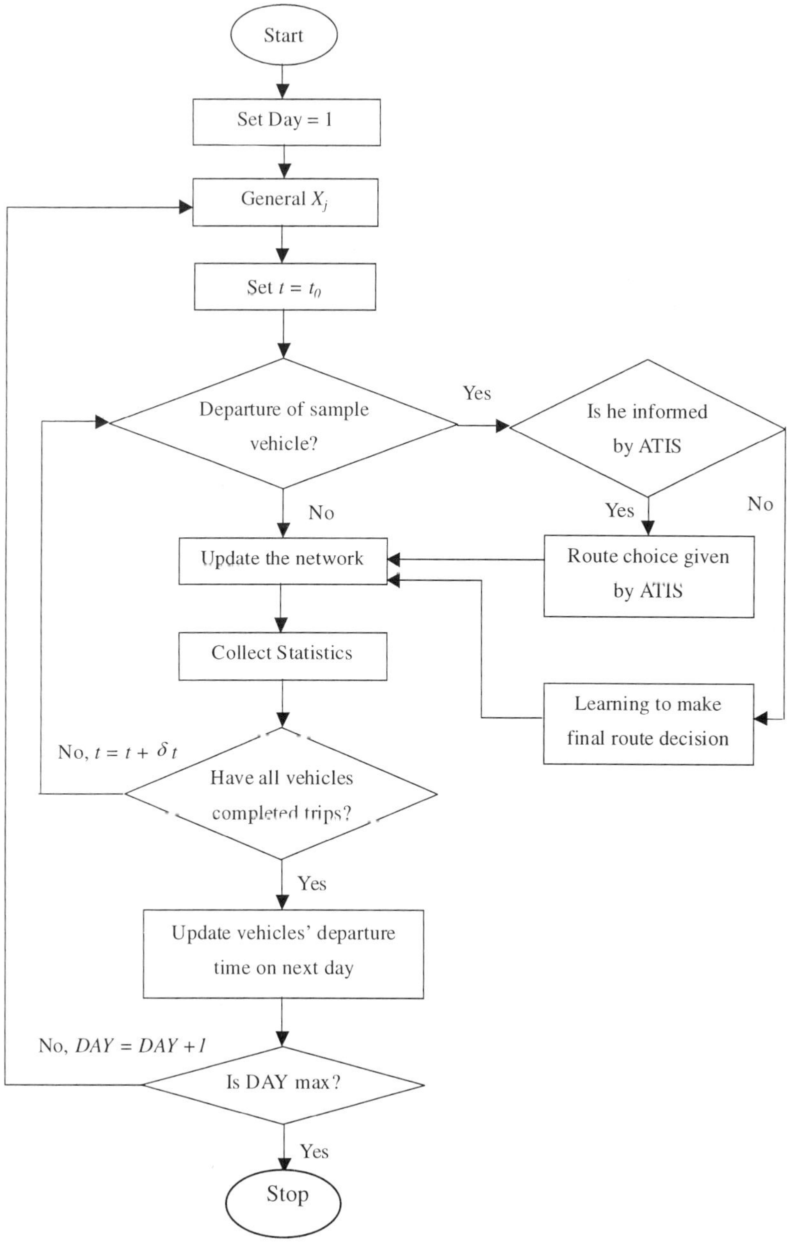

Figure 2 Flow chart of simulation procedure

DEPARTURE TIME AND ROUTE CHOICE

Calculation of Route Travel Time

By the time each driver departs, the ATIS will provide travel time to equipped drivers on each of the two routes at node 1. In our simulation experiment, predictive information is only used. Let $Q_a(t)$ and s_a denote the number of queuing vehicles at bottleneck a and at time t and s_a the capacity of bottleneck a $(a = 1,2,3,4)$. For calculation of predictive travel time on each route, it is recognized that the time to clear bottleneck 2 on route A is not $Q_2(t)/s_2$ (the travel time computed at time t), but rather that Q_2 has to be computed at time $t + Q_1(t)/s_1$, which represents the arrival time at the second bottleneck of that particular sample vehicle joining the queue of bottleneck 1 at time t. This type of information is realistic since actual expected travel time is informed. Now, in this case, the travel time for a driver arriving at node 1 at time t and following route A is:

$$T_A^P(t) = Q_1(t)/s_1 + Q_2(t + Q_1(t)/s_1)/s_2 \tag{1a}$$

and similarly for route B:

$$T_B^P(t) = Q_3(t)/s_3 + Q_4(t + Q_3(t)/s_3)/s_4 \tag{1b}$$

For those uninformed drivers (without ATIS), they will try to choose the route of which the travel time is perceived to be smaller based on their past experiences. The route travel time updating algorithm is actually considering a conditional probability. With past experiences, we assume that each driver will get the statistics of the travel time from the first day to the last day. On a particular day, if an uninformed driver departs, the corresponding past experience of his or her route choice is then recalled. The mean as well as standard deviation of choosing each route (i.e. route A and route B) is calculated. The concept of Bayesian updating is applied in this calculation. The Bayesian mean (μ'') and standard deviation (σ'') is determined by the following equations:

$$\mu'' = \frac{\left[\bar{x}/\left(\sigma/\sqrt{n}\right)^2\right] + \left[\mu'/(\sigma')^2\right]}{\left[1/\left(\sigma/\sqrt{n}\right)^2\right] + \left[1/(\sigma')^2\right]} = \frac{\bar{x}(\sigma')^2 + \mu'(\sigma^2/n)}{(\sigma')^2 + (\sigma^2/n)} \tag{2}$$

$$\sigma'' = \sqrt{\frac{(\sigma')^2(\sigma^2/n)}{(\sigma')^2 + (\sigma^2/n)}} \tag{3}$$

where μ' and σ' are the predefined mean and standard deviation of travel time; and $\bar{x}$ and σ are the mean and standard deviation of travel time accumulated in that particular category in past experience respectively; and n is the accumulated number of cases fell into that particular category. For each route choice, we can combine the corresponding probability distribution with the cost function of the driver, and the expected travel cost $EC(i,k)$ caused by choosing the route r for each driver i on day k can be determined by integration:

$$EC(r,i,k)=\int_{T}^{\infty} C(i,k)\cdot\frac{1}{\sqrt{2\pi}}\cdot\exp\left\{-\left[\frac{(t-T)-\mu^{''}}{\sigma^{''}}\right]^{2}\right\}dt \tag{4}$$

where T is the departure time of driver i on day k, $C(i,k)$ is the cost function of driver i on day k, r is the route choice of either A or B. Then a final route choice will be determined by comparing the cost induced. By expect utility theory, the one with less cost will be chosen. For drivers equipped with ATIS, we suppose predictive travel information is provided and drivers will follow the advice provided.

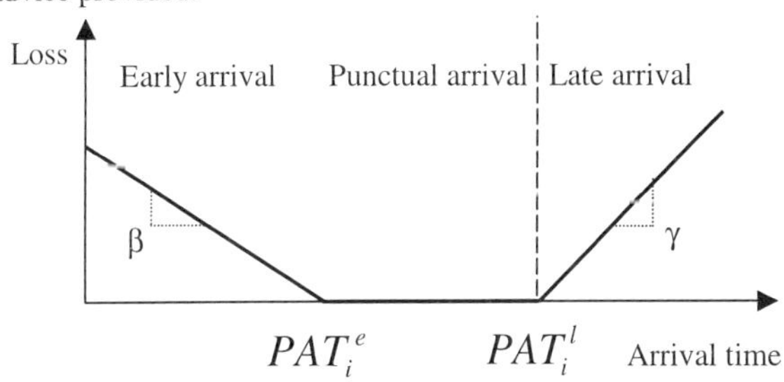

Figure 3 Graphical representation of early and late arrival penalty.

Adjustment of Day-to-Day Departure Time

On a particular day of our simulation, when all sample vehicles have reached their destinations, their travel costs will be calculated and they will make decisions of whether their departure times on next day will change according to their individual experiences. For each sample driver i, his travel cost consists of three components, travel time (or waiting time), early arrival schedule delay and late arrival schedule delay. Analytically, travel cost $C(i,k)$ of traveler i on day k is calculated below.

$$C(i,k)=\alpha\cdot(T_Q)_{i,k}+\delta_{i,k}\cdot\beta\cdot(SDE)_{i,k}+(\delta_{i,k}-1)\cdot\gamma\cdot(SDL)_{i,k} \tag{5}$$

where $\delta_{i,k}$ is 1 for early arrival and 0 for late arrival, $(T_Q)_{i,k}$ is the travel time (or queuing time) for driver i on day k, α,β and γ are the parameters representing the unit costs of travel time, early arrival schedule delay and late arrival schedule delay, respectively.

The schedule delay for early and late arrival for driver i on day k are respectively given by

$$(SDE)_{i,k} = PAT_i^e - AT_{i,k} \quad (6a)$$

$$(SDL)_{i,k} = PAT_i^l - AT_{i,k} \quad (6b)$$

where $AT_{i,k}$ is the actual arrival time of driver i on day k, PAT_i^e is the upper limit of the range of preferred arrival time and PAT_i^l is the lower limit of the range of preferred arrival time. The value of γ must be larger than that of β as the late arrival penalty must be larger than the early arrival penalty. PAT_i^e and PAT_i^l define the preferred arrival time window, as shown in Figure 3, in which no penalty will be imposed on the punctual drivers. A discrete rather than continuous time approach is used to model the impact of pre-trip departure time information and drivers' departure time choice. The range of possible departure times of all drivers is divided into M small time intervals of equal length. Each driver tries to choose a departure time such that his expected experienced travel cost will probably be minimized on a specific day.

We suppose ATIS can provide the actual travel costs for all departure time intervals on previous day (note that a driver may just know his experienced cost for his selected departure time interval on previous day, and may not be aware of the actual travel costs for other unselected departure time intervals). Based on the historically accumulated travel cost and the newly available actual travel cost on the last day, each driver will form his perceived cost for all departure time intervals on current day. One can again use the Bayesian updating formula to calculate the perceived cost, here the following linear convex combination formula is used to determine the predicted or perceived travel cost for each departure time interval by each driver.

$$PC(i,m,k) = (1-\phi)AC(m,k-1) + \phi BC(i,m,k-1) \quad (7)$$

$PC(i,m,k)$ is the perceived cost for departure time interval m by driver i on current day k; $AC(m,k-1)$ is the day $(k-1)$ or previous day's actual travel costs for departure time interval m experienced by driver i or other drivers i; $BC(i,m,k-1)$ is the Bayesian updated travel cost obtained by historical information from first day to previous day for departure time interval m for driver i in each departure time interval. ϕ can be considered to be an experience factor that reflects the relative importance of the newly available travel information and historical information in forming a driver's current perceived travel cost. The range of our predefined probable departure times is divided into a number of intervals, and historical travel costs from first day to previous day are stored in the corresponding departure time interval. For each interval, Bayesian updating is applied to obtain a Bayesian updated travel cost owing to past experiences. Note that eqn.(7) is applied to drivers equipped with ATIS because those drivers are assumed to be informed of the actual travel costs in all departure time intervals on previous day. If a driver is not equipped with an ATIS, his or her perceived cost for other departure time intervals on current day will remain to be $BC(i,m,k-1)$ only.

Once the current perceived costs for all departure time intervals are calculated, a logit model is used to determine the probability of choice of each departure time interval as follows:

$$P(i,m,k) = \frac{\exp\left(-\theta PC(i,m,k)\right)}{\sum_n \exp\left(-\theta PC(i,n,k)\right)} \tag{8}$$

where θ is a parameter reflecting the impact of the uncertainty of perceived cost on driver's choice of departure time, clearly, as $\theta \to +\infty$, driver will choice a departure time interval with least perceived cost.

RESULTS OF A NUMERICAL EXAMPLE

Based on the algorithm mentioned in the previous section, a simulation experiment is carried out to investigate the travel time reliability for two types of commuters: one type with ATIS and another type without ATIS.

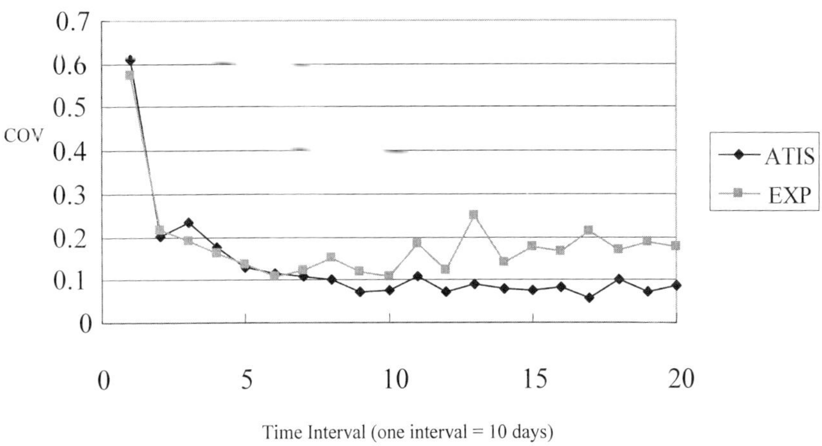

Figure 4 COV of day-to-day average generalised travel cost

Owing to the fact that accumulated traveling experience is one of important factors for both ATIS and non-ATIS (uninformed) commuters, the period within which people get the experience from network is regarded as 'warm up' period. After this period, it is expected that the variation of travel cost should be reduced to a certain level. Figure 4 shows the coefficients of variation (COV) of day-to-day average generalized travel cost. This figure is based on the case without side flow variation and the market penetration of the ATIS is 50%. The points plotted in the figure are the coefficients of variation obtained by taking the mean and standard deviation of the resultant generalized travel costs for the previous ten days for both ATIS and non-ATIS commuters. e.g. Interval 2 considers the travel statistics from Day 11 to Day 20 in our simulation and etc. The coefficient of variation for the ATIS commuters is smaller than that for the non-ATIS commuters and the trends of both lines become stable within a small range of value. This means the degree of fluctuations of the day-to-day generalized travel cost for the ATIS commuters is much smaller than that of the non-ATIS commuters. One reason is that the information given by the ATIS is more reliable than the commuters' individual past experience.

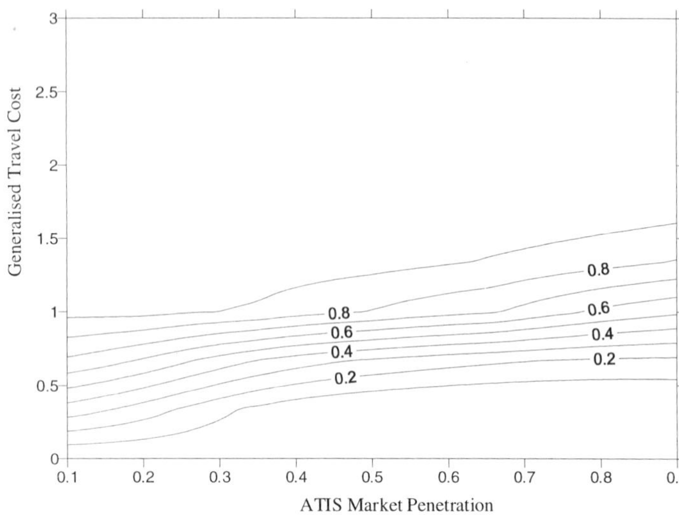

Figure 5a Travel cost reliability contours for commuters equipped with ATIS (the case without side flow variation)

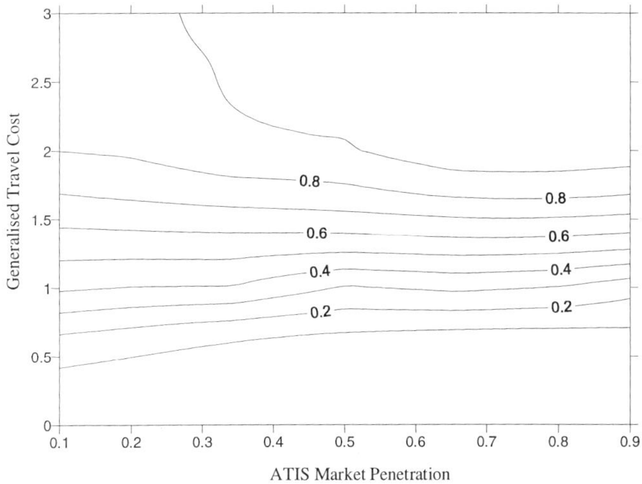

Figure 5b Travel cost reliability contours for commuters without equipped with ATIS (the case without side flow variation)

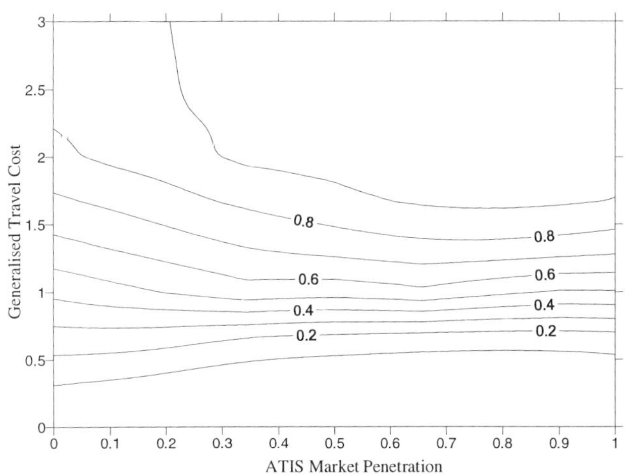

Figure 5c Travel cost reliability contours for all commuters in the simulation (the case without side flow variation)

Figure 5a shows the travel cost reliability contours for the ATIS commuters for various market penetrations in the network. The value of market penetration means the proportion of commuters equipped with ATIS. It is observed that when the market penetration increases, the reliability of travel cost of ATIS commuters steadily decreases and the travel cost generally increases. The reason is due to an occurrence of over-reaction to the information provided by ATIS. The resultant probability of having generalized travel cost less than a particular value will become lower with increasing ATIS market penetration. Note that this observation from the simulation results seems to be contradictory with the expected aim of ATIS, and might well depend on the manner in which the simulation scenario is set and the information is provided. Indeed, as observed in earlier studies (Yang, 1999; Yang *et al.*, 1999; Yang and Meng, 2001), ATIS should be designed carefully particularly in terms of market penetration in order to achieve better system and individual driver performance. From Figures 5b, it seems that, for non-ATIS commuters, the increase in the ATIS market penetration does not affect their resulting travel cost reliabilities so much as ATIS commuters. The main reason is that the route choices of the non-ATIS commuters are based on their past experience while those of the ATIS commuters are based on the ATIS route guidance only. The performance of ATIS commuters is much better than non-ATIS commuters according to their travel cost reliabilities. It depicts the travel cost reliability can be improved by equipping ATIS.

Figure 5c shows the travel cost reliability contours for both ATIS and non-ATIS commuters involved in our simulation. When the ATIS market penetration is small, the gaps among the contours are large; when the ATIS market penetration increases, the gaps among the contours gradually become narrow. This figure shows the inter-relationship between the generalized travel cost and the market penetration in the whole system, and it is useful for traffic engineers to design an optimal market penetration, for example, through appropriate ATIS service charge (Yang and Huang, 2002).

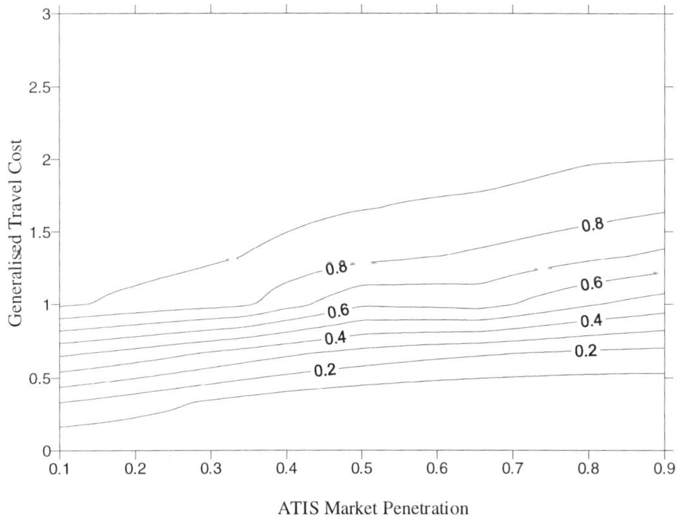

Figure 6a Travel cost reliability contours for commuters equipped with ATIS (the case with side flow variation)

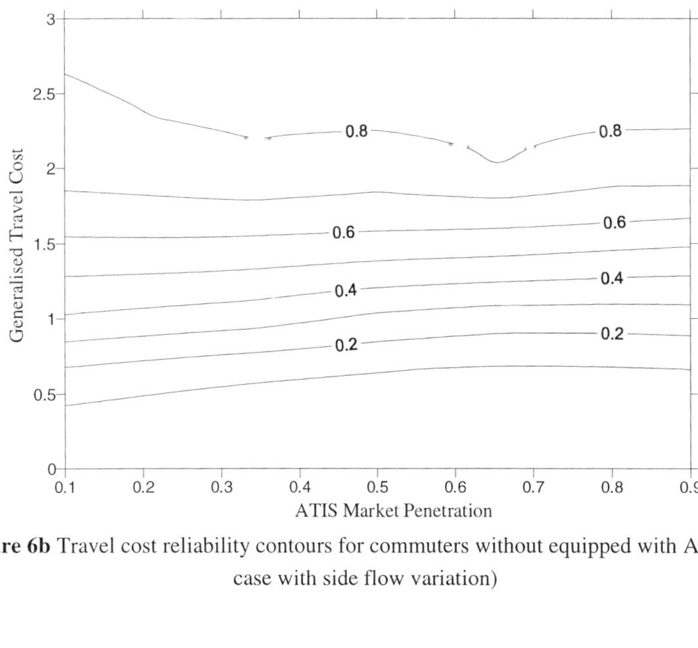

Figure 6b Travel cost reliability contours for commuters without equipped with ATIS (the case with side flow variation)

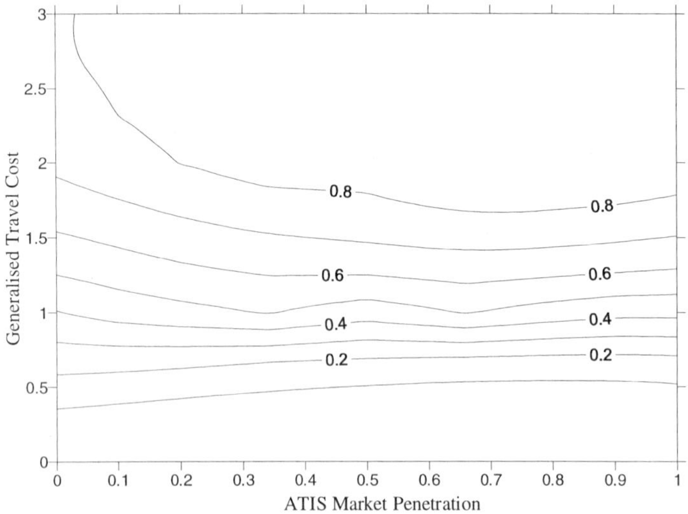

Figure 6c Travel cost reliability contours for all commuters in the simulation (the case with side flow variation)

Figures 6a, 6b, and 6c show the same details as those shown in Figures 5a, 5b and 5c respectively except that there exists a side flow variation in the simulation. Side flows from node 2 and node 3 vary with a normal distribution with the coefficients of variation are equal to 0.25 for both nodes. Due to uncertainties induced by side flow variation; the performance of ATIS commuters in the case without a side flow variation is better than that in the case with a side flow variation. Side flow variation mainly influences commuters to predict their departure times. Nevertheless, ATIS commuters can still get the current information for selecting their routes. Similarly, travel performance of the non-ATIS commuters in the case with side flow variations is worse than that in the case without side flow variations. This is because not only the departure time choices of non-ATIS commuters but also their route choices would be influenced by the variation of side flows. In the case without side flow variation, non-ATIS commuters can estimate the value of side flow for both nodes by their own experience. However, their experience is unreliable for making their route choice decisions with existence of side flow variation. The general pattern of the reliability contours shown in Figure 6c looks similar to that shown in Figure 5c. However the overall performance for all ATIS and non-ATIS commuters in our simulation in the case with constant side flows is better than that in the case with day-to-day side flow variation. It is believed that the side flow variation is one of the factors to influence the performance of system for the same generalized travel cost and a certain market penetration. This in turn suggests that precise prediction of flow variations is critical for successful implementation of ATIS. Implementation of side flow variation in our experiments indeed gives more realistic results.

CONCLUSION

Travel time reliability is an important measure of a road network especially when describing the performance of network infrastructures by considering travel time of drivers only. However, travel time is not the only factor that commuters consider. Usually, commuters consider their total utility of traveling instead of travel time only. For a given network subjected to a heavily traffic congestion and side flow variation, ATIS is introduced to improve travel cost reliability. It is proved that the performance of commuters is much better when they equip with ATIS. It is useful especially when there exist uncertainties in the road network. However, the results of our simulation depict that although ATIS can improve the network performance, the benefits to informed commuters decrease with increasing market penetration. A two-dimensional travel cost and ATIS market penetration are represented by iso-reliability contours. Our graphical representation allows for intuitive perception of which ATIS market penetration such that a required travel cost reliability can be achieved. Our approach established as efficient procedure for reliability based comprehensive performance assessment of a road network.

ACKNOWLEDGMENTS

This research described in this paper was substantially supported by a grant from the Research Grants Council of the Hong Kong Special Administrative Region, P. R. China (Project No. 6208/99E).

REFERENCES

Asakura, Y. (2000) Reliability measures of an origin and destination pair in a deteriorated road network with variable flows. In: *Transportation Networks: Recent Methodological Advances* (M.G.H. Bell, ed.), pp.273-287. Pergamon.

Ang, A.H.S. and Tang, W.H. (1990) *Probability Concepts in Engineering Planning and Design*: Vol.1. *Basic Principles*, Vol.2. *Decision, Risk and Reliability*. John Wiley & Sons, New York.

Bell, M.G.H. (1999) Measuring network reliability: a game theoretic approach. *Journal of Advanced Transportation*, **33**, 135-146.

Chen, A., Yang, H., Lo, K.K., Tang, W.H. (1999) A capacity related reliability for transportation networks. *Journal of Advanced Transportation*, **33**, 183-200.

Chen, A., Yang, H., Lo, H.K. and Tang, H. W. (2002) Capacity reliability of a road network: an assessment methodology and numerical results. *Transportation Research*, **36B**, 225-252.

Du, Z.P. and Nicholson, A. (1997) Degradable transportation systems: sensitivity and reliability analysis. *Transportation Research*, **31B**, 225-237.

Iida, Y. (1999) Basic concepts and future directions of road network reliability analysis. *Journal of Advanced Transportation*, **33**, 125-134.

Wakabayashi, H. and Iida, Y. (1992) Upper and lower bounds of terminal reliability of road networks: an efficient method with Boolean algebra. *Journal of Natural Disaster Science*, **14**, 29-44.

Yang, H. and Huang, H.J. (2002) Modeling user adoption of advanced traveler information systems: A control theoretical approach for optimal endogenous growth. *Transportation Research* Part C (in press).

Yang, H. and Meng, Q. (2001) Modeling user adoption of advanced traveler information systems: dynamic evolution and stationary equilibrium. *Transportation Research*, **35A**, 895-912.

Yang, H., Ma, O.S. and Wong, S.C. (1999) New observations on the benefit evaluation of advanced traveler information systems. *ITS Journal*, **5**, 251-274.

Yang, H. (1999) Evaluating the benefits of a combined road pricing and route guidance system in a traffic network with recurrent congestion. *Transportation*, **26**, 299-322.

20

ESTIMATING BENEFITS OF TRAVEL DEMAND MANAGEMENT SCHEMES WITH UNCERTAINTY OF TRAVEL TIMES

Eiichi Taniguchi, Department of Civil Engineering Systems, Kyoto University, Kyoto, Japan
Kensuke Ikeda, Toyota Motors Ltd., Toyota, Japan

ABSTRACT

This paper presents models for evaluating effects of implementing TDM schemes on road traffic with the consideration of uncertainty of travel times. The models are composed of drivers' behavioural model for identifying the optimal starting time and route with the designated time for arrival at the destination as well as dynamic traffic simulation model for estimating the change of travel times. A case study for applying the models on a test road network showed that TDM schemes of park and ride systems and road pricing enabled drivers to obtain benefits by improving the reliability of travel times as well as benefits by reducing travel times.

INTRODUCTION

Recent increase in transport demands has generated traffic congestion and environmental problems in urban areas. Some travel demand management (TDM) schemes including park and ride systems and road pricing have been proposed and implemented in many cities around the world. However, it is required to evaluate effects of these schemes before the implementation. Network modelling becomes a key issue in the evaluation of effects of these schemes on traffic conditions. In particular the uncertainty of travel times within networks plays an important role for modelling, as the traffic conditions on the network vary with time.

Traffic congestion increases the uncertainty of travel times as well as the travel times themselves. It is expected that implementing TDM schemes will reduce the variation of travel times and the amount of travel times. If the reliability of travel times is improved, drivers can cut the time margin of travelling and take later starting time for a trip. Moreover, providing alternatives of travel modes and routes by TDM schemes allows traffic congestion to be alleviated and the reliability on road network to be improved.

The reliability on road network has become more important, given the ongoing increases in the time value of people involved in business and other activities. There are three types of reliability of road network; connectivity reliability, time reliability and capacity reliability (Bell et al., 2001). This paper focuses on the time reliability of networks that is related to the uncertainty of travel times within the network.

Normally constructing new roads or implementing TDM schemes can reduce travel times. The benefits generated by the reduction of travel times are estimated in the traffic planning. However, the measures can contribute to decrease the variation of travel times as well as the reduction of mean travel times. This leads to the reduction of risk of arriving late at the destination, which is beneficial to drivers. This type of benefits should be incorporated in estimating economic benefits by implementing some measures for improving road traffic conditions (Taniguchi *et al.,* 2001). It becomes essential in an information-oriented society, since the trips on the road network are then increasingly associated with a designated arrival time. A typical example can be seen in the Just-In-Time delivery system for freight transport.

There have been intensive studies on drivers' choice behaviour on starting time for a trip. Many of them assume drivers behave by maximising their utility (or minimising their disutility) under the uncertainty of travel times that can be represented by a probabilistic distribution. Hall (1983) presented a model for identifying the starting time for a trip of drivers by minimising the loss function of drivers. Chang *et al.* (1988) analysed drivers' choice behaviour on starting times for daily commuting by an experimental approach. Ben-Akiva (1986) presented a model for simultaneously identifying the starting time and route by applying the maximisation of drivers' utility.

This study presents a simulation model that incorporates drivers' behaviour under the uncertainty of travel times. After applying the model to a test network, benefits by reducing travel times and improving the reliability of travel times can be estimated for commuting drivers by means of TDM schemes. As well the changes in traffic conditions on the network can be evaluated after implementing TDM schemes.

MODELS

Drivers' Behavioural Model

Models for adjusting starting time. The drivers' behaviour of adjusting the starting time for commuting by car under the uncertainty of travel times can be formulated as given below. A driver who is expected to arrive at his/her office at a designated time identifies the starting time by minimising the expected disutility taking into account the early arrival and delay penalties as shown in Figure 1. If a driver arrives at the office later than the designated time t_d, he/she needs to pay the penalty that is proportional to delay time. If he/she arrives earlier than the designated time t_d, he/she needs to wait for the work starting time and incurs the time costs. Multiplying the distribution of arrival time at the office with the penalty function in Figure 1 gives expected disutility of early arrival and delay.

The expected disutility of using car can be given as the sum of disutility of early arrival and delay penalty, time costs of moving by car and costs incurred by the road pricing scheme. It can be formulated as:

$$EDU_c(t_0) = C_p(t_0) + C_m(t_0) + \delta \cdot C_r \tag{1}$$

where,

$$C_p(t_0) = \int_{-\infty}^{t_d - t_0} c_f \{t_d - (t_0 + t)\} f(t)dt + \int_{t_d - t_0}^{\infty} c_d \{(t_0 + t) - t_d\} f(t)dt \tag{2}$$

$$C_m(t_0) = \int_{-\infty}^{+\infty} c_m \cdot t\, f(t)dt = c_m \cdot E(f(t)) \tag{3}$$

t_0: starting time

t_d: designated time to be arrived

$f(t)$: probability density function of travel time t

c_f: coefficient for time cost due to early arrival (yen/min.)

c_d: coefficient for delay penalty (yen/min.)

c_m: coefficient for time cost due to the increase of travel time (yen/min.)

δ: = 1; if vehicle passes cordon line, = 0; otherwise

C_r: cost for road pricing (yen)

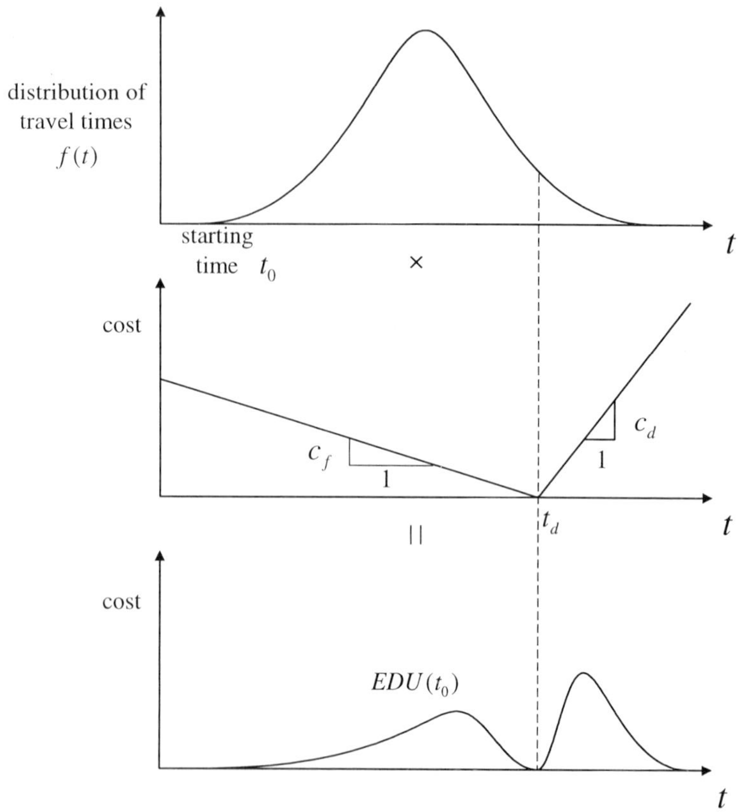

Figure 1 Identification of expected disutility using the distribution of travel times and penalty function

Let us consider park and ride systems in which travellers use trains after leaving their cars at car park attached to the railway station. The expected disutility of using park and ride systems is the sum of the disutility of early arrival and delay and time cost of moving by car and train and cost for car park and train ticket. It can be formulated:

$$EDU_p(t_0) = C'_p(t_0) + C_m(t_0) + c_t \cdot t_t + C_c + C_t \tag{4}$$

where,

$$\begin{aligned} C'_p(t_0) = & \int_{-\infty}^{t_d - t_i} \{t_i \cdot c_f + c_w(t_d - t_i - t)\} f(t - t_0)dt + \int_{t_d - t_i}^{t_d} c_w(t_d - t) \cdot f(t - t_0)dt \\ & + \int_{t_d}^{\infty} [t_i \cdot c_d + c_w\{t - (t_d + t_i)\}] \cdot f(t - t_0)dt \end{aligned} \tag{5}$$

t_i : time interval of train operation (min.)

c_w : coefficient for time cost due to waiting at station (yen/min.)

c_t : coefficient for time cost for taking train (yen/min.)

t_t : time for riding train (min.)

C_c : cost for car park (yen)

C_t : cost for train ticket (yen)

Equation (5) incorporates the time cost for waiting for trains at station. Here, it is assumed that each train is operated at the same interval of t_i. Drivers try to determine the optimal route and starting time by choosing the latest starting time that can be given by

$$t_0^* = \min_{t_0} EDU(t_0) \tag{6}$$

where, $EDU(t_0)$ indicates $EDU_c(t_0)$ or $EDU_p(t_0)$

The model formulated here can also be applied for describing travel behaviour under other TDM schemes than park and ride systems and road pricing.

Models for choosing transport mode. Drivers are to choose optimal transport mode and route by minimising the expected disutility. Then the probability of choosing park and ride systems based on the equation (6) can be given by

$$P_r = \frac{1/EDU_p}{1/EDU_p + 1/EDU_c} \tag{7}$$

Note that drivers can update travel times based on their experience on travelling in the previous day. The model used in this study only allows travel times on a route that a driver actually used to be updated.

Dynamic Traffic Simulation Model

This study used a dynamic traffic simulation model based on BOX model (Fujii *et al.*, 1994). The BOX model is essentially a macroscopic model but because the origin and destination of each vehicle is defined, it is actually a hybrid macroscopic/microscopic model. Vehicles are assumed to choose the shortest path when they arrive at a node using an estimated average travel time. The BOX model consists of two components, flow simulation and route choice simulation as shown in Figure 2. A sequence of boxes is used to represent each link. Groups of vehicles flowing out of a box and into the next box during the scanning interval represent the flow on links. The BOX model was combined with the drivers' behavioural model that was described in the previous section to estimate benefits of reducing travel times and improving the reliability of travel times as well as traffic conditions.

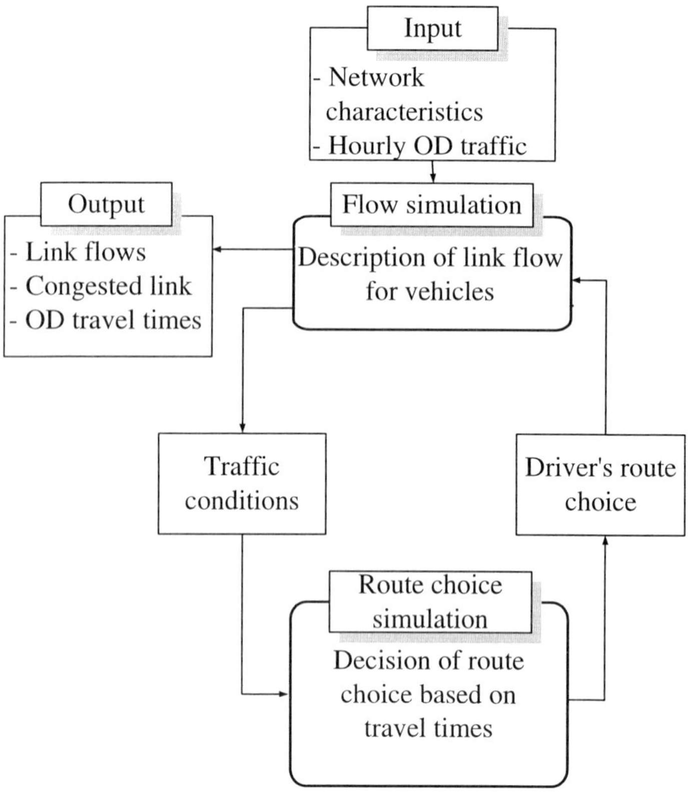

Figure 2 Dynamic traffic simulation (BOX model)

CASE STUDY ON A TEST NETWORK

Test Conditions

The model was applied to a test network with 25 nodes and 80 links. A railway line with 5 stations that facilitate park and ride car parks was located along a central axis of the network as shown in Figure 3. The study considered commuting drivers who made trips from their home to office by cars in the morning time. It is assumed for simplicity that homes of commuters are randomly located in the suburb (nodes 1-20 in Figure 3) and their offices are located in the city centre (node 23 in Figure 3). Each commuting driver can make a decision each day whether to use car or the park and ride system based on historical data of travel times on the route in which the driver has travelled. All commuting drivers are assumed to have a designated time to arrive at their offices. The free running speed of all links is 60 km/h. The capacity of link is 45 vehicle/min. The daily traffic volume for whole network is 110,000 vehicle/day. Figure 4 shows the hourly generation ratio of traffic that was based on the survey of traffic on trunk roads in Kyoto-Osaka area, Japan.

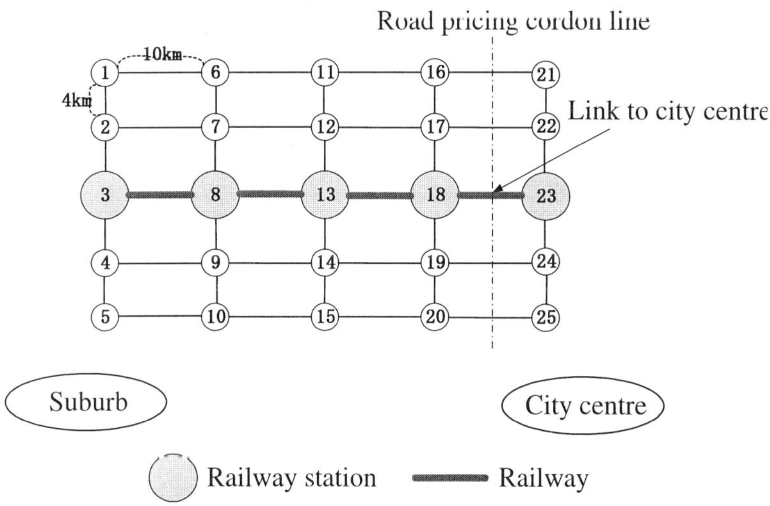

Figure 3 Road and railway network as used in the case study analysis

Some parameters required for analysis are assumed to be:

(a) Designated arrival time: 8:30, 9:00 and 9:30 a.m.

(b) Coefficient for
- Delay penalty: 300 yen/min
- Time cost of early arrival: 30 yen/min
- Time cost of trip by car: 65.5 yen/min
- Time cost of trip by train: 80 yen/min
- Time cost of waiting for train: 90 yen/min

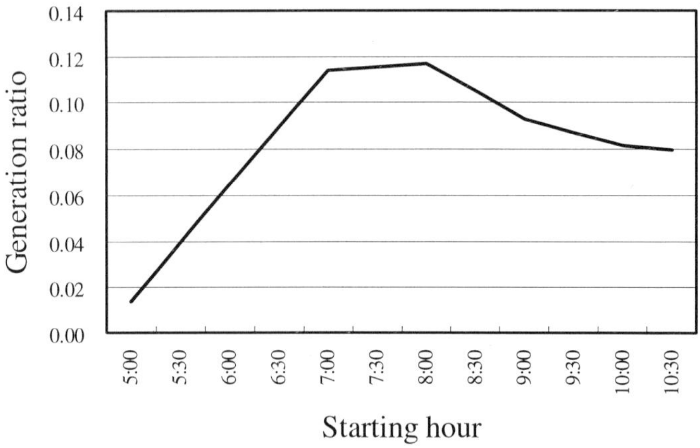

Figure 4 Generation ratio of hourly traffic

The delay penalty was assumed to be 10 times higher than the time cost of early arrival. The time cost of trip by train was assumed to be 1.6 times of the time cost of trip by car, since the disutility of travelling by train was estimated 1.6 times higher than that by car. The time cost of waiting for train was set 1.8 times of the time cost of trip by car.

It is assumed that travellers who use park and ride systems will choose the nearest station, since car parks at stations are to have enough capacity for all users. Commuting drivers can choose a route with the minimum disutility on the previous day and follow the route on current day. On the other hand other drivers will follow the route choice procedure in BOX model based on the updated average travel times.

Drivers who leave home in the period 7:00 – 9:00 a.m. for their offices (5,543 vehicles) are taken for this study. This study compares the changes of traffic behaviour in cases with and without TDM schemes. We considered following cases:

Case 0 : without any TDM schemes
Case 1 : Park and ride systems are implemented
Case 2 : Park and ride systems and road pricing (300 yen /vehicle) are implemented
Case 3 : Park and ride systems are implemented and employers pay the train fare
Case 4 : Park and ride systems are not implemented and level of congestion on roads is increased
Case 4' : Park and ride systems are implemented and level of congestion on roads is increased

Case 5 : Park and ride systems are not implemented and generation of traffic varies with +10% and -10%

Case 5' : Park and ride systems are implemented and generation of traffic varies with +10% and -10%

For Case 2 the price of 300 yen is imposed on all the vehicles that pass the cordon line shown in Figure 3. For cases 4 and 4' the free running speed was decreased to 50 km/h to get more congested traffic conditions.

Results

Simulation was performed for 365 days in which drivers commuted based on their learning process of travel times. Figures 5 and 6 show the change of conversion rate to park and ride systems of drivers and the average standard deviation of travel times during 365 days. The average standard deviation of travel times is defined as the average of standard deviation for cognitive travel times of the commuting drivers. Both the conversion rate to park and ride systems and the average standard deviation of travel times become almost stable after about 110 days of iteration. Then the average value of the last 10 days of 365 day iteration was taken for the analysis.

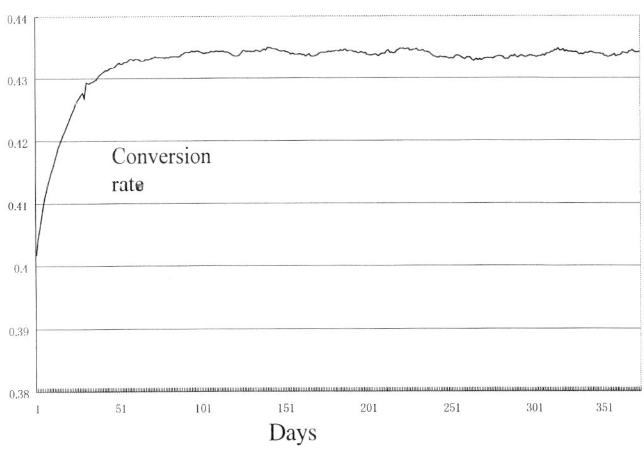

Figure 5 Change of conversion rate to park and ride systems (Case 1)

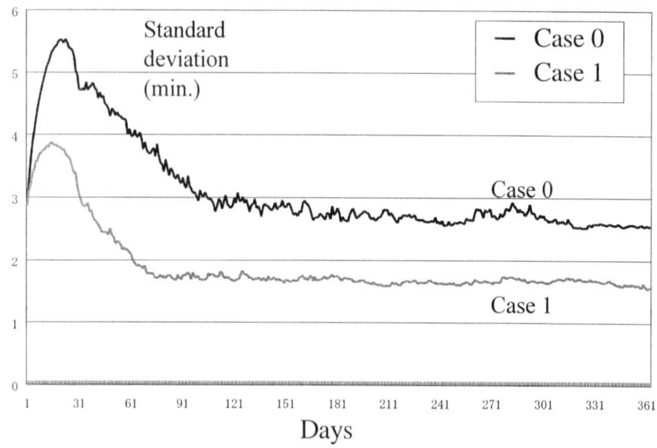

Figure 6 Change of average standard deviation of travel times between case 0 and case 1

Figure 7 and Table 1 show results of simulation. The benefits by reducing travel times are generated by the reduction of actual travel time costs. The benefits by improving reliability of travel times are generated by the reduction of actual disutility of early arrival and delay at the destination. Figure 7 and Table 1 indicate that each TDM scheme can generate benefits by improving reliability of travel times as well as benefits by reducing travel times. Note that benefit by improving reliability of travel times simply means benefit by decreasing the standard deviation of travel times. The conversion rate to park and ride systems of drivers who may use the systems was 42.9% for Case 1. The number of vehicles that changed to park and ride systems was about 2% of whole traffic. The conversion rate increased to 48.8% when road pricing of 300 yen was introduced in Case 2, since the additional price is charged to drivers who come into the city centre.

It is also interesting that the conversion rate to park and ride systems increased when the reliability of travel times by car became lower in Cases 4' and 5'. Therefore, drivers tend to prefer park and ride systems under the conditions of low reliable travel times by car. In both Cases 4' and 5'the percentage of benefits by improving reliability of travel times to benefits by reducing travel times, increased compared with Case 1. It means that drivers can obtain more benefits by improving the reliability of travel times in higher level of congestion.

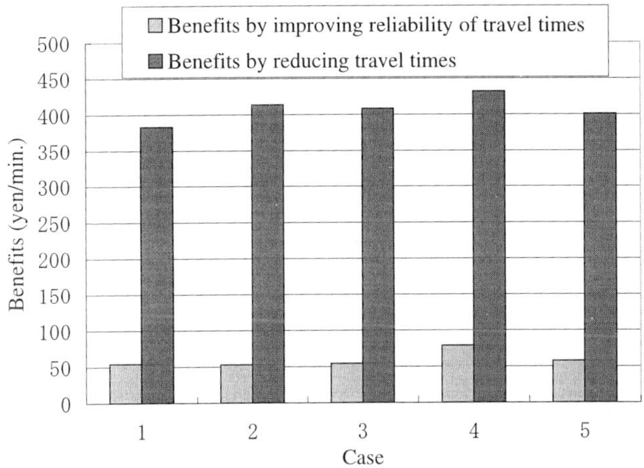

Figure 7 Comparison of benefits for a commuting driver by implementing TDM scheme

Table 1 Effects of TDM schemes on benefits

Case	0	1	2	3	4	4'	5	5'
Benefits by reducing travel times (yen /person)	-	395.5	417.3	403.3	-	493.4	-	426.8
Benefits by improving reliability	-	56.1	60.5	61.7	-	88.1	-	66.8
Change in loss of travel times (%)		-15.7	-16.6	-16.0	-	-15.6	-	-16.7
Change in expected disutility of early arrival and delay (%)	-	-33.2	-35.8	-36.5	-	-45.0	-	-35.6
Real penalty of early arrival and delay (yen/person)	169	169	169	169	196	108	188	121
Conversion rate to park and ride system (%)	-	42.9	48.8	45.6	-	48.4	-	43.3

Table 2 Effects of TDM schemes on traffic conditions

Case	0	1	2	3	4	5	5'
Variance of travel times (min^2)	6.620	2.519	2.541	2.471	2.673	8.857	2.471
Percentage of delay (%)	3.6	2.1	0.81	2.6	8.1	3.2	0.93
Real disutility of car commuters (yen/person)	2686	2231	2421	2220	2790	2739	2246
Travel time costs (thousand yen)	178,060	162,560	162,130	162,270	204,170	182,090	161,090
Vehicle hours of travel (hour)	45,167	41,333	41,333	41,333	52,000	46,333	41,000
Average travel speed at link to city centre (km/h)	31.5	44.3	44.4	45.0	26.9	33.6	41.0

Table 2 shows effects of TDM schemes on traffic conditions. Introducing park and ride systems for Case 1 considerably improved the road traffic conditions compared with Case 0 in terms of variance of travel times, percentage of delay, real disutility of car commuters, travel time costs, vehicle hours of travel and average travel speed at the link to city centre. The vehicle hours of travel for Case 1 was decreased by 8.5% compared with Case 0. It is remarkable that changing mode of about 2% of whole vehicles to park and ride systems compared with Case 0 made substantial reduction of 8.5% in terms of vehicle hours of travel. The decrease in variance of travel times for Case 1 is also an important effect of the park and ride systems from viewpoint of travel time reliability in the road network. For Case 2 in which road pricing as well as park and ride systems is introduced, the effects on traffic conditions are almost same level in many of indicators as Case 1. But the percentage of delay for Case 3 is better than for Case 2. This would be beneficial for car commuters. For Case 5 in which the traffic demands fluctuated by 10%, park and ride systems are effective to reduce the variance of travel times and vehicle hours of travel and to increase the average travel speed at the link to city centre. Therefore, TDM schemes are effective to alleviate traffic congestion even for the case that traffic demands vary. Note that the variation of travel demands in Case 5 is not explicitly related to the level of congestion on road network. It leads to generate benefits not only to commuting drivers but also to other general drivers.

Table 3 Expected disutility of early arrival and delay

	Time cost for early arrival	Delay penalty	Time cost for waiting for train
Car	69.2	23.4	-
Park & ride	0.6	9.1	63.2

Table 3 illustrates expected disutility of commuters by car and park and ride systems on early arrival and delay at the office. This table indicates that expected disutility due to early arrival and delay for commuters by park and ride systems is much smaller than that for commuters by car. Most of expected disutility of commuters by park and ride systems comprises the time cost for waiting for train at the railway station. These commuters have chosen the behaviour of waiting for a train at the platform for avoiding any delay at the office. On the other hand commuters by car have taken the expected disutility of early arrival to avoid delay.

Table 4 Starting time for a trip of commuters

Case	0	1	2	3	4'	5'
Car	7:25	7:34	7:36	7:34	7:22	7:33
Park & ride	-	7:39	7:39	7:39	7:39	7:40

Table 4 compares the starting time for a trip of commuters who live in the suburb 40 km away from the city centre and the designated arrival time at the office is 8:30 a.m. This table indicates that commuters by park and ride systems can shift the starting time by 14-15 minutes later in Cases 1-5' compared with Case 0. In addition, commuters by car also shifted the starting time by 8-11 minutes later in Cases 1, 2, 3 and 5'. This is attributed to the fact that the implementation of TDM schemes reduced travel times and improved the reliability of travel times.

CONCLUSIONS

This paper presented models for evaluating the effects of TDM schemes on road traffic with the consideration of the uncertainty of travel times. The following concluding remarks were drawn from the application of the models to a test network.

(a) Results indicated that TDM schemes of park and ride systems and road pricing enabled drivers to obtain benefits by improving the reliability of travel times as well as benefits by reducing travel times. The percentage of benefits by improving the reliability of travel times to benefits by reducing travel times increased for the case with a higher level of uncertainty of travel times.

(b) The considerable reduction in the variance of travel times and the vehicle hours of travel was observed by implementing park and ride systems and road pricing. The average travel speed was also increased. This means that these TDM schemes are effective in alleviating traffic congestion.

These results are only from a test network. We need further research on the application of models to more realistic large-scale network.

REFERENCES

Bell, M.G.H. and Cassir, C. (eds.) (2001). Reliability of transport networks. Research Studies Press.

Ben-Akiva, M. (1986). Dynamic model of peak period traffic congestion with elastic arrival rates, *Transportation Science* **20(2)**, 164-181.

Chang, G. and Mahmassani, H.S. (1988). Travel time prediction and departure time adjustment behavior dynamics in a congested traffic system, *Transportation Research* **22B(3)**, 217-232.

Fujii, S., Iida, Y. and Uchida, T. (1994). Dynamic simulation to evaluate vehicle navigation. *Vehicle Navigation & Information Systems Conference Proceedings*, pp.239-244.

Hall, R.W. (1983). Travel outcome and performance: The effect of uncertainty on accessibility, *Transportation Research* **17B (4)**, 275-290.

Taniguchi, E. and Yonetani, K. (2001). Estimating benefits by improving the reliability on travel times, In: *Urban Transport and the Environment for the 21st Century VII,* WIT Press, pp.271-280.

21

SIMULATING ROAD TRAFFIC INTERRUPTIONS – DOES IT MATTER WHAT MODEL WE USE?

Katja Berdica, Royal Institute of Technology, Stockholm, Sweden
Zarko Andjic, Gabites Porter Consultants, Christchurch, New Zealand
Alan J Nicholson, University of Canterbury, Christchurch, New Zealand

ABSTRACT

Computer models are often used for studying the effects of changing conditions in the road network. State-of-the-art macroscopic models generally take some kind of network equilibrium approach and therefore have difficulties in appropriately representing short-term capacity reductions, probably resulting in too low estimates of delays. Recently developed dynamic models may be more promising. The purpose of this paper is to investigate the implications of model choice further, as well as the possibilities to study effects of short-term incidents. Three different computer programs were used: TRACKS, SATURN, and Paramics. The results show that microsimulation is a feasible tool for studying short-term disturbances in the road transportation system.

INTRODUCTION

Road transports are tightly linked to everyday life, both private and business, and problems in the road network can have wide spread repercussions. Extreme weather conditions and other natural phenomena can make large parts of the road network impassable. More rare events like collapsing bridges, as well as every day incidents like traffic accidents or closures due to maintenance activities, can reduce the capacity of the traffic system. The resulting delays cause serious losses in terms of travel time, as well as other costs in the case of disrupted "just in time" deliveries etc. The road network is also one of our most important lifeline systems – crucial for assisting, maintaining, and repairing other infrastructures and for evacuation in a disaster. All this put together makes reliability in the road transportation system a highly important subject in terms of public economy and welfare (Berdica, 2002).

Since controlled, real life, field experiments are somewhat difficult to perform, computer models are often used for studying the effects of changing conditions in the road network. Recently, the EMMA model (the everyday name for the Swedish implementation of the traffic analysis package emme/2) was used in a case study of the road network in Stockholm, Sweden, to study how closures and capacity variations influence the users' travel time, trip length, etc (Berdica, 2000). EMMA belongs to the group of state-of-the-art models, in which traffic flow is handled on the approximate, macroscopic level. These generally take some kind of network equilibrium approach, in which the road-user is assumed to have perfect information on the system status (i.e. they can be said to simulate the effects of long-term disturbances). Once an equilibrium route choice pattern has been determined, the travellers do not deviate from it. This may be appropriate when modelling a bridge failure, say, but many of the every day incidents causing reductions in road serviceability are of short duration and a new equilibrium is never reached; the road-user has to revise his/her course of action continuously during the journey, as he/she encounters unexpected events and changing conditions. Using a network equilibrium approach could underestimate the effect of the disturbance since this transient state is not captured (see Figure 1). Dynamic models, where changing conditions like congestion are taken into account by regular "updating" of route choice, may be more promising for studying short-term capacity reductions. In their most disaggregate form, traffic is modelled by microsimulation of individual vehicles in the road network.

The purpose of the study presented here is twofold: to investigate the implications of model choice in further detail, as well as the practicability of studying the effects of short-term incidents. The first objective is achieved by simulating the same range of closures, in the same road network, for the same traffic demand, by means of three different computer programs: TRACKS, SATURN, and Paramics. These represent macroscopic, mesoscopic, and microscopic modelling respectively. Running and analysing a number of scenarios with short-term incidents in Paramics achieves the second objective.

The three models are described in the second section, followed by the study design, analyses of the results, and the summarised conclusions.

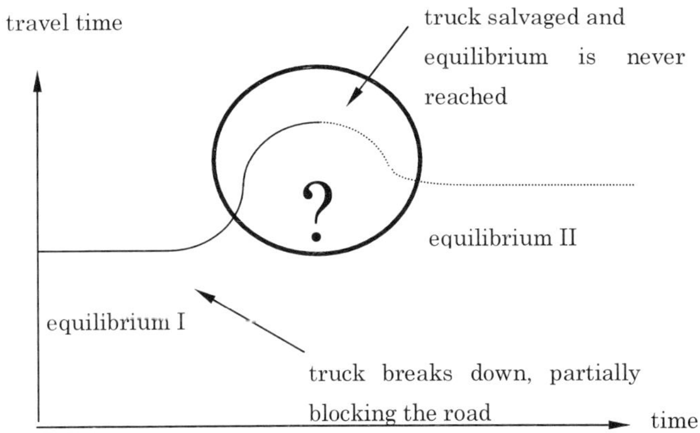

Figure 1 Illustration of the problem of the "adjustment process" between two equilibrium states (Source: Berdica, 2000)

MODEL CHARACTERISTICS

The programs TRACKS, SATURN and Paramics can be said to represent three different levels of detail for modelling road traffic. They have all been implemented and calibrated for the road network around the University of Canterbury campus in Christchurch, New Zealand, over the last few years (Cameron, 1996; Laird and Nicholson, 2000; Andjic, 2000). This provides an opportunity for a comparative study not often encountered. The coded network, with the simulation network used for the comparison being the portion within the polygon, is shown in Figure 2.

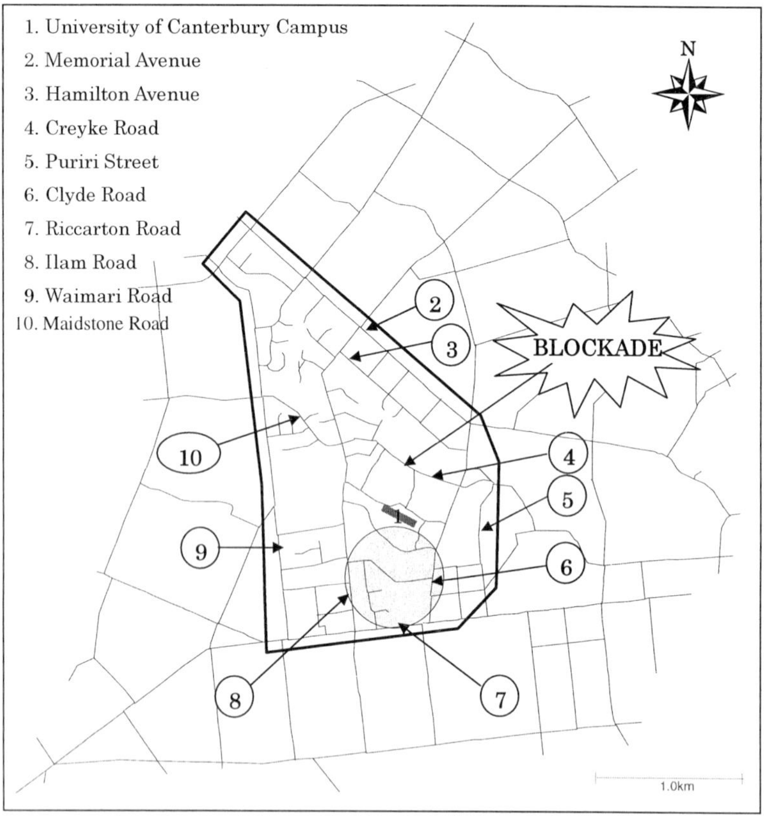

Figure 2 Coded road network: simulation network within the marked polygon (Adapted from: Andjic, 2000)

TRACKS is a suite of programs developed by Transportation and Traffic Systems Ltd in New Zealand over the years since the late 1970's. It consists of a complete set of programs for performing the traditional four-stage transport planning process of trip generation, distribution, mode choice and assignment, but only the assignment part was used in this study. TRACKS was chosen to represent the traditional macroscopic models. These are generally based on the Wardrop condition of user equilibrium, which means that individuals choose routes in order to minimise their generalised travel costs and in an equilibrium situation all used routes between any origin-destination pair have equal costs, not greater than those on unused routes. However, for this application the TRACKS option of a time-dependent incremental assignment was chosen, rather than the true equilibrium assignment. This involves dividing the total trip matrix into fractional matrices that are assigned one at a time until the whole travel demand has been loaded onto the network, with re-calculation of optimum paths between increments. The results can be interpreted as the build up of congestion for the peak period. One might expect that the smaller the increments the closer the final solution will be to the equilibrium solution, although Ortuzar and Willumsen (1990) state that the "algorithm does not necessarily converge to Wardrop's equilibrium solution even if the number of fractions... is large and the size of the increments... is small". The basis for the route choice is a function of travel time and travel distance summed over links and intersections. Each link in the network is assigned one of twenty link types, for which a predetermined speed-flow relationship describes the average speed and response to flow variations (i.e. the rate of increase in travel time as a function of traffic flow assigned to the link). For intersections, the delay is estimated for signals and roundabouts using the algorithm in SIDRA (Akcelik, 1986) and a modification of Tanner's queuing theory (Fisk & Tan, 1986; Gabites Porter, 1991) for priority intersections, for the calculated flow rates and a number of user-specified parameters.

SATURN (Simulation and Assignment of Traffic in Urban Road Networks) has been developed at the Institute for Transport Studies, University of Leeds (UK) since the late 1970's. The model is based on an iterative loop between a detailed simulation model to estimate delays at junctions, and an assignment model to identify the routes taken given such junction delays. The simulation model represents traffic by so called "cyclic flow profiles". A more detailed account of the process can be found in Hall et al. (1980). SATURN uses an equilibrium-based assignment, but it is more of a mesoscopic model thanks to the more detailed modelling of junctions. It is hence essentially node-based (as opposed to TRACKS, which is primarily link-based and uses link speed-flow relationships), with the philosophy being that the relationship between flow and delay is largely determined by the interaction of queues at intersections. One very important feature is the blocking-back facility in SATURN (i.e. modelling of excessive queues). This makes it possible to take into account when a queue on a link extends back as far as the upstream junction, reducing the flow of vehicles entering the link. This is very different from TRACKS, in which all traffic is let through with the only consequence being a higher calculated delay, and is an important consideration when assessing the effects of closures in the road network.

Paramics is an acronym derived from Microscopic Simulation on Parallel Computers. This microsimulation traffic model is the result of a 6-year collaboration between Quadstone and SIAS in Edinburgh (UK), although these two firms have developed their own Paramics software independent of each other since 1998. Each individual vehicle is modelled for its entire trip through the road network. Neither the flow-delay relationships commonly used as a proxy for driver behaviour in traditional models, nor travel times or junction capacities, are defined. Speed is instead a result of the interaction between all vehicles, with their movements being governed by models for vehicle following, gap acceptance and lane changing, within the constraints formed by the physical properties or layout of the road network (i.e. lane arrangement, on-street parking, traffic signal timings, etc.). Individual driver characteristics include the level of aggression (used for determining the critical gap for lane changing) and the level of awareness, which distinguishes between drivers who are familiar or unfamiliar with the road network. The latter are essentially confined to the main roads, while the former can use minor roads as well. In addition, familiar drivers can at certain intervals be presented with feedback information on the level of congestion in the network and can re-route accordingly, giving a dynamic assignment.

STUDY DESIGN

The incident simulated was a complete closure about halfway up Creyke Road (see Figure 2), which was chosen because of its major importance to traffic both to/from and through the area, and its location in the centre of the network. The modelled time period is the 1999 morning peak hour (8-9 AM), with an average travel demand of 10 796 trips. Apart from a base scenario of normal operation, closures of 10, 20, 30, 40, 50 and 60 minutes were modelled. In Paramics, the length and time for occurrence of the desired incidents are specified and can be seen graphically during the simulation as vehicles "breaking down" to block the road. The incident was set symmetrically around the middle of the morning peak hour, i.e. the 20-minute closure was coded to start at 8:20 and end 20 minutes later. The simulation was set to run beyond 9 AM to allow for the effects of the incidents to fully disperse. Also, because of the stochastic nature of Paramics, it is necessary to do multiple runs with different seed values for randomisation of trip release so the results presented are the average of five runs.

For a 1-hour closure in SATURN, cars were simply prohibited from using that particular link during that model run, while to simulate shorter closures the method chosen was to reduce the capacity; halving the saturation flow of a dummy node in the middle of Creyke Road corresponds to an incident blocking the road for 30 minutes out of the modelled hour. This rather crude method was chosen rather than the time-slice method available in SATURN, to retain the equilibrium-based assignment that was essential for our comparison. Also, time slicing would have involved substantial model alterations and recalibration would have been needed. In TRACKS, the already available option of an incremental assignment procedure makes it possible to code the particular link as missing for certain increments only, to simulate a closure of shorter duration.

The input data relating to the travel supply (i.e. physical properties/layout of the road network) and travel demand (i.e. trip matrix) were as consistent as possible. There are differences in the travel supply input requirements for the different models, and whenever there was an overlap in the input data requirements, the same values were used. If an item of data was required by only one model, then the most appropriate value (to describe accurately e.g. the physical properties of the network) was input.

RESULTS

As was mentioned before, the Paramics simulation was set to run for up to three hours, to allow for the effects of the incidents to fully disperse. However, for the 50 minutes and 1-hour closures the network became so congested that many trips were not completed by the end of this period. This is seen in the constant increase in accumulated travel time in Figure 3 for these two scenarios, which were therefore left out in the subsequent comparisons and analyses. Also, since the traffic system is not really congested (i.e. not operating close to its capacity limit during normal conditions) even the half-hour closure in SATURN did not show any substantial effects, so SATURN results for closures shorter than 30 min are therefore omitted.

Model Comparison

The comparison is made for net effects at the system level and concentrates on the time spent and distance travelled in the road network. These represent two important areas in terms of travel costs and thereby also possible savings: operating cost savings and user-time savings. The effect of closure duration on total travel time and total travel distance is shown in Figures 4 and 5 respectively.

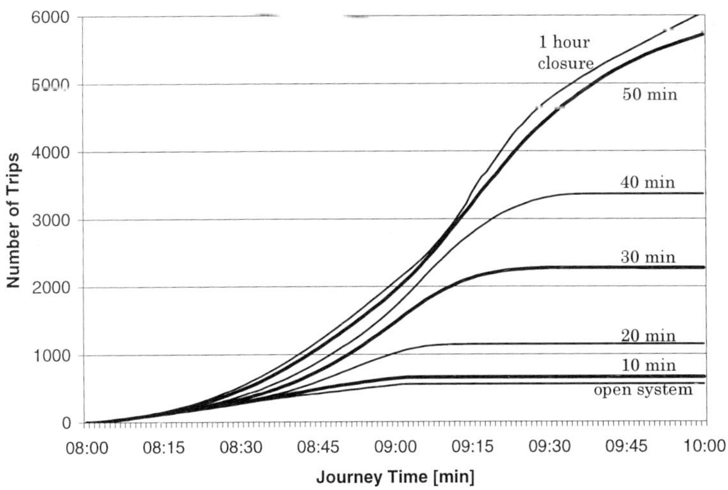

Figure 3 Total system travel time for different closures in Paramics

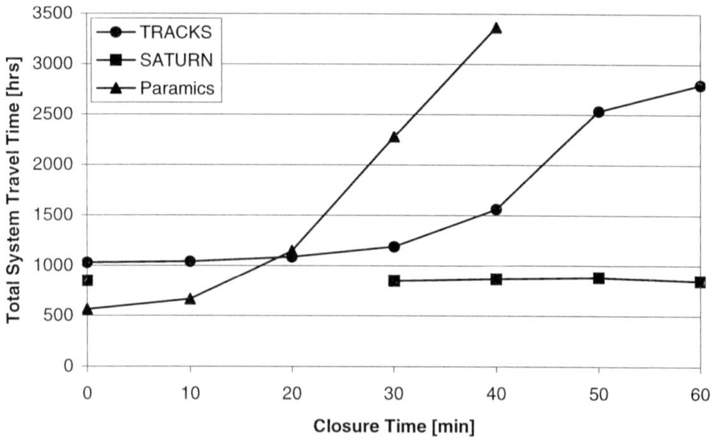

Figure 4 Model comparison of system travel time for different closure durations

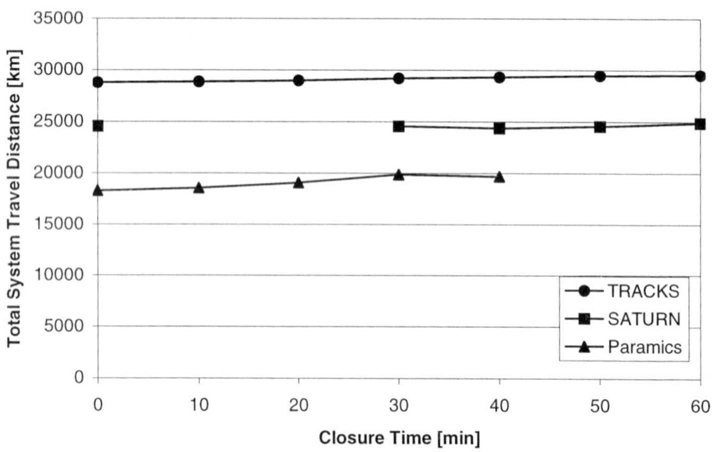

Figure 5 Model comparison of system travel distance for different closure durations

It should be noted that there are differences between the three models for the base scenario (i.e. no closure). The difference in travel distance is mainly explained by the way that trips enter/exit the network from/to the origin/destination zones. TRACKS connects zone centroids to dummy nodes located somewhere in the middle of the links. Each centroid connector has its length specified and travel on it is included in the trip length calculations. In SATURN, centroid connectors are connected to links and trips into/out of the zone are assumed to enter/exit at the intersections. Neither travel on the links in question, nor travel on the centroid connectors, is recorded in the calculations. Paramics simply loads trips from a zone onto the links in that zone in proportion to their length, the vehicles appearing at the upstream end of the link in question when there is a suitable gap in the traffic flow (if not, vehicles form a "virtual queue" in the zone until there is a suitable gap). The above also explains part of the difference in total travel time, but the main reason for this is most likely that the three models use different methods for assessing intersection delays. The models have been calibrated to give a good match between observed and predicted flows but it is not known precisely how well they predict travel time. Part of the differences may therefore also be explained by initial calibration discrepancies, but this has not been investigated further as model calibration was not part of this study.

From Figure 4 it is clear that Paramics is more sensitive to disturbances. Already a closure of 10 minutes increases system travel time by 18%, compared to a 1% increase in TRACKS. Closures shorter than half an hour in SATURN do not (as mentioned above) result in any noticeable difference. The microsimulation model is also more sensitive to incident duration, as shown by a higher rate of increase for longer closures. An interesting result is that the possibility of coding the link closure for only certain increments in the TRACKS assignment apparently makes this model more sensitive in terms of changes in travel time than SATURN, which is otherwise more detailed in its formulation. As for the total distance travelled the intuitive notion is that trip lengths should generally increase compared to the base scenario, due to detours caused by the closures. This is true for TRACKS and Paramics, although the effect isn't very big, and Paramics is again the most sensitive. One might expect a longer closure to generally lead to longer detours, as in the TRACKS results, but this seems to be true only to a certain point. A longer closure may well cause such an extensive spread of queues that further re-routing is useless and the only option is to wait. This is especially true for small networks, such as the one in this study, in which there is only a limited number of alternative routes. Also, route choice was based upon minimising travel time, which means that the chosen route is not necessarily the shortest. As traffic conditions change in the network, a previously shorter but slower route may well become the fastest alternative.

Paramics is clearly a more sensitive model when simulating short-term road interruptions. It probably presents a more realistic picture of the resulting consequences, as travellers in the microsimulation are faced with the closures and the subsequent changes in congestion conditions en-route and can change their route choice accordingly. In the equilibrium-based models, the status of the road network is assumed to be known beforehand and route choice does not change en-route, which clearly fails to correspond with reality. All three models have a weakness when considering real traffic behaviour, in that they do not allow vehicles to simply make a U-turn to avoid a queue (something that quite often occurs in a real life situation). Paramics can allow U-turns at discrete locations, where dummy nodes allowing U-turns have been coded. This was deemed impractical since such dummy nodes would have to be put in at a large number of discrete locations, to allow for dynamic changes in queue lengths. Paramics may therefore over-estimate the effects of incidents to some extent.

Short-term Incidents in Paramics

In Paramics, the average travel time during normal operation is around 3 min/trip, increasing drastically with increasing closure time, and for a 40-minute incident it is over 18 min/trip (Figure 6). The effect on average trip distance on the other hand is not as great. From 1.68 km/trip during normal conditions, it reaches a maximum of 1.83 km/trip for a half-hour closure. However, this is reasonable considering the size of the network, as was mentioned above. Figure 7 shows the average travel time distribution for trips with respect to their start time during the simulation and it is clear that trips released before and after the incidents aren't affected as much as trips released during the closure itself. It is interesting to notice the resemblance to the hypothetical illustration in Figure 1. The graph does in fact depict that problematic adjustment phase previously discussed, to some extent.

Another way of illustrating the effect on overall travel conditions in the network from various closures is the total travel time distribution (Figure 8). For shorter closures a majority of the trips are completed within some 10 minutes but this proportion quickly becomes smaller as the closure time increases. For the 40-minute closure half of the trips take longer than this, and as much as 30% have travel times exceeding half an hour. A very rough estimate of the economic impact for the system as a whole, based on user travel time and vehicle operating costs (i.e. distance) only, is presented in Figure 9. The values used are 27 cents/min and 21 cents/veh-km, in accordance with the New Zealand standard procedure for economic appraisal of urban travel (Transfund New Zealand, 1999). Expressed at the level of the individual, the average cost for travel during the morning rush hour is $1.35 and $5.40 per trip during an incident closing Creyke Road for 10 and 40 minutes respectively, compared to $1.20 during normal conditions.

It was noted above that the two longest closures were not successfully modelled in Paramics. The main reason was the occurrence of "routing errors" during the simulation, due to vehicles waiting for more than an arbitrary time to make a turn, which was then deemed impossible, leaving the driver without a route to the destination. It is also likely that the small size of the network contributes to the severe congestion and routing problems. For longer closures, it is likely that drivers would in reality use more of the surrounding street network to avoid congestion. Since these more peripheral routes are not included in the current simulation network, the system simply comes to a stand still and never clears.

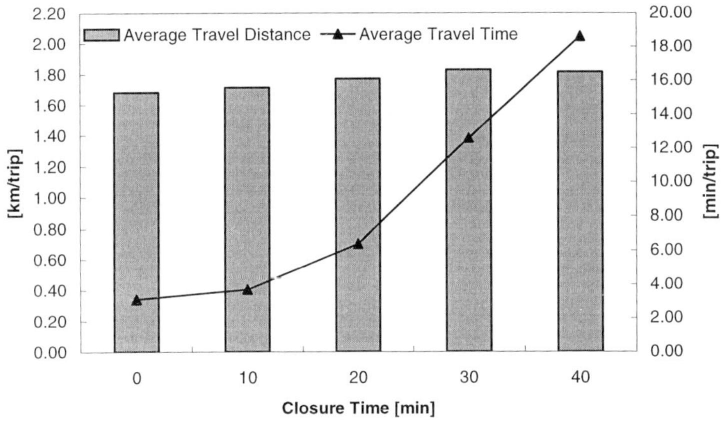

Figure 6 Average travel times and distances for different closures in Paramics

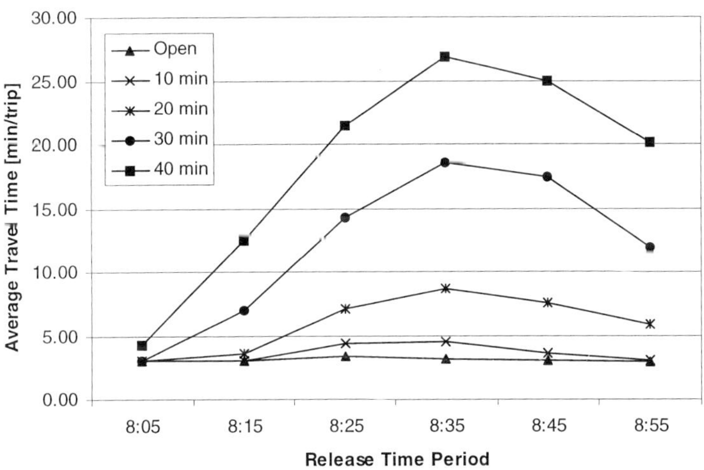

Figure 7 Average travel times for different closures in Paramics

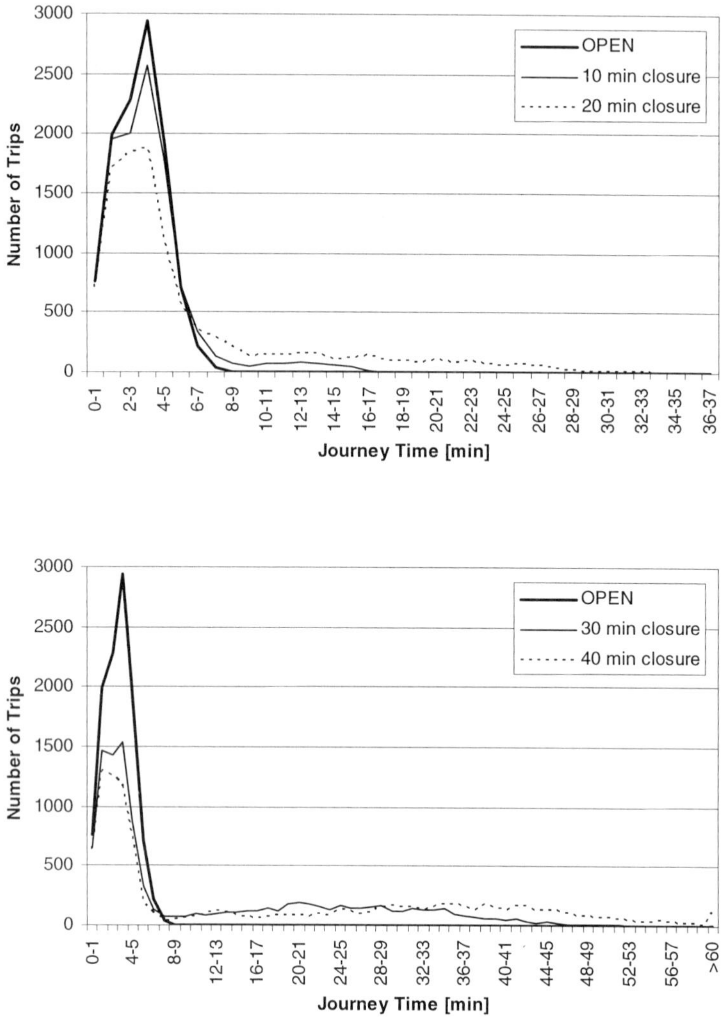

Figure 8 Travel time frequencies for different closures in Paramics

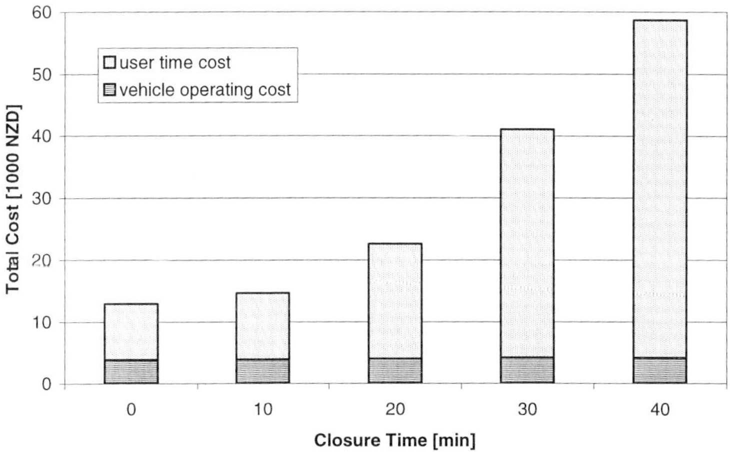

Figure 9 Cost estimation of different closures in Paramics

Rather than going through the cumbersome process of increasing the size of the coded road network for the Paramics model, maybe the problem mentioned above can be ameliorated by changing the frequency of information supply and the weight attached to new information. The travel cost information given to familiar drivers as a basis for their route choice in the next period is a combination of the current cost (V_{now}) and the value in the previous period (V_{old}). That is $V_{new} = a \cdot V_{now} + (1\text{-}a) \cdot V_{old}$. The feedback factor 'a' determines the amount of smoothing, and it determines how much account the drivers take of prior knowledge and new information in making their next decision. The feedback interval is then the frequency with which new information is provided. When the feedback interval for familiar drivers was reduced from the default value of 5 minutes to 3 minutes, the system actually cleared within the simulation run time of three hours for the 50-minute closure for one seed value. It could be that, if left for longer than three hours, the other four runs would also have finished successfully. A quick sensitivity analysis shows negligible differences in results for varying feedback intervals in the base scenario. For a closure of half an hour, total system travel time (compared to that for a 5-minute feedback time) is roughly 8% and 13% lower for intervals of 3 and 1 minute(s) respectively. A similar sensitivity analysis was performed varying the feedback factor, and again no significant difference could be seen for the base scenario. For the 30-minute closure, increasing or decreasing 'a' from the default value of 0.5 resulted in reductions in total system travel time of 7% (a = 0.75) and 5% (a = 0.25). Judging from the small effects in the base scenario, it may be that changes to these parameters do not influence initial calibration to any appreciable extent, but further studies should be carried out to verify this.

CONCLUSIONS

The results show that Paramics is not only more sensitive to a disturbance as such, but is also more sensitive to the disturbance duration, than are SATURN and TRACKS as used in this study. This supports the hypothesis that the use of equilibrium-based models for studying short-term incidents underestimates the effects of the disturbances. However, it should be noted that there is a possibility of Paramics over-estimating the effects because of its present inability to allow vehicles to simply make a U-turn to get out of a queue. Thanks to the incremental assignment procedure in TRACKS, this model seems to be more responsive to incidents than is SATURN as used here. However, it should be noted that we have not made use of the time-slice feature available in SATURN, a choice that to some extent puts this model in an unfavourable light. It was, however, the equilibrium property of the existing SATURN model that we wanted to focus upon. Hence, the focus of the study has been assessing different modelling approaches, rather than comparing particular models. Problems and possibilities involved in traffic model comparisons are discussed further in Nicholson et al. (2001).

The different types of models are usually used for different sized networks but here all three were used for a small network. As network size increases the scope for traffic to find alternative routes vill generally increase, unless the system as a whole is severely congested. With growing network size, however, it becomes less practical to use a microscopic model (and, to a lesser extent, a mesoscopic model) due to the greater level of detail in network coding and the computer capacity needed for running the program. We need to identify the separate effects of network size and modelling assumptions, of which we have looked only at the latter so far. One hypothesis that could and should be tested is that the effect of different modelling assumptions (e.g. using a static equilibrium assignment method rather than a dynamic method as in Paramics) will perhaps diminish as the network size increases, especially for large duration closures, but will remain important.

During this study we identified a number of principal features desirable in a program for modelling short-term disruptions in road networks. The ability to vary network characteristics during the model period is of course essential, as is the possibility of dynamic adjustment to varying conditions, especially re-routing of vehicles en-route. Detailed interaction of vehicles on the individual level, together with randomness in their release onto the network, is also important when trying to match the real life situation. It seems that Paramics possesses more of these features than the other models used, and microsimulation thus seems the most suited to studying the effect of short-term incidents in the road transportation system. Future work should also include a closer look into what, if any, are the advantages of adjusting the feedback interval and/or the feedback factor.

On a more general note, none of the models used today seem to be designed with reliability studies of this kind in mind. The current development in urban areas all over the world, however, makes it necessary to get the most out of our existing road systems through better traffic management etc. It is also important to remember that although a model may be well calibrated for normal conditions, there is no guarantee that it will also predict abnormal conditions correctly. It would be good to have empirical data about actual traffic behaviour during a closure, to enable calibration of the model for closure conditions. It would also allow a comparison of the effects predicted by each model with the observed effects, rather than just comparing the results from one model with those from another model. Further development of modelling tools is clearly needed to increase our knowledge about the status of the road network, the expected effects when incidents do happen, and possible mitigation measures.

ACKNOWLEDGEMENT

Our deepest gratitude goes to Axel Wilke and Jonathan Harrington, for helping out with running the Paramics program. Special thanks also go to James Laird for his advice on how Paramics works. Last but not least, thanks to the Erik Philip Foundation at the Royal Institute of Technology for their financial support.

REFERENCES

Akçelik, R. (1986). *Traffic Signals: Capacity and Timing Analysis.* Research Report 123, ARRB Transport Research Ltd, Victoria, Australia.

Andjic, Z. (2000). *University of Canterbury Traffic Study: Development of SATURN and TRACKS Models.* Master of Engineering Thesis, Dept. of Civil Engineering, University of Canterbury, Christchurch, New Zealand.

Berdica, K. (2000). *Analysing Vulnerability in the Road Transportation System – Putting Theory into Practice Using Sweden as an Example.* TRITA-IP FR 00-76, Dept. of Infrastructure and Planning, Royal Institute of Technology, Stockholm, Sweden.

Berdica, K. (2002). An introduction to vulnerability - what has been done, is done and should be done. *Transport Policy*, **vol. 9**, no. 2, pp 117-127.

Cameron, J. L. (1996). *A Traffic Model for the University of Canterbury and its Environs.* Master of Engineering Thesis, Dept. of Civil Engineering, University of Canterbury, Christchurch, New Zealand.

Fisk, C. S. and Tan, H. H. (1989). *Delay Analysis for Priority Intersections.* Dept. of Civil Engineering, University of Auckland, New Zealand.

Gabites Porter (NZ) Ltd (1991). *Performance Analysis of Priority Intersections - Practitioner's Guide.* Transit New Zealand, Wellington, New Zealand.

Hall, M. D., Van Vliet, D. and Willumsen, L. G. (1980). SATURN – A Simulation-Assignment Model for the Evaluation of Traffic Management Schemes. *Traffic Engineering & Control* , **vol 21**, n 4, 168-176.

Nicholson, A. J., Berdica, K. and Andjic, Z. (2001). Comparing Traffic Models: Two Case Studies. *Transport Engineering in Australia*, **vol 7**, n 1/2, 65-76.

Laird, J. J. and Nicholson, A. J. (2000). *Application of Microsimulation Traffic Modelling to a New Zealand Road Network.* Research Report 2000-10, Dept. of Civil Engineering, University of Canterbury, Christchurch, New Zealand.

Ortuzar, J. D. and Willumsen, L. G. (1990). *Modelling Transport.* John Wiley & Sons, University College, London, Great Britain.

Transfund New Zealand (1999). *Project Evaluation Manual.* Transfund New Zealand, Wellington, New Zealand.

22

TRAFFIC FLOW ANALYSIS ON INTER-CITY EXPRESSWAY FOR EMERGENCY SITUATION

Keiichi Ogawa, Department of Environment Systems Engineering, Ritsumeikan University, Shiga, Japan
Takamasa Akiyama, Department of Civil Engineering, Gifu University, Gifu, Japan

ABSTRACT

Traffic accident on inter-city expressways might cause large-scale traffic congestion. It might increase travel times of many drivers and it produces a large social loss. This study aims to estimate the social loss of travel time of drivers caused by traffic accidents on inter-city expressway using traffic simulation model, and to evaluate the effects of outflow recommendations when an accident occurs on the expressway. The traffic simulation model on Tomei Expressway is constructed to estimate the dynamic traffic congestion. Travel time losses of drivers are estimated by the simulation results with hypothetical traffic accidents. It is understood that the total losses of travel times are depending on the positions of accident spots and the occurrence times of accidents, because it might influence to congestion at the bottlenecks of flow capacity. Moreover, the effect of traffic control in emergency situation is discussed. The influences of outflow recommendations for drivers are estimated using the traffic simulation model.

INTRODUCTION

Traffic accident on inter-city expressways might cause large-scale traffic congestion. It might increase travel times of many drivers and it produces a large social loss. This study aims to estimate the social loss of travel time of drivers caused by traffic accidents on inter-city expressway using traffic simulation model, and to evaluate the effects of outflow recommendations when an accident occurs on the expressway. Traffic simulation model is based on the assumptions that traffic flow on the expressway can be approximated as a compressive fluid flow. It aims to evaluate the social losses of traffic accidents based on the travel times of drivers and to evaluate the effects of traffic control in emergency situation. Then the effect of traffic control in emergency situation is discussed based on the differences of the losses of travel time of drivers using the constructed traffic simulation model.

BASIC STRUCTURE OF THE SIMULATION MODEL

Basic Assumptions in the Modelling

Basic estimation of dynamic traffic congestion is based on the macroscopic traffic simulation model. The basic structure of the model was originally developed for dynamic traffic flow estimation on urban expressways (e.g. Inoue *et al.*, 1988; Tsuchida *et al.*, 1999; Ogawa and Akiyama, 2001). Generally, traffic flow is assumed as compressive fluid flow in the model. The basic assumptions of the model structure are as follows.

(1) Objective district is divided into several sections. Traffic condition in each section is represented by traffic density.

(2) The density-speed relationships are assumed for all sections. The running speed of the vehicles in each section is estimated by using these relationships.

(3) The number of vehicles in a section at time $t+dt$ is calculated with using the number of vehicles in the section at time t, the traffic flow from upstream section between time t and $t+dt$, and the traffic flow to downstream section between time t and $t+dt$.

(4) The traffic flow in the simulation must satisfy the continuity condition.

Therefore, it is necessary to formulate the sequence of traffic flow between two sections in each time interval dt. Traffic flow sequence in the model is illustrated in Figure 1. It is formulated as follows. Notations in the following description are shown in Table 1.

Standard Sequence between Two Sections

The standard sequence between two sections is formulated as follows. Traffic flow from upstream section i_0 to downstream section i_1 between time t and $t+dt$ is represented as a minimum value of desired outflow of the section i_0 and acceptable inflow of the section i_1.

$$Q(i_0,t) = \min\left(QD(i_0,t), QM(i_1,t)\right) \tag{1}$$

where $Q(i_0, t)$ is traffic flow from section i_0 to section i_1 between time t and $t+dt$, $QD(i_0, t)$ is desired outflow of the section i_0, and $QM(i_1, t)$ is acceptable inflow of the section i_1.

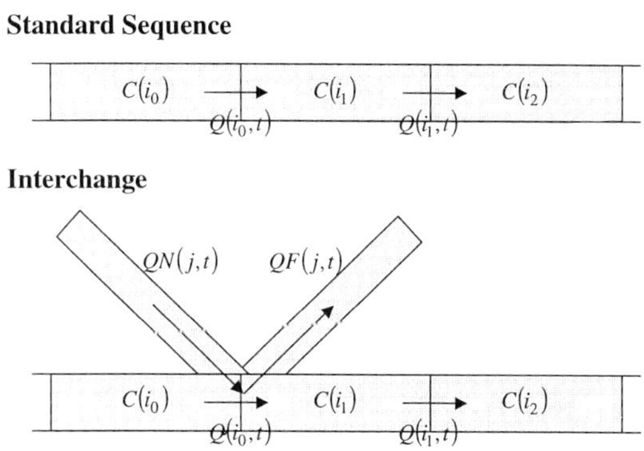

Figure 1 Traffic Flow Sequence in the Model

Desired outflow $QD(i_0, t)$ is defined as a maximum number of vehicles those are able to flowing out from section i_0 between time t and $t+dt$. It depends on the traffic situation on the section i_0 at the time t. Concretely, it is determined by the current traffic flow of section i_0 if traffic flow on section i_0 belongs to free flow and it is determined by the flowing capacity of section i_0 if traffic flow on section i_0 belongs to congested flow. Flowing capacity is defined to be a maximum flow of section i_0.

On the other hands, acceptable inflow $QM(i_1, t)$ is defined as a maximum number of vehicles those are able to flowing into the section i_1 between time t and $t+dt$. It depends on the traffic situation on the section i_1 at the time t. Concretely, it is given by the flowing capacity of section i_1 if traffic flow on section i_1 is free flow and it is given by the current traffic flow of section i_1 if traffic flow on section i_1 is congested flow. Flowing capacity is defined to be a maximum flow of section i_1.

These definitions are represented as follows.

$$QD(i_0,t)=\begin{cases} K(i_0,t)\cdot V(i_0,t)\cdot N(i_0)\cdot dt, & \text{if } K(i_0,t)\le KC(i_0) \\ C(i_0)\cdot dt & \text{if } K(i_0,t)\ge KC(i_0) \end{cases} \tag{2}$$

$$QM(i_1,t)=\begin{cases} C(i_1)\cdot dt & \text{if } K(i_1,t)\le KC(i_1) \\ K(i_1,t)\cdot V(i_1,t)\cdot N(i_1)\cdot dt, & \text{if } K(i_1,t)\ge KC(i_1) \end{cases} \tag{3}$$

where $K(i, t)$ and $V(i, t)$ are traffic density and average speed on section i at the time t. $C(i)$, $KC(i)$ and $N(i)$ are flowing capacity, critical density and number of lanes of the section i respectively.

Based on the above estimation, number of existing vehicles $IM(i_1, t+dt)$ and traffic density $K(i_1, t+dt)$ on section i_1 at time $t+dt$ are calculated with as follows.

Table 1 Notation in the Simulation Model

Standard sequence	
$K(i,t)$	Density on section i at time t
$V(i,t)$	Average speed on section i at time t
$IM(i,t)$	Number of vehicles on section i at time t
$QD(i,t)$	Desired outflow from section i to the downstream section at time t
$QM(i,t)$	Acceptable inflow of section i at time t
$Q(i,t)$	Traffic flow from section i to the downstream section at time t
$KC(i)$	Critical density of section i
$N(i)$	Number of lanes of section i
$L(i)$	Length of section i
$C(i)$	Flowing capacity of section i
Sections with interchange (on-ramp)	
$QDN(j,t)$	Desired outflow from on-ramp j to the mainline at time t
$QN(j,t)$	Traffic flow from on-ramp j to the mainline at time t
$QB(j,t)$	Traffic flow on tollbooths of on-ramp j at time t
$D(j,t)$	Traffic demand of on-ramp j at time t
$W(j,t)$	Number of waiting vehicles on tollbooths of on-ramp j at time t
$X(j,t)$	Number of waiting vehicles between on-ramp j and the mainline at time t
$CN(j)$	Capacity of tollbooths of on-ramp j
$T(j)$	Service time of tollbooths of on-ramp j
$NB(j)$	Number of tollbooths of on-ramp j
$CX(j)$	Capacity of the section between on-ramp j and the mainline
$Y(j)$	Maximum number of waiting vehicles between on-ramp j and the mainline
Sections with interchange (off-ramp)	
$QDF(k,t)$	Desired outflow from the mainline to off-ramp k at time t
$QF(k,t)$	Traffic flow from the mainline to off-ramp k
$P(k)$	Diversion ratio of off-ramp k
$W(k,t)$	Number of waiting vehicles on tollbooths of off-ramp k at time t
$CN(k)$	Capacity of tollbooths of off-ramp k
$T(k)$	Service time of tollbooths of off-ramp k
$NB(k)$	Number of tollbooths of off-ramp k

$$IM(i_1, t+dt) = IM(i_1, t) + Q(i_0, t) - Q(i_1, t) \tag{4}$$

$$K(i_1, t+dt) = \frac{IM(i_1, t+dt)}{L(i_1) \cdot N(i_1)} \tag{5}$$

where $L(i_1)$ and $N(i_1)$ are length and flowing capacity of section i_1 respectively.

The Sections Connected with Interchanges

The interchanges are supposed to be located at the boundary of sections. An interchange consists of two ramp-ways, on-ramp *j* and off-ramp *k*, and they are supposed to be connected with mainline at the boundary of sections. If an interchange is located between section i_0 and section i_1, traffic flow sequence at the boundary of sections in the model is illustrated in Figure 1. The flow sequence is formulated as follows.

Acceptable inflow $QM(i_1, t)$ of section i_1 is split into the acceptable inflow for the section i_0 and that for the on-ramp *j*. For this modification, the merging ratio from each section must be defined. In this model, it is determined as a ratio of acceptable inflows of the section i_0 and the on-ramp *j*; $QM(i_0, t)$ and $CX(j)$. Traffic flow from each section is determined as a desired outflow from the section if total desired outflow from both sections is less than the acceptable inflow of section i_1. On the contrary, traffic flow from each section is determined by the minimum value of desired outflow and acceptable inflow respectively if total desired outflow is more than the total acceptable inflow.

Moreover, acceptable inflow for the section i_0 must be increased to include the difference of outflow and inflow for on-ramp *j* if desired outflow from section i_0 is more than the acceptable inflow and that from on-ramp *j* is less than the acceptable inflow. Contrarily, acceptable inflow for on-ramp *j* is increased if desired outflow from on-ramp *j* is more than the acceptable inflow and that from section i_0 is less than the acceptable inflow.

Similarly, desired outflow $QD(i_0, t)$ from section i_0 is split into the outflow to the section i_1 and that to the off-ramp *k*. The diversion ratio $\Gamma(k)$ to each section must be defined. Traffic flow to section i_1 and that to off-ramp *k* are determined by the minimum value of desired outflow and acceptable inflow for each section respectively.

Moreover, flow capacities of tollbooths; $CN(j)$ and $CN(k)$; are determined for all ramp-ways. They are based on the average service times; $T(j)$ and $T(k)$; and the numbers of tollbooths; $CN(j)$ and $CN(k)$; on the ramp-ways.

MODELLING OF THE OBJECTIVE ROUTE

The Objective District

The Tomei Expressway is one of the inter-city expressways connecting Tokyo and Nagoya that is having the heaviest traffic demands in Japan. The objective district is an up-bound route of the district between Kikugawa I.C. and Shizuoka I.C. The length of district is 40.0 kilometres.

Two lanes are operated for one direction in the objective district. The outline of the objective district is shown in Figure 2.

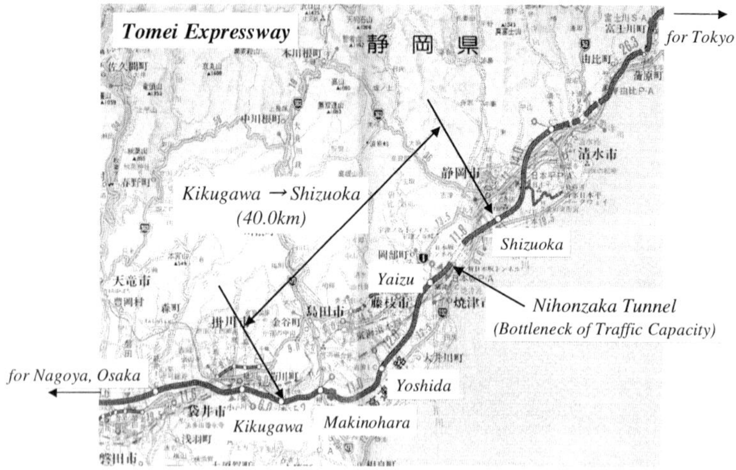

Figure 2 The Objective District in Tomei Expressway

The objective district includes Nihonzaka Tunnel that is a bottleneck of flow capacity on Tomei Expressway. Generally, traffic congestion might occur frequently at the entrance of tunnel on the expressways, because many drivers reduce the speed caused by narrow and dark field of vision, and frictions of narrow margin of roadways. Nihonzaka Tunnel is one of the capacity bottlenecks on Tomei Expressway and heavy traffic congestion might occur frequently during peak seasons in every years.

The objective district is divided into 40 sections in the model. The length of each section is 1.0 kilometre. Time interval *dt* of flow calculation is 20 seconds. They are determined to satisfy the continuity condition of traffic flow. Three interchanges; Sagara-Makinohara I.C., Yoshida I.C. and Yaizu I.C.; are included in the objective district. The ramp-ways of each interchange are supposed to be connected with the mainline at the boundary of two sections.

Parameter Calibration of the Model

The vehicle detectors are installed at the intervals of 2 kilometres on the mainline of Tomei Expressway and they measure traffic volumes, time occupancies, and average speeds every 5 minutes. The density-speed relationship of each section is determined based on the detected data. Figure 3 shows the density-speed relationship of the detected data on 177.38 [kp], which is one of the sections located at the upstream of the capacity bottleneck. It is having an ordinary geometric condition.

The density-speed relationship can be assumed as a linear function as shown in Figure 3. The parameters of the model, free flow speeds and maximum densities, are calibrated to make accurately the estimation of traffic congestion at the bottleneck. The result of the parameter calibration of density-speed relationship is shown in Table 2.

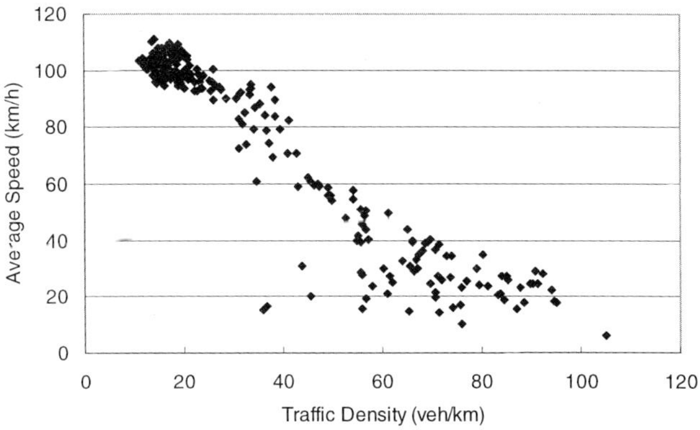

Figure 3 Density-Speed Relationship (177.38 [kp])

Table 2 Parameters of Density-Speed Relationships

	Free flow speed [km/h]	Maximum density [veh/km]
203 - 198 [kp]	110	81
198 - 195 [kp]	95	77
195 - 179 [kp]	110	81
179 - 178 [kp]	110	77
178 - 171 [kp]	110	81
171 - 170 [kp]	60	72
170 - 169 [kp]	95	77
169 - 161 [kp]	110	81

The sections located between 171 [kp] and 169 [kp] are corresponding to the entrance of Nihonzaka Tunnel, that is a bottleneck of flow capacity. The sections located between 198 [kp] and 195 [kp] are corresponding to the mountainous district near the Sagara-Makinohara I.C. Table 2 shows that both of free flow speeds and maximum densities at the bottlenecks are less than those of other ordinary sections. Therefore, flow capacity at the entrance of Nihonzaka Tunnel is determined to be almost half of ordinary sections. It becomes to be able to estimate traffic congestion at the entrance of Nihonzaka Tunnel and Sagara-Makinohara I.C.; those are bottlenecks of flow capacity.

Estimation Performance of the Model

The estimation performance of the model is examined by vehicle detectors data on August 12 and 16, 1994. There are heavy traffic demands on Tomei Expressway on the surveying dates, and heavy traffic congestion had occurred at the entrance of Nihonzaka Tunnel. In this season the Buddhist Festival, called "Obon", is held in Japan and many people living in metropolitan area visit their hometown to attend the Buddhist Festival. Therefore, there are heavy traffic congestions on inter-city expressways in every year.

Estimation performance of the model is shown in Figure 4. The sections that the average speed is less than 40 [km/h] are determined as congested sections. It is corresponding with the definition of traffic congestion using for information provisions to the drivers on inter-city expressways. In this figure, the sign "*" indicates the corresponding sections of estimated traffic congestion with observed traffic congestion. The sign "+" indicates the sections in estimated congestion. The sign "-" indicates the sections in observed congestion.

The hit ratio of the traffic congestion is 0.970 for all sections during the objective periods; from 8:00 a.m. to 2:00 p.m. It is known that the traffic congestion caused by the capacity bottleneck at the entrance of Nihonzaka Tunnel can be estimated enough accurately in this simulation model.

Traffic Flow Estimation in Emergency Situation

Hypothetical Traffic Accident in the Model

Traffic accident on inter-city expressway often causes heavy traffic congestion on the mainline. It makes increasing travel times of many drivers and it produces a large external diseconomy for the drivers. The flow capacity at the accident spot might be decreased because traffic flow is obstructed by lane closing or narrow margin of roadway during the accident processing time. The dynamic traffic congestion can be estimated by using the traffic simulation model if the traffic condition at the accident spot is determined appropriately in the model.

In this study, a hypothetical traffic accident is determined in the model. It is assumed that one lane is closed at the accident spot and free flow speed is reduced to be 70 [km/h] during the processing time. The density-speed relationship is changed as shown in Figure 5. It means the significant reduction of traffic capacity at the accident spot in the model. This assumption is based on the speed regulations on the district of the inter-city expressway which is having only one lane for one direction. However, of course traffic condition at the accident spot are different by the situation of traffic accidents. Therefore, it is necessary to determine the traffic condition at the accident spot more particularly for practical evaluation of traffic accidents on the expressway in detail.

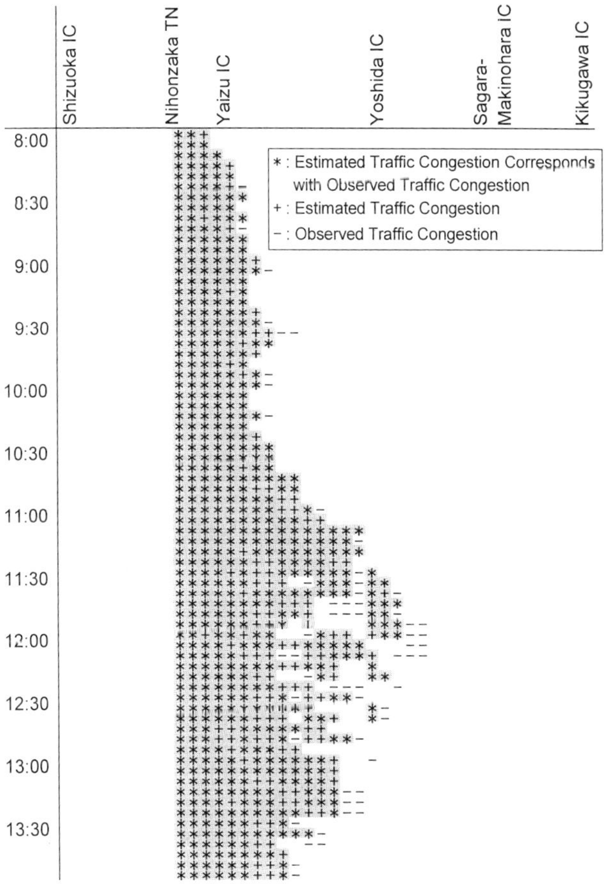

Figure 4 Estimation Performance of the Simulation Model

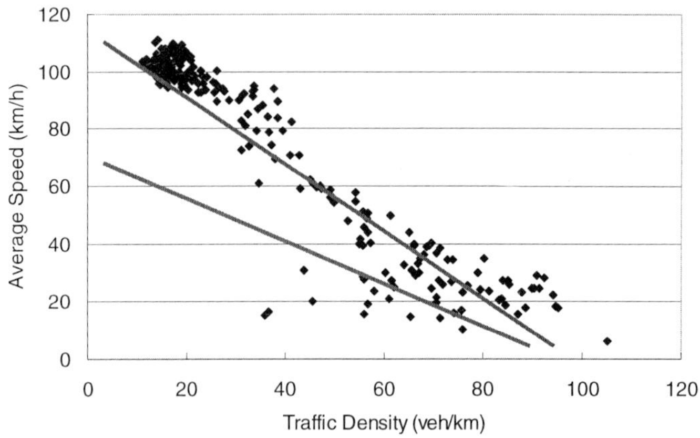

Figure 5 Density-Speed Relationships of the Accident Spot

Result of Traffic Flow Estimation

Figure 6 shows the traffic congestion estimated by the simulation in case that a traffic accident occurs on the mainline. It is supposed that the accident spot is 187 [kp] and that the accident occurs at 9:00 a.m. One lane is closed at the accident spot and free flow speed might be reduced to be 70 [km/h] during the processing time of accident. The length of processing time is supposed to be 60 minutes. Lane closing would be over at 10:00 a.m. In this figure, the sign "#" indicates the sections in lane closing at the accident spot.

It is understood that the traffic accident may cause heavy traffic congestion at the accident spot during the processing time. The congestion is extended to Kikugawa I.C. in this case. Moreover, comparing with Figure 4, congestion has occurred at Sagara-Makinohara I.C. after the processing time is over. It is because that the flow capacity of the section at Sagara-Makinohara I.C. is less than that of the other sections. The vehicles that have been in the congested section during the processing time cannot pass smoothly even after the processing time has been over.

Furthermore, it impacts to the traffic congestion at the entrance of Nihonzaka Tunnel in downstream sections. The length of traffic congestion decreases at the Nihonzaka Tunnel during the processing time. It is because the number of flowing vehicles is reduced caused by the flow obstruction at the accident spot. However, the traffic congestion increases after the processing time is over. It is because the number of vehicles flowing from the accident spot upstream is increased after the processing time is over. Therefore, it must be considered the influence to other congestions to evaluate the impacts of the traffic accidents.

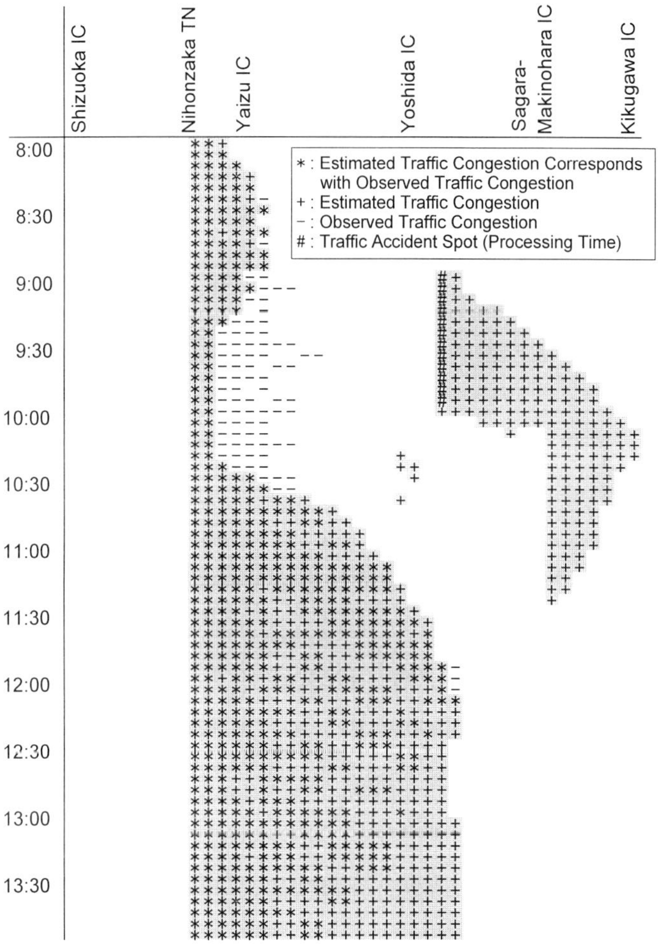

Figure 6 Estimated Traffic Congestion Caused by the Traffic Accident

Travel Time Loss of Drivers

Travel time loss of drivers flowing between Kikugawa I.C. and Shizuoka I.C. also can be estimated by the simulation model. Figure 7 shows the comparison of the estimated travel times of drivers in the accidental case and that in the non-accidental case. The traffic accident is supposed to be same as the case shown in Figure 6. The total loss of travel time of all drivers is defined as a difference between the total travel time of all drivers for the accidental case and that for the non-accidental case.

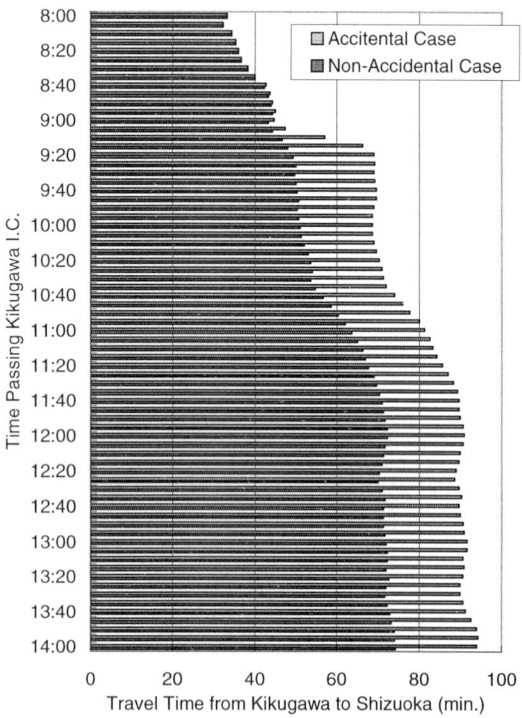

Figure 7 Comparison of the Estimated Travel Times of Drivers

Comparison of the Impacts of Traffic Accidents

Figure 8 and Figure 9 show the estimated total volumes of traffic congestion and the estimated total travel time losses in several cases respectively. It is known that they are depending on the position of accident spot and the occurrence time of accident. For examples, the traffic accidents that occur at the sections downstream of the bottleneck (195 [kp] and 187 [kp]) make traffic congestions and travel time losses larger compared with the accidents at the sections upstream of the bottleneck (172 [kp] and 165 [kp]). The traffic accidents that occur before the peak of traffic demand (e.g. 8:00 a.m.) make traffic congestions and travel time losses larger compared with the accidents after the congestion has started (e.g. 11:00 a.m.).

These differences of volumes of traffic congestion and travel time losses are caused by the indirect influences of the traffic accident to congestions. It is because the traffic accidents might influence to congestions at the capacity bottlenecks. It must be considered the influence to other congestions to evaluate the impacts of the traffic accidents.

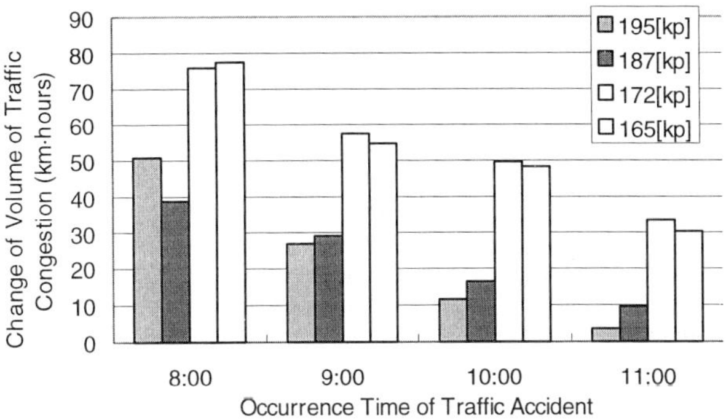

Figure 8 Change of Volumes of Traffic Congestion Caused by Traffic Accident

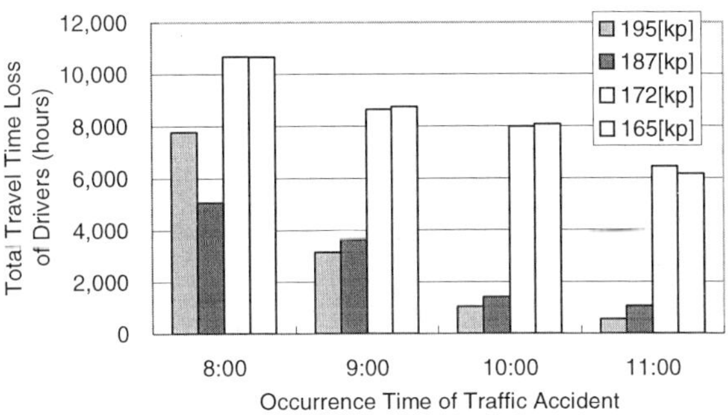

Figure 9 Total Travel Time Losses of Drivers Caused by Traffic Accident

TRAFFIC CONTROL IN EMERGENCY SITUATION

Outflow Recommendation for Drivers

It is important to control traffic flow to keep traffic condition more smoothly and safely in emergency situation. It is considered that the outflow recommendation for drivers at the upstream interchanges using traffic information devices is one of the feasible methods to reduce the traffic congestion at the accident spot.

Of course it is important to find the practical and feasible methods to recommend the drivers to change their route to other roads, such as information providing for the drivers. Today due to the progress of information and communication technologies, many types of traffic information systems becomes in practical use. It is considered that such information devices; as on-road information boards, in-vehicle navigation systems and so on; can be used for the recommendation for the drivers. Of course it is necessary to analysing and modelling the route choice behaviours of drivers and to estimate behaviours of the drivers under the outflow recommendation. However, in this study, behaviour of drivers is supposed to be given exogenously.

Figure 10 shows the estimated traffic congestion on the mainline if outflow recommendation is curried out at Sagara-Makinohara I.C., which is located at the upstream section of the hypothetical accident spot. The traffic accident is supposed to be same as the case shown in Figure 6. In this case, it is supposed that 20 % of vehicles passing through Sagara-Makinohara I.C. between 9:15 a.m. and 10:00 a.m. should change their route to the surface road. The time delay for the action of traffic control is supposed to be 15 minutes. It is considered as a time to identify the traffic accidents on the mainline. Compared with Figure 6, it is known that the traffic congestion at the accident spot is significantly reduced.

Effects of Outflow Recommendation

Figure 11 shows the relationship between outflow rates of vehicles at Sagara-Makinohara I.C. and estimated total travel time loss of drivers. The traffic accident is supposed to be same as the case shown in Figure 6. The incident identification time is fixed to be 15 minutes in all cases. It is understood that of course traffic congestion on the mainline might be reduced significantly if outflow rate of vehicles is increased.

Figure 12 shows the relationship between incident identification times and estimated total travel time loss of drivers. The traffic accident is supposed to be same as the case shown in Figure 6. The outflow rate of vehicles is fixed to be 20 % in all cases. It is understood that traffic congestion might be reduced if the incident identification time is reduced. Therefore, it is important to identify the traffic incident and inform it for the drivers as soon as possible for the traffic control in emergency situation.

Concluding Remarks

In this paper, the macroscopic traffic simulation model on inter-city expressway was constructed and traffic flow on inter-city expressway in emergency situation was analysed. Travel times of drivers flowing between Kikugawa I.C. and Shizuoka I.C. on Tomei Expressway were estimated by the simulation model. It is understood that the total travel time losses are depending on the positions of accident spot and the occurrence times of accident. The differences of travel time losses are caused by the indirect influences of the traffic accident to other congested sections. Moreover, it is understood that the outflow recommendation for drivers at the upstream interchanges is effective to reduce the traffic congestion at the accident spot in emergency situation.

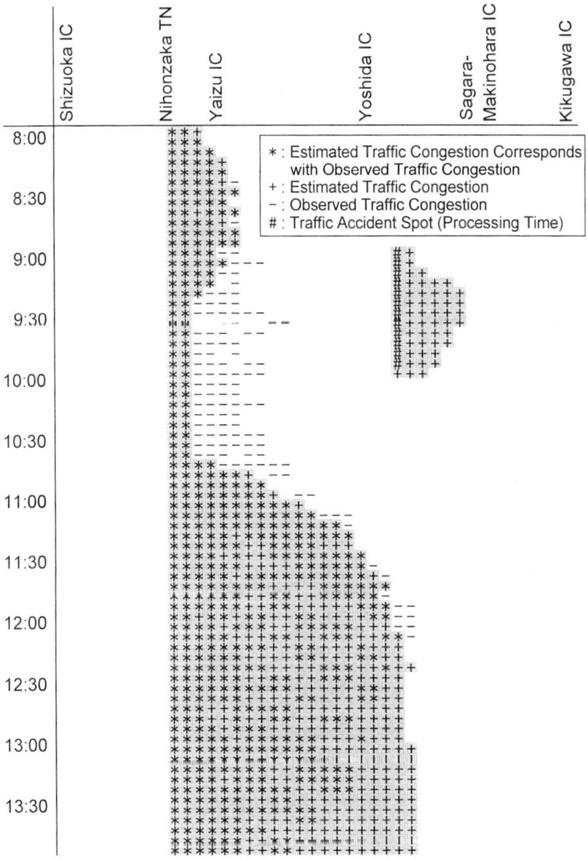

Figure 10 Influences of the Outflow Recommendation for the Drivers

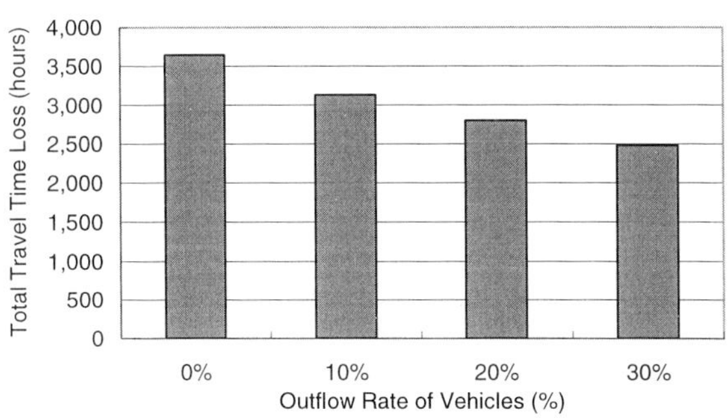

Figure 11 Relationship of Outflow Rate and Travel Time Loss

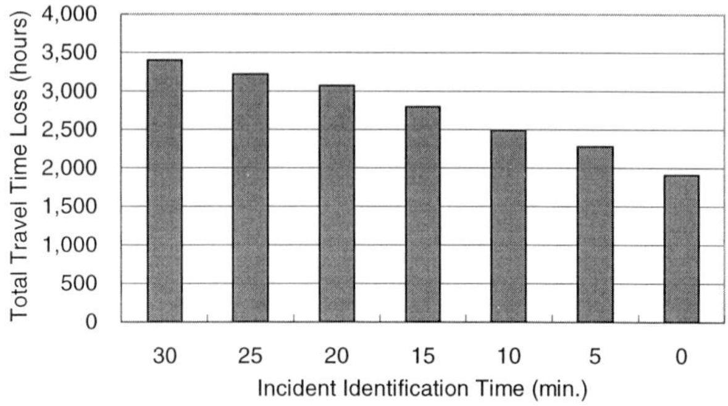

Figure 12 Relationship of Incident Identification Time and Travel Time Loss

ACKNOWLEDGEMENTS

The authors would like to thank Mr. Masaru Akao, who was a student of Gifu University, for the helping for programming and calculating.

REFERENCES

Inoue, N., Hasegawa, T. and Matsuo, T. (1988). Traffic Control System on the Hanshin Expressway - Further Developments -, *Management and Control of Urban Traffic System, Engineering Foundation*, pp.149-160.

Tsuchida, T., Yokoyama, T. and Akiyama, T. (1999). Construction of the Traffic Management Support System with Traffic Simulation, *Infrastructure Planning Review*, **16**, pp.879-886. (In Japanese)

Ogawa, K. and Akiyama, T. (2001). Traffic Simulation Model on Urban Expressway with Individual Vehicle Description, *Proceedings of the 8th World Congress on Intelligent Transport Systems*, CD-ROM, No.412.

23

TRAVEL TIME VARIABILITY AFTER A SHOCK: THE CASE OF THE TWIN CITIES RAMP METER SHUT OFF

David Levinson, Department of Civil Engineering, University of Minnesota, USA
Lei Zhang, Department of Civil Engineering, University of Minnesota, USA

ABSTRACT

Ramp meters in the Twin Cities were turned off for 8 weeks in the Fall of 2000. This paper analyzes traffic data collected in this experiment on travel time variability with and without ramp metering for several representative freeways during the afternoon peak period. Travel time variability is generally reduced with metering. However, it is found that ramp meters are particularly helpful for long trips relative to short trips. The annual benefits from reducing travel time variability with meters are estimated to be $33.1 million, compared to the annual ramp metering costs of $2.6 million in the Twin Cities metro area. Thus, the impact on travel time variability should be captured in future ramp metering benefit/cost analysis.

INTRODUCTION

Ramp meters have been depicted as ITS devices being able to prevent freeway breakdown and hence maximize the throughput at freeway bottlenecks. Reduced travel delay, improved safety, fuel consumption savings and emission reduction are reported frequently by researchers and practitioners after deployment of ramp metering systems (e.g. Kang and Gillen, 1999). It is well recognized that ramp metering decreases the variability of freeway travel speed and so the freeways become more reliable. However, the additional delay at on-ramps increases the overall trip travel time variability. Therefore, whether ramp metering makes travel more reliable is not clear under these two opposite forces. Previous benefit/cost studies on ramp metering are unable to take travel reliability change into account because of a lack of both theoretical and empirical work in the field. As reported by many researchers (Banks, 1991; Hall and Agyemang-Duah, 1991; Persaud and Hurdle, 1991; Cassidy and Bertini, 1999), the reduced freeway capacity after breakdown ranges from 0% to 8%. A majority of papers conclude that the capacity reduction when queue is present is negligible. Thus the importance of meters may not lie in reducing total travel time, as it mostly just shifts the time from freeways to meters. Rather it could be in reducing the variability in travel time by taking noise out of the system. If ramp metering does improve travel reliability, the benefits resulting from it should be considered.

In what may be the single most comprehensive experiment in the history of surface transportation, ramp meters in the Twin Cities of Minneapolis and St. Paul were the subject of a recent test of their effectiveness. By turning the meters off for 8 weeks in October, November, and December 2000, it has been possible to determine the effects of ramp metering on travel times and travel time variability. This paper presents the detailed results of an analysis of observed data on several representative freeways, I494 outer loop (eastbound and southbound), I494 inner loop (westbound and northbound), TH169 northbound and TH62 westbound (see Figure 1), during the afternoon peak period, and considers travel time variability with and without ramp meters to control freeway traffic in the Twin Cities area.

The first ramp meter was installed in the Twin Cities in 1970 on southbound I- 35E north of downtown Saint Paul. Now after 30 years of evolution, meters are standard on many freeways. There are currently 443 meters regulating ramps throughout the metropolitan area. Since the first installation of ramp meters, which operated as an isolated system, there has been sustained improvement in the system. Now most of the ramp meters are controlled centrally in real-time. Also, initially there were single lane ramp meters, but to better utilize the system so that it does not affect the local arterial or connecting roads, the usual practice now is to have two-lane ramp meters, which increase the storage capacity of ramps. Ramp meters in the Twin Cities were intended "to optimize flow in metro area freeway corridors by making efficient use of available transportation facilities" (Mn/DOT 1996a). The Minnesota Department of Transportation made this goal operational through a control strategy that divided freeways into zones terminating at bottlenecks. The number of vehicles in each zone at a given time was constrained by the capacity of the bottleneck. Ramp metering was used to limit those vehicles. The metering zone equation can be expressed as:

$$A + U + M + F = X + B + S \qquad (1)$$

s.t. incident override and occupancy override

Where:

A	upstream mainline volume (measured variable);
U	sum of unmetered entrance ramp volumes (measured variable);
M	sum of metered local access ramp volumes (controlled variable);
F	sum of metered freeway-to-freeway access ramp volumes (controlled);
X	sum of exit ramp volumes (measured variable);
B	downstream bottleneck volume at capacity (constant);
S	space available within the zone (computed volume using the difference between capacity of the zone at the threshold density (usually 20 veh/lane-km), and the actual number of vehicles presently in the zone).

(For more details on the Minnesota algorithm, interested readers may refer to (Bogenberger and May, 1999) and (Mn/DOT 1998)).

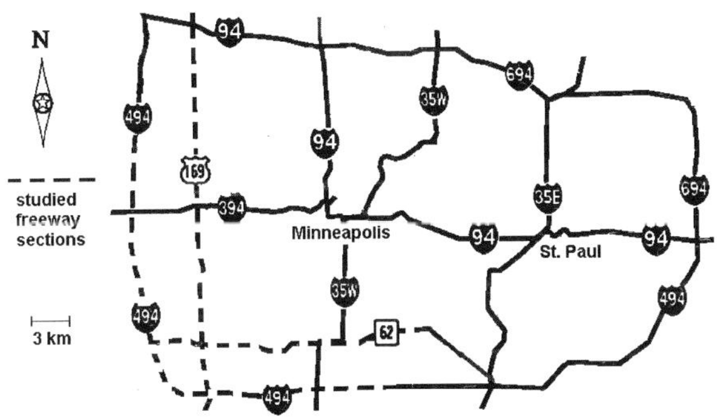

Figure 1 Freeway segment location map

Controversy arose when a State Senator from rural Minnesota challenged the strategy of the state Department of Transportation. The long delays at some ramps (at times, though not generally, in excess of 20 minutes), designed to ensure that the freeway remained free-flowing, drew the ire of some commuters, who believed the system was at best inefficiently managed. The state legislature passed a bill in Spring 2000 requiring a ramp meter shut off experiment. This paper is an analysis of data collected with and without metering.

The next section details how data were chosen from raw volume and occupancy counts collected by inductive loop detectors with the aim that impacts on travel time from factors other than ramp meters can be minimized. The following section shows the method used to measure travel times on ramps and freeway segments from the data available. Then inter-day and intra-day travel time variability are defined for this case study and results of variability with and without ramp metering control are shown and discussed. Recommendations and conclusions are delivered at the end of this paper.

DATA

Data were collected on six representative freeway sections in workday PM peaks (14:30 to 19:00 pm). These sections are: I494 outer loop (eastbound and southbound), I494 inner loop (westbound and northbound), TH169 northbound and TH62 westbound. We followed two criterion in selecting freeway sections and the study period:

- The time space domain must be free of congestion at the beginning and end of study. The flow conditions within those freeway sections will not be affected by further downstream queues.
- Data must be complete. There is no idle detector in the studied freeway sections.

We expect that ramp meters affect freeway travel time reliability. However besides ramp meter control, many other supply and demand side factors can also lead to travel time variation. For the purpose of this research, it is important to appropriately control for these other factors in the analysis.

Seasonal demand fluctuation. It is well known that demand on freeways changes seasonally. The duration of the Mn/DOT ramp metering shut off experiment was eight weeks from the last week of October to the first week of December 2000. To minimize the impacts from seasonal demand change, comparable weeks in October to December 1999 were chosen in our analysis as days with ramp meter control.

Weather. Although impacts of weather on freeway travel time have often been ignored in quantitative studies because it is discontinuous and difficult to plan or respond to, no evidence shows that weather can just be neglected. A study performed by the Mn/DOT TMC (traffic management center) in the late 1970s (Ries 1980) shows that precipitation, temperature, sky cover and wind speed all affect freeway capacity and hence travel time. In our study, days with more than 0.4 inches of hourly precipitation (rain, snow) in any PM peak period (14:30 to 19:00pm) are excluded.

Crashes. It has been shown by many studies (and also in this ramp metering shut off experiment) that ramp metering can reduce freeways crashes. So, crashes are considered endogenous when comparing travel time variation with and without ramp metering. Days both with and without crashes are included.

MEASURING TRAVEL TIMES

This section summarizes the calculation methodology required to measure travel times (and speeds) on entrance ramps, freeway segments, and O-D pairs on a highway with and without ramp meters.

The data collected for entrance ramps come in two types of paired data:

- Departure rate (Q_k), arrival rate (q_k) pair in each time interval (k) obtained from volume detectors.
- Departure rate (Q_k), queue length (n_k) pair for each time interval obtained from volume detectors and periodic visual observation of queue length by remote cameras.

The second type data can be transformed to the first type by equation 2:

$$q_k = Q_k + n_k - n_{k-1} \tag{2}$$

Where:

q_k the arrival rate in time interval k (vehicles/hour);
Q_k the departure rate in time interval k (vehicles/hour);
n_k the queue length in time interval k (number of vehicles).

Throughout the studied peak periods, all ramp upstream detectors have low occupancy readings and videotapes (at ramps without upstream detectors) do not show any queue spill-over effects to local connecting streets. This assures that the delays at on-ramps represent total delays caused by ramp meters.

Using the I/O queuing diagram shown in Figure 2, it is possible to find the total travel time every individual vehicle spends at ramps: (t_2-t_1). This travel time contains two parts, the free flow travel time from ramp upstream detector to the departure detector and ramp delay. Since ramps are short in distance and the free flow travel time at ramps can be neglected, the time duration (t_2-t_1) will be just called ramp delay for the remainder of this paper (A discussion on distinguishing "delay" and "waiting time in a queue" can be found in Lawson et al. (1997) and Lovell and Windover (1999) among others). The same method can be used to obtain arrival time, but assuming a uniform arrival at the back of the queue is not as accurate as assuming a uniform departure from the front of the metered queue, and is discussed more below.

The data are collected in 5 minutes intervals (*5* minutes = *300* seconds). So, the average time headway for time interval k equals $300/Q_k$ (sec). Then the departure time of the first vehicle is t_1 which is equal to h_1. The departure time of the second vehicle is t_2 which is equal to (h_1+h_2). As the time headway is accumulated, the departure time (T_v) for every individual vehicle can be calculated. The arrival time (t_v) can also be obtained by the same methods. Then the delay (d_v) of every vehicle can be calculated by equation 3:

$$d_v = T_v - t_v \tag{3}$$

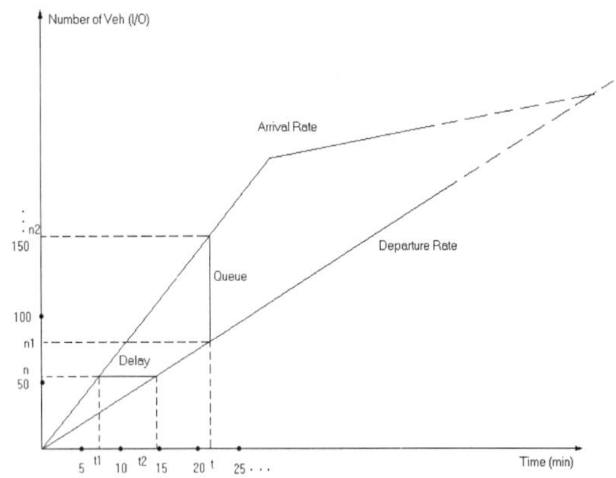

Figure 2 Queuing diagram

Where:

- t_v arrival time for vehicle v;
- T_v departure time of vehicle v;
- d_v delay of vehicle v (sec).

The departure rate within the window is still assumed to be uniform due to the presence of ramp meters. An assumption must be made about the arrival rate of vehicles at the back of the queue. Observed data is only available in five minute traffic counts. A uniform arrival rate would give a lower bound for delay estimation, a more reasonable assumption is to use an Poisson arrival process (the headways are exponentially distributed), which allows for bunching of vehicles. If the number of vehicles arriving at a queuing system has a Poisson distribution with a mean of q_k customers per unit of time, the time between arrivals has an exponential distribution with a mean of $1/q_k$. With the data collected by departure rate detectors, arrival rate detectors and queue length cameras at these ramps, the ramp delay for each vehicle and the average delay in each 5-minute time interval are obtained. The headway between vehicles is simulated using an exponential distribution 50 times, computing the delays using equation 3, and then averaging the delay.

Freeway loop detectors provide the volume and occupancy information in 5 minutes intervals, useful for computing traffic flow on freeway segments. Based on these data, the space mean speeds in every time interval can be computed as in equation 4:

$$\bar{u}_{s,k_1} = 2.76\times10^{-5}\left(\frac{Q_{k_1}L_v}{K_{k_1}}\right) \tag{4}$$

Where:

$\bar{u}_{s\,k_1}$ space mean speed of detector 1 in time interval k (km/sec);

Q_{k1} volume of detector 1 in time interval k (vehicles/hour);

L_v average vehicle length plus the length of the loop detector (m);

K_{k1} time occupancy of the detector 1 in time interval k (%).

Average vehicle length plus detector length, commonly known as the effective vehicle length is a crucial factor in estimating speed from single inductive loop detector flow and occupancy readings. Our average vehicle length estimates were taken from the Mn/DOT Traffic Management Center effective loop detector length calibration/normalization study. Because freeways have multiple lanes, there are more than one loop detectors at each station, there are multiple space mean speeds for one station derived from equation 4 (see figure 3). The weighted mean of all lanes will be used as the speed at each station, which is, e.g. for a two-lane freeway section:

$$\bar{u}_{s,k} = \frac{K_{k_1}\bar{u}_{s,k_1} + K_{k_2}\bar{u}_{s,k_2}}{K_{k_1} + K_{k_2}} \tag{5}$$

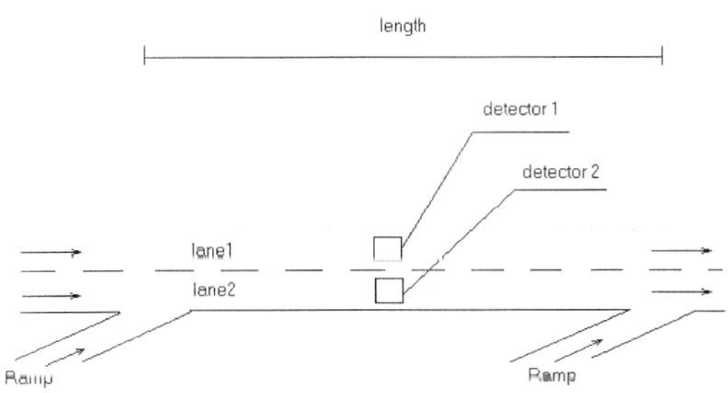

Figure 3 Illustration of freeway detectors

Assume that this speed is the average speed within this segment (that is, speeds within a segment are uniform). If there are no detectors in a segment, the speed of the nearest station will be assumed as the average speed in this segment (though this kind of situation rarely happens). Then the travel time of every time interval can be obtained:

$$\tau_k = L / \bar{u}_{s,k} \tag{6}$$

Where:

τ_k travel time in time interval k,

L Length of the section.

Once the average delay on ramps and the average travel time in freeway segments in each time intervals are obtained, it is possible to build the travel time O-D matrices. However, this requires that the data be synchronized (that is we cannot simply add up the travel times of the entrance ramp and all freeway segments between an OD pair in time interval k to obtain the total OD trips travel time in this interval, because a trip may take longer than 5 minutes and hence the travel time on one or more freeway segments for this trip may not be τ_k but τ_{k+1} or τ_{k+2} or …). Firstly, a database which records the delay and travel time data for each 5-minute period is built. Then the travel time for each O-D pair (from ramp i to ramp j) will be calculated. Similarly, a delay matrix can be constructed. The trip travel time is calculated with equation 7 (see also figure 4):

$$\tau_{i,j,k} = d_{r,k} + \tau_{i,k+x1} + \tau_{i+1,k+x2} + \ldots + \tau_{j-1,k+x(j-i)} \qquad (7)$$

Where:

$\tau_{i,j,k}$ travel time from origin i to destination j departing in time interval k;
$d_{r,k}$ delay on ramp r in time interval k;
τ_{k+x1} travel time in the freeway segment i in time interval $(k+x_1)$;
$x_1, x_2, \ldots, x_{j-i}$: synchronization coefficient, x is equal to the integer part of the result of the travel time from the origin to the beginning of this segment divided by the time interval (300sec).

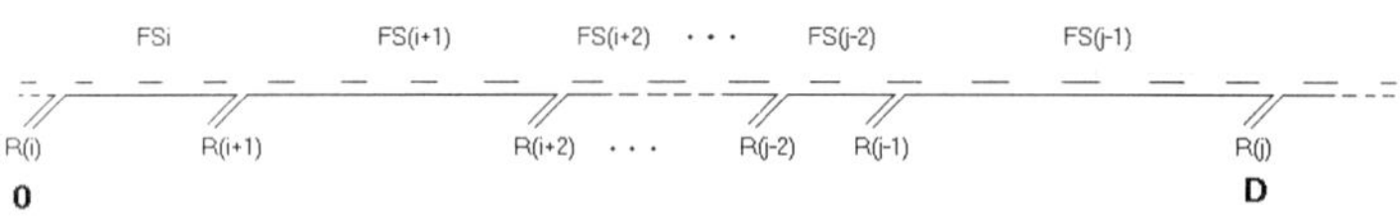

Figure 4 Building trip travel times

With the method developed above, we are able to calculate travel times for all freeway OD pairs (where Os are entrance ramps and Ds are exit ramps) in 5-minute time intervals. To illustrate, we will use one trip group (trips with the same origin and destination) as an example in the following section and the computed travel times for this trip group can be summaries in one matrix (see table 1).

Table 1 Travel time matrix (days × time intervals) for one OD pair

Time Interval 1999 On 2000 Off	Day 1 (10/25) (10/23)	Day 2 (10/26) (10/24)	...	Day n –1 (12/04) (12/02)	Day n (12/05) (12/03)	Inter-Day Travel Time Variation
1 (14:30)	τ_{11}	τ_{12}	...	$\tau_{1(n-1)}$	τ_{1n}	$V_{1,on/off}$
2 (14:35)	τ_{21}	τ_{22}	...	$\tau_{2(n-1)}$	τ_{2n}	$V_{2,on/off}$
3 (14:40)	τ_{31}	τ_{32}	...	$\tau_{3(n-1)}$	τ_{3n}	$V_{3,on/off}$
...	...	...	...	...	...	...
t – 1 (18:25)	$\tau_{(t-1)1}$	$\tau_{(t-1)2}$	...	$\tau_{(t-1)(n-1)}$	$\tau_{(t-1)n}$	$V_{t-1,on/off}$
t (18:30)	τ_{t1}	τ_{t2}	...	$\tau_{t(n-1)}$	$\tau_{t\,n}$	$V_{t,\,on/off}$
Intra-Day Travel Time Variation	$v_{1,on/off}$	$v_{2,on/off}$	...	$v_{(n-1),on/off}$	$v_{n,\,on/off}$	$\bar{v}_{on/off}$

On: with ramp metering control; *Off: without ramp metering control;*
V: inter-day travel time variability; *v: intra-day travel time variability;*
$\bar{v}$*: the average of intra-day travel time variability*

TRAVEL TIME VARIATION

There is considerable literature discussing measures for assessing travel time uncertainties (Asakura and Kashiwadani, 1991; Bell et al., 1999; Small, 1992). A recent review work on the subject was done by Bates *et al.* (2001). Given a series of travel times, several different summary measures can be derived indicating their deviation. Currently, two measures dominate the field: *travel time reliability* and *travel time variation (or variability). Travel time reliability* is defined as the probability that a trip can be made within a specific duration of time or alternatively, just a percentile of travel times (the former one sets a critical travel time while the latter sets a particular probability). Travel time variation is simply the standard deviation of travel times. Although theoretically those two summary statistics can be converted to each other if the distribution of the travel times are also known, its better to treat them as two different measures because at the current stage we are not clear about the distribution of travel times and how people perceive this distribution. Traditional travel demand theory prefers a measure that can be easily interpreted in terms of scheduling convenience. In evaluating value of travel uncertainty, a recent study (Lam and Small 2001) suggests the difference between the 90 percentile and the median of travel times.

From the freeway control perspective, a good measure of travel uncertainty must be calculable from routinely collected traffic data and can be easily incorporated into a benefit/cost analysis on control alternatives. The methodology we developed in the above section enables computation of both *travel time reliability* and *variation* from loop detector data. However, there is currently no conclusive answer to how travel uncertainty should be evaluated. It is noted by Bates *et al.* (2001) that both theory and recent empirical work suggests that the valuation of reducing travel uncertainty can be explained by travellers' scheduling consideration. But, since we know neither travellers' preferred schedule nor the flexibility of their actual schedule, it is impossible to aggregate scheduling convenience. Small *et al.* (1999) fit an econometric model which treats scheduling considerations implicitly using stated preference survey data and obtain an value of reliability in terms of reduction in *travel time variation*. This way, the difficulty of understanding travellers' true scheduling behaviour can be avoided. We will follow this line and evaluate the benefits of ramp metering on reducing *travel time variation*.

Here we look at two distinct types of travel time variation: *inter-day* and *intra-day*. For instance, *inter-day* travel time variation $V_{1,off}$ is the standard deviation of travel times of all trips with this OD (see table 1) starting between 14:30 and 14:35 pm from Oct. 25, 1999 to Dec. 5, 1999 excluding selected days noted in the Data section. The equation for calculating V_t is simply given by:

$$V_t = std(\tau_{t,1}, \tau_{t,2}, \quad \ldots, \quad \tau_{t,n}) \tag{8}$$

where:

V_t	Inter-day travel time variation of trips starting at time interval *t*;
std()	Standard deviation function;
$\tau_{t,n}$	Travel time for trips starting at time interval t in day *n*.

On the other hand, v_l is the *intra-day* travel time deviation of all PM peak trips with this OD in day 1 (Oct. 25, 1999). That is:

$$v_n = std(\tau_{1,n}, \tau_{2,n}, \quad \ldots, \quad \tau_{t,n}) \tag{9}$$

where:

v_n	Intra-day travel time variation of trips in day *n*;
std()	Standard deviation function;
$\tau_{t,n}$	Travel time for trips starting at time interval *t* in day *n*.

Inter-day travel time variation affects long-term scheduling decisions on non-discretionary tips (e.g. work trips) which occur daily. Intra-day travel time variation more likely influences discretionary trips (e.g. a shopping trip) or one-shot trips (e.g. a trip to visit a friend) that have flexibility in the time-of-day that they are scheduled. However, it is not obvious that the driver knows the variation in travel time across the peak period. These two types of variation may work interactively to shape travellers' overall perception of travel uncertainty. That perception arises from travellers' own experience, and determines their final scheduling decisions.

Previous evaluations of travel reliability focus on work trips, hence their results more likely reflect the value of *inter-day* travel variation.

To compare the metering-on case with the metering-off case, the inter-day variation difference (D_t) is computed by subtracting metering-on values from metering-off values:

$$D_t = V_{t,off} - V_{t,on} \tag{10}$$

For trips with this OD pair, we acquire a vector of *inter-day* travel time variation differences across different time intervals during the PM peak: D_1, D_2, ... , D_t. These results can be presented by a range-median diagram in which the range (lower/upper bound) and median value for this vector of *inter-day* travel time variation can be clearly identified (see Figure 5 for an example of this graphical form).

Because of data processing techniques introduced in the above section, we couldn't get a perfect one-to-one match for *intra-day* travel time variation (i.e. the dates excluded from analysis due to weather etc. with ramp metering in 1999 may not be excluded without ramp metering in 2000 and vice versa). Rather than showing travel time variation differences between metering-on and metering-off cases directly (as equation 10 for inter-day travel time variation), the averages of travel time variation are computed separately for metering-on and metering-off cases. Then by comparing these two average values ($\bar{v}$), we ascertain whether ramp metering control reduces or increases intra-day travel time variation.

$$\bar{v} = average(v_1, v_2, ..., v_n) \tag{11}$$

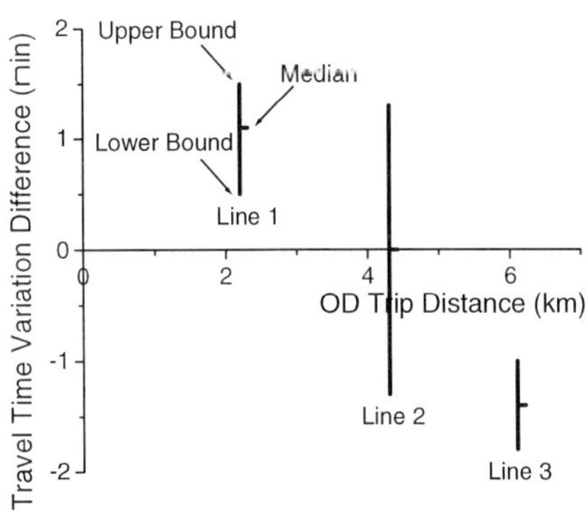

Three vertical lines in Figure 5 graphically describe three possible results of the range and median of a vector of inter-day travel time variation differences (metering off value - metering on value). If in one time interval, ramp meter control decreases (increases) OD travel time variation, this would be represented by line 1 (line 3). If ramp meter control doesn't affect travel time variation at all, line 2 is the case. From the x-axis, one can read the trip OD distance (total travel distance).

Figure 5 Possible results of inter-day travel time variation difference

RESULTS

Inter-day travel time variation results for four freeways are shown graphically in Figure 6. It is obvious that for most OD pairs (103 out of 127 in which 26/45 OD pairs ≤ 5 kms, 77/82 OD pairs > 5 kms), inter-day travel time variability is reduced by implementation of ramp metering (t tests of $V_{off} - V_{on} > 0$ are statistically significant at level 0.01). Freeway peak hour travel reliability (including ramp delays) increases. The average reduction in inter-day travel time variation for short trips (≤ 5 kms) is 0.17 min and for long trips (> 5 kms) 1.91 min. One can find that for extremely short trips (≤ 3 kms), it is hard to say whether ramp meters improve travel time variations.

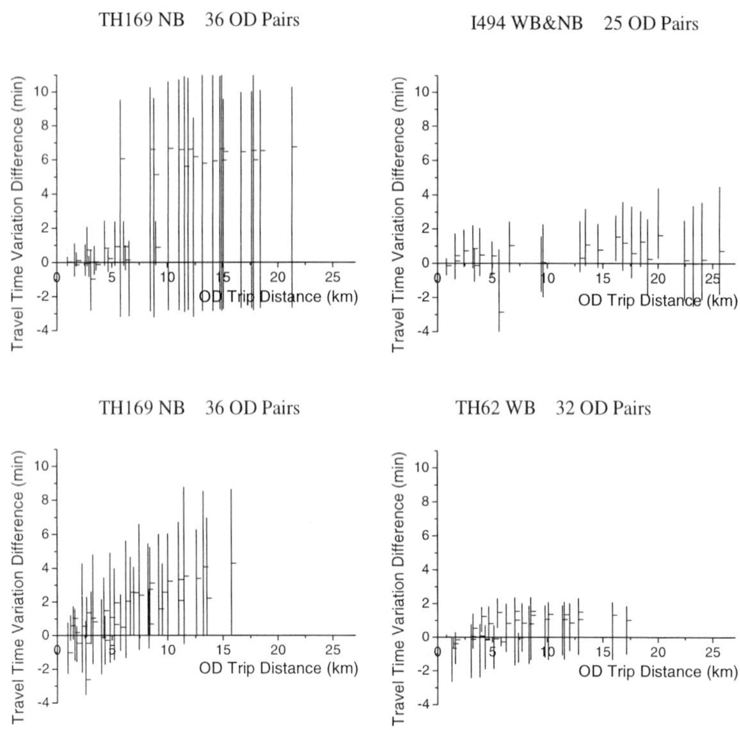

Figure 6 Inter-day travel time variation difference of each freeway without ramp meters – with ramp meters ($V_{off} - V_{on}$)

Figure 7 illustrates intra-day travel time variation results with two curves representing metering-on and metering-off cases. Although the intra-day travel time variation reductions caused by ramp meters differ by freeway segment, it is clear that ramp meters play a positive role in reducing intra-day travel time variation. Just as the inter-day results, intra-day travel time variation of long trips (> 5 kms) is reduced more significantly (on average 3.33 min) than those of short trips (≤ 5 kms, 0.01 min).

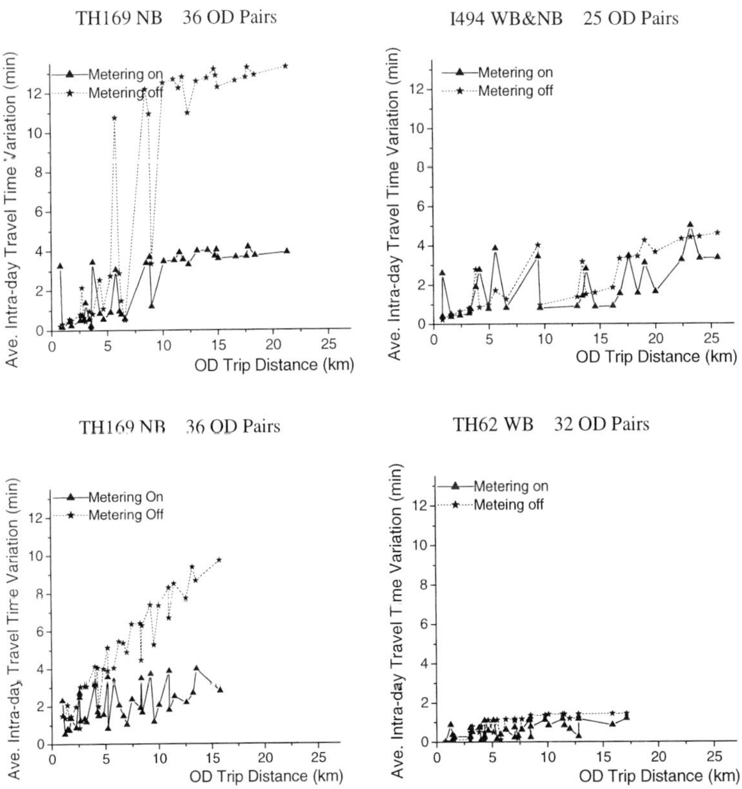

Figure 7 Intra-day travel time variation: with/without ramp metering control

Figure 8 compares the following three parameters of 36 OD trip pairs on TH 169 northbound: travel time saving per kilometer of travel; inter-day travel time variation reduction per kilometer of travel; ramp delay per kilometer of travel. Several points are worth to be noted from this limited comparison:

- For long trips, the Minnesota metering algorithm aiming at minimizing total travel time also significantly reduce congestion-related inter-day travel time variation, which can be seen at the right half of Figure 8 where both travel time saving and variation reduction become more and more constant.

- For shorter trips, since the weight of ramp delay become larger, both of the two objectives are not fulfilled satisfactorily and their relationships are not clear.

If day-to-day ramp delay is perfectly constant (i.e. the travel time variation with ramp metering is only due to freeway mainline travel time variation), the overall inter-day travel time variation could be additionally reduced by more than 50% for many OD pairs on the studied freeways (30/45 short trips, 15/82 long trips). Therefore, a metering algorithm able to achieve more stable ramp delay can significantly improve travel reliability, especially for short trips. Eliminating extremely long ramp delay is one practical improvement in this regard that can be made toward the existing or proposed metering algorithm in order to reduce ramp delay variability. Because of the non-linear nature of the value of ramp waiting time, this may also significantly reduce the perceived total travel time.

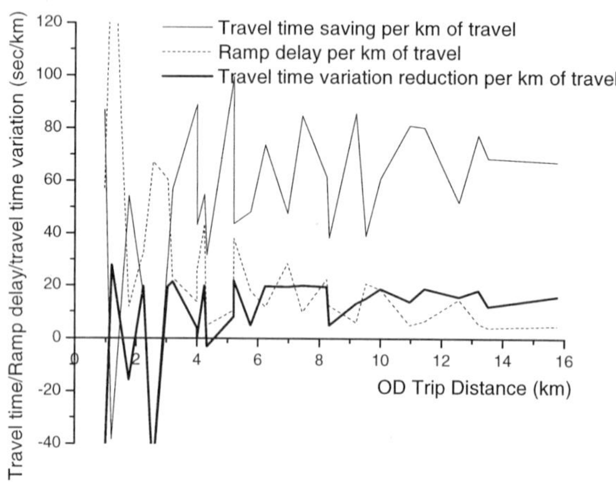

Figure 8 Travel time saving/ramp delay/inter-day travel time variation reduction Per kilometer of travel with ramp metering

VALUATION

Black and Towriss (1993), Small (1995) and Small et al. (1999) estimated a "reliability ratio", defined as the ratio of cost of standard deviation to cost of mean travel time when scheduling costs are not separately considered (the ratio was estimated from a model where absolute travel time, travel time variation and monetary cost all appear as explanatory variables). They obtained consensus results of 0.7, 1.27 and 1.3. Applying a 1.3 reliability ratio on $0.164/min ($9.8/hour) value of time which is currently used by Mn/DOT, the value of per minute reduction in travel time variation would be $0.213/min. The overall average inter-day travel time variation difference for all OD pairs in this study can be obtained, which is 1.5 minutes[1]. In other words, this means on average, implementation of ramp metering control reduces 1.5 minutes of inter-day standard deviation of travel time per trip. That would be $0.32 savings per trip in monetary value (*1.5* min × *$0.213*/min). This is still a conservative estimate because the benefits from reduced intra-day travel time variation are not taken into account. The total annual morning peak (6:00am to 9:00am) work trips can also be derived from detector counts[2] and for year 2000, there are 103 million morning peak work trips. Therefore, the annual benefits brought by ramp meters in terms of reducing travel time variation for morning work trips alone would be $33.1 million (*$0.32* /trips × *103* million trips). Compared to the $2.6 million annual ramp metering costs in the Twin Cities (Cambridge Systematics, Inc. 2001), this is definitely a huge benefit. Table 2 summarizes the valuation process.

Table 2 Annual benefits from travel time variation reduction

Value of travel time (VOT)	$0.164/min
Reliability Ratio (RR)	1.3
Value of travel time variation (RR*VOT)	$0.213/min (1)
Inter-day travel time variation reduction per trip	1.5 min (2)
Annual total number of home-to-work trips	103 million (3)
Annual Benefits (1)*(2)*(3)	$33.1 million

RECOMMENDATIONS AND DISCUSSION

Ramp metering was designed to improve freeway traffic flow and safety. While it generally does both, it also has the affect of improving travel time reliability. This benefit should be captured in the analysis of ramp metering benefits. Further research is needed to develop a better way to monetize the value of reliability, however it is clearly positive.

Most of the existing and proposed ramp metering algorithms aim to minimize total travel time. Our results suggest that the benefits from reduced travel uncertainty outweigh the benefits from absolute travel time saving. Should we still aim to minimize absolute travel time and consider the benefits from reduced travel uncertainty as a positive outcome, or should we reframe our rationale for ramp metering? This is not an easy question because we do not know the inherent relationships between these two objectives, minimizing total travel time vs. minimizing travel time variation. They may be achieved simultaneously since a less congested freeway can reduce both absolute travel time and congestion-related travel time variation. But they may trade-off with each other. For instance, a metering algorithm which operates freeway at relatively high risk of breakdown could eventually save more travel time than another algorithm operating at low risk, but the latter control algorithm may achieve a higher travel reliability. Future studies should look at the theoretical relationships between those two objectives in more details.

ACKNOWLEDGMENTS

This research was part of the project *Measuring the Equity and Efficiency of Ramp Meters* funded by the Minnesota Department of Transportation. The authors want to thank Pat Otto, Dave Berg, John Hourdakis, James Aswegan, Rich Lau, John Bieniek, Atif Sheikh, Shantanu Das, and Frank Lilja for their assistance. The opinions and errors remain those of the authors.

REFERENCES

Asakura, Y. and M. Kashiwadani (1991). Road Network Reliability Caused by Daily Fluctuation of Traffic Flow. *Proceedings of the 19th PTRC Summer Annual Meeting*, pp. 73-84

Banks, J.H. (1991). Two Capacity Phenomenon at Freeway Bottlenecks: A Basis for Ramp Metering? *Transportation Research Record* **1320**, pp. 83-90

Bates, J., M. Dix and May A. (1987). Travel Time Variability and its Effect on Time of Day Choice for the Journey to Work. Transportation Planning Methods, *Proceedings of Seminar C, PTRC*, pp. 299–311

Bates, J., J. Polak, Jones P. and A. Cook (2001). The Valuation of Reliability for Personal Travel. *Transportation Research Part E* **Vol. 37**, pp. 191-229

Bell, M.G.H., C. Cassir, Iida, Y. and W.H.K. Lam (1999). A Sensitivity Approach to Reliability Assessment. *Proceedings of the 14th International Symposium on Transportation and Traffic Theory*, July 20-23, pp. 283-300

Black, I.G. and J.G. Towriss (1993). Demand Effects of Travel Time Reliability. Center for Logistics and Transportation, Cranfield Institute of Technology

Bogenberger, K. and A. D. May (1999). Advanced Coordinated Traffic Responsive Ramp Metering Strategies. California PATH working paper UCB-ITS-PWP-99-19

Cambridge Systematics, Inc (2001) Twin Cities Ramp Metering Evaluation Report. Prepared for Minnesota Department of Transportation pursuant to laws 2000: Chapter 479, HF2891

Cassidy, M.J. and R.L. Bertini (1999). Some Traffic Features at Freeway Bottlenecks. *Transportation Research Part B* **Vol. 33**, pp. 25-42

Hall, F.L. and K. Agyemang-Duah (1991). Freeway Capacity Drop and the Definition of Capacity. *Transportation Research Record* **1320**, pp. 91-98

Kang, S. and D. Gillen (1999). Assessing the Benefits and Costs of Intelligent Transportation System: Ramp Meters. California PATH Research Report, UCB-ITS-PRR-99-19

Lam, T. and K. Small (2001). The Value of Time and Reliability: Measurement from a Value Pricing Experiment. *Transportation Research Part E* **Vol. 37**, pp. 231-251

Lawson, T., W., Lovell, David J. and C. Daganzo (1997). Using the Input-Output Diagram to Determine the Spatial and Temporal Extents of a Queue Upstream of a Bottleneck. *Transportation Research Record* **1572**

Lovell, D. J. and Windover, J. R. (1999). Analysing Freeway Traffic under Congestion: Traffic Dynamics Approach — Discussion by D.J. Lovell and J.R Windover. J*ournal of Transportation Engineering*, July/August 1999

Levinson, D., L. Zhang, Das, S. and M. Sheikh (2001). Ramp Meters on Trial: the Case of Twin Cities Ramp Metering Shut Off. To be presented at TRB 81st Annual Meeting, Washington, D.C., Jan. 13-17, 2002

Minnesota Department of Transportation (1996a). Traffic Management Program Overview: Twin Cities Metro Area. Report Number TMC07043-0196 February 1996.

Minnesota Department of Transportation (1996b). Freeway Operations Section Business Plan. Feb. 1996.

Minnesota Department of Transportation, Traffic Management Center (1998). Ramp Metering by Zone – The Minnesota Algorithm. Working Report

Minnesota Department of Transportation, Traffic Management Center (1999, 2000). All Detector Report.

Minnesota Department of Transportation (2000). website http://www.dot.state.mn.us Mn/DOT Metro Division, Traffic Management Center

Persaud, B.N. and V.F. Hurdle (1991). Freeway Capacity: Definition and Measurement Issues. *Proceedings of the International Symposium of Highway Capacity*, A.A. Balkema press, Germany, pp.289-307

Small, K. A. (1992). Urban Transport Economics. *Fundamentals of Pure and Applied Economics series* **Vol. 51**, Harwood Academic Publishers, New York.

Small, K, R. Noland, and P.M. Koskenoja (1995). Socio-economic Attributes and Impacts of Travel Reliability: A Stated Preference Approach. Draft PATH Report, MOU-117

Small, K., R. Noland, Chu, X. and D. Lewis (1999). Variation of Travel-Time Savings and Predictability in Congested Conditions for Highway User-Cost Estimation. NCHRP Report 431

Ries, G. (1980). Impact of Weather on Freeway Capacity. Mn/DOT Metro Division, Traffic Management Center research project report (unpublished)

ENDNOTES

[1] This average is computed by assuming that all OD pairs have the same number of trips. This may not be a bad assumption because we are work on a total number of 105 OD pairs and errors may be cancelled out to a large extent. Also the average OD distance for all OD pairs are 8.4 km which should be close to the actual average freeway travel distance.

[2] The basic idea is that we sum up morning peak (6:00am to 9:00am) volume across all boundary detectors of the managed freeway system (freeways controlled by ramp metering). The selected detectors are either located at mainline entrance stations or entrance ramps (freeway to freeway entrance ramps are excluded). In case that a loop detector was not functioning, we look for substitutive detectors (e.g. if an entrance ramp detector was not functioning but both mainline detection stations immediately before and after this entrance ramp worked well, these mainline detectors can be the substitutes for this entrance ramp detector). We also assume that 80% of the morning commutes are work trips.

24

EFFECTS OF DEMAND VARIATION ON THE LOCATION OF OPTIMAL ROAD PRICING CORDONS

Agachai Sumalee, Institute for Transport Studies, University of Leeds, Leeds, UK
Anthony D. May, Institute for Transport Studies, University of Leeds, Leeds, UK
Simon P. Shepherd, Institute for Transport Studies, University of Leeds, Leeds, UK

ABSTRACT

This study reports on the effect of demand variation on the optimal location of road-pricing cordons. The optimal road-pricing cordon, in this study, aims to maximise the social welfare function. This optimisation program is categorised as Bi level optimisation programming which is a NP hard problem. The paper first describes the method developed to solve the optimal toll problem for a given set of chargeable links. The tests were carried out with a small toy network and a larger scale network. For the small network, four single user class demand characteristics were varied individually; these were the elasticity of trip generation with respect to increases in travel cost, value of travel time, volume of traffic, and traffic distribution pattern. For the larger scale network, only elasticity, value of time, and trip volume were tested. The results of the larger scale network are also analysed by including the cost per toll point. The tests with the larger scale network were modified so that the constraint of uniform charge is applied. The results showed that demand variation could influence the best location of toll points. This finding raises the question of whether the implementation of the same cordon all day in an urban traffic network is the optimal approach under the existence of demand variations by time of day, and also whether the evaluation process of the cordon location should consider the effect of different time periods together.

INTRODUCTION

The main objective of this research is to test the influence of demand characteristics on the location of the optimal road pricing cordon. In this paper, the term "optimal road pricing cordon" refers to the optimal location of the cordon and the charge level at each toll point. The variation of traffic demand in an urban network exists. The impact of this variation on the design and operation of the traffic management tools is significant. Particularly in the case of road pricing, the optimal design of road pricing depends on the demand characteristics, and on the other hand the demand characteristics also change in response to the implementation of the pricing measures. The problem of designing the optimal road pricing system is, thus, bilevel optimisation programming problem.

The main emphasis in previous research into road pricing has been concentrated on charge levels. Morden (1996) described the system of State Route 91 in Southern California offering value pricing toll rates adjusted according to the time of day. Recently, Swenson et al (1999) described the employment of time-of-day variable toll pricing to change trip making from peak to off-peak hours leading to a reduction in congestion on two toll bridges in Lee County, Florida. The Singapore electronic road pricing (ERP) also implements different charge levels in different periods of the day (Menon, 2000).

However, the awareness of the variation of the location of the charge cordon due to the variation of the traffic demand characteristics has gained less attention. This may be due to the difficulties in implementation in real life. However, the current and future technology of satellite based and GPS based systems may give rise to a more flexible location of charging boundaries. This paper is organised into four further sections. The second section presents the approach to solve the second-best optimal toll problem based on Verhoef (2000). Thrid explains the description of the small network and the results of the tests. Forth section explains the detail of larger scale network and the test results. Final section draws the conclusions found from the tests.

SECOND-BEST PRICING WITH AN UNIMODAL NETWORK

The idea that efficient use of roads requires users to pay for the traffic congestion costs that they impose on urban road networks is long established (Pigou, 1920; Walters, 1961). However, there are also some problems obstructing the optimised use of urban roads following simplistic economic theory. Some economists for example, Sharp (1966) have long argued that the basic idea of road pricing is too simplistic. Road pricing, by its nature, is a first-best solution to the congestion problem and, as such, implicitly assumes full marginal cost pricing is the norm throughout the economy. If this is not so then adjustments to the framework are necessary to define second-best criteria (Button, 1998).

Following Verhoef (1998, 2000), Shepherd et al (2001) developed the mathematical approach associated with SATURN (Van Vliet, 1982) to find the optimal road pricing cordon under the second-best condition in which not all links in the network can be tolled. SATURN is a steady-state equilibrium assignment model which predicts route choice and traffic flows on a road network based on the generalised cost of travel. However, in this early stage the method still can find only the optimal charge. The programming method developed by Shepherd et al (2001) termed CORDON will be used in this paper with the small network to find the optimal toll levels and also total benefit, defined as economic efficiency. More precisely, the objective function of the optimal road pricing design is as follows:

$$Max\ W = \sum_{i=1}^{I} \int_{0}^{T_i} D_i(x_i)dx_i \; - \; \sum_{j=1}^{J} V_j c_j(V_j) \tag{1}$$

given the notation in Appendix A ,which equals to user benefit (UB) – social cost (SC) (Yang and Huang, 1998). However, in this paper we are interested in the social welfare improvement of the do-something compared to do-nothing schemes. Thus, the benefit of the scheme shown in the results throughout the paper can be defined as:

$$W = \sum_{i=1}^{I} \left(\int_{0}^{T_i^1} D_i(x_i)dx_i - \int_{0}^{T_i^0} D_i(x_i)\,dx_i \right) - \left(\sum_{j=1}^{J} \left(V_j^1 c_j(V_j^1) - V_j^0 c_j(V_j^0) \right) \right) \tag{2}$$

where T_i^0, T_i^1 are the total demands between O-D pair i in do-nothing and do-something cases respectively, and similarly V_j^0, V_j^1 are the total link flows in do-nothing and do-something cases respectively. It should be noted that the objective function of the optimal road pricing design in this paper is purely related to economic efficiency and does not include the other benefits, e.g. environmental improvement.

For a larger network, due to the running time of the program, the test was modified to use uniform charges which will be explained in the forth section. Those who are interested in the mathematical detail and the implementation with a traffic assignment model should consult Verhoef (2000) and Shepherd et al (2001).

SENSITIVITY TESTS WITH THE SMALL NETWORK

The characteristics of demand are categorised into four main groups, i.e. elasticity, value of time, distribution of the traffic over the network, and volume of the traffic in the network. This section discusses the results of varying each parameter. There are two test sets, single tolled links and pairs of tolled links. In the single tolled link tests, each link will be taken in turn and use the CORDON process described in the previous section to find its optimal charge and the associated benefits. The second test set uses the CORDON process to generate solutions for each possible combination of pairs of links.

Treatment of Demand

Four characteristics of traffic demand have been analysed, i.e. elasticity of demand, value of time (single user class in this research), total trips (volume), and trip matrix distribution. It is assumed that all users are homogeneous in this research. The values used in the sensitivity analysis of these variables are reported in Table 1.

Table 1 Tested level of each demand characteristic (highlighted value is the base value)

Demand Characteristics	Tested range
Elasticity	0.25, 0.35, 0.43, 0.53, 0.80
Value of time (Pence per minute)	3.63, 4.63, 5.63, 6.63, 7.63, 8.63, 9.63
Trip matrix volume (PCU/hr)	2000, 3000, 4000, 5000, 6000
Trip distribution pattern	See appendix B

In this research, we refer to the elasticity of the change of the number of car trips with respect to the change of travel costs. The demand function assumes constant elasticity and is determined by equation (3). The range of elasticity was specified as 0.25-0.8 following the previous research (Fowkes, *et al* 1993; Goodwin, 1992).

$$C = C_0 \left(\frac{T}{T_0} \right)^{\frac{-1}{\beta}} \tag{3}$$

where T is the traffic demand between an OD pair and C is the cost of travelling between that OD pair; β is the elasticity. The subscripts 0 represent the base conditions (or equilibrium with no tolls in this case).

Network Topology

Figure 1 shows the five-link network used in this paper. We assume only two origin and destination pairs are used by travellers, i.e. A-C and B-C. All links are one-way links, and thus the network is not symmetrical. It should be stressed that this is only a "toy" network constructed for the tests.

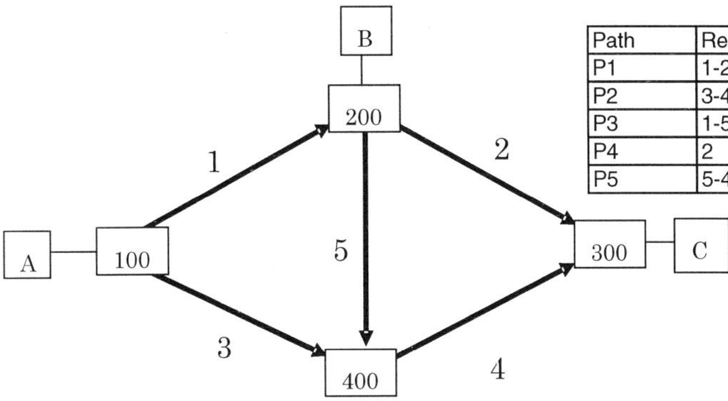

Path	Relevant links
P1	1-2
P2	3-4
P3	1-5-4
P4	2
P5	5-4

Figure 1 Five-link network

100, 200, 300, and 400 represent nodes in the network where as mentioned A, B, C represent zones in the network. There are in total five paths in the network between the two relevant O-D pairs.

Results of the Sensitivity Tests

Elasticity of demand. Each link in the network is taken in turn as a single toll point and the cordon program is used to derive its optimal toll and associated benefit. Figure 2 shows the results of the sensitivity test by changing the elasticity of demand. The horizontal axis represents values of elasticity in each test and the vertical axis represents the total benefit in generalised time (seconds). The optimal total benefit for each link is plotted on the graph.

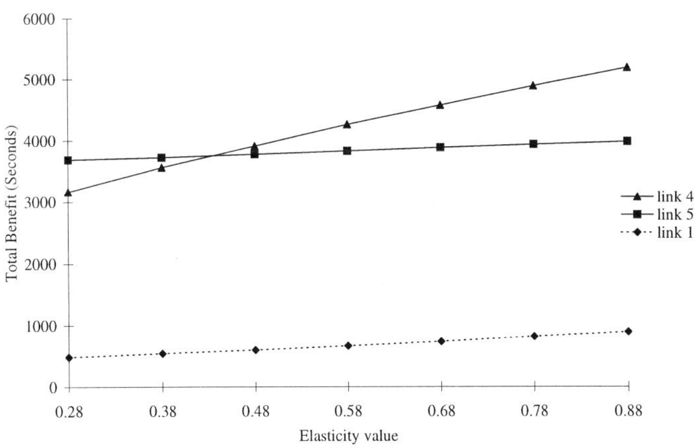

Figure 2 Results of sensitivity test of elasticity with single tolled link

Links 2 and 3 gave negative benefits for charging positive tolls on these links. In fact, the cordon program predicted negative tolls, which means the travellers should be subsidised to travel on these links. Thus, the results for these two links are not included in the graph. From figure 2, link 5 is the best link to be tolled under the elasticity value of 0.28 and 0.38 whereas link 4 is the best link with the higher values of elasticity. The total benefit generated by tolling link 4 is more sensitive to changes in elasticity compared to that generated by tolling on link 5. In practice, the result implies that with uncertain information regarding elasticity of traffic demand, one should consider tolling on link 5 rather than link 4. The other implication is the location of toll point during peak and off-peak periods. The operator, according to this result, should toll link 5 during peak period (assuming that the elasticity during peak period is low) whereas link 4 should be tolled during off-peak (elasticity is high). However, the consideration of implementing the different toll location should also take into account of costs per toll point.

The charge on link 1 is imposed on the demand between A-C on paths P1 and P3, but not on the demand between B-C. Thus, tolling on this link gives lower benefits compared to tolling on the link that can control the demand between both O-D pairs. The interesting point is the variation of the optimal tolled link between link 4 and 5. The reason is not because of the demand between B-C since the tolls on both links are imposed on the demand between B-C on the same path (P5). The concentration, thus, should be on the demand between A-C and its related paths. When the elasticity is low tolling on link 5 is better than tolling on link 4. The reason is that the low elastic demand condition approaches the fixed demand case. The demand is not very sensitive to the change of travel costs, thus the allocation of the traffic pattern (route choice) on the network is a better approach to optimise the benefit compared to the suppression of demand. Charging on link 5 allows trips between AC to avoid the charge by using paths P1 and P2 without increasing the congestion on links 2 and 4 too much. On the other hand when the elasticity of demand is high, the suppression of traffic can produce higher benefits for the network. This is due to the decrease in congestion and consequently the total travel cost in the network. Link 4 is the best link to be tolled in this case as it covers more paths and any re-routing imposes higher costs on link 2 which in turn helps to suppress the demand.

The second test is the test with the combination of two tolled links. Figure 3 shows the results of the Cordon program for all combinations. There is no variation of the first and second best combinations (links 2 and 4, and links 1 and 4 respectively) regardless of the elasticity value. It is noteworthy that the pair (2,4) can capture all demand on all paths, which brings it close to the first-best condition. Therefore, it is not a surprise that this combination appears to be the best combination in all tests. There appears to be a variation of the third, fourth, and fifth best tolled pairs with a change in elasticity value.

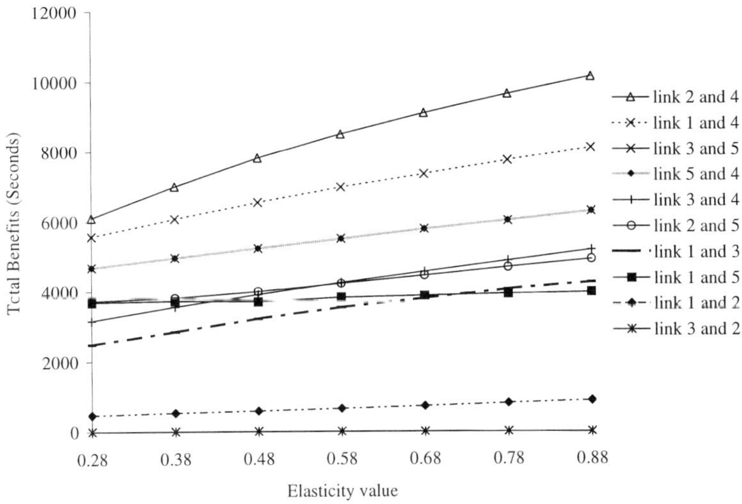

Figure 3 Results of sensitivity test of elasticity with two tolled links

Value of time. Figure 4 presents the results of the sensitivity test with the value of time for single tolled links. Similar to the sensitivity test with elasticity, putting positive tolls on links 2 or 3 could not raise any benefit for the network. However, there is no variation of the best link to be tolled in the range of value of time tested. Links 4 and 5 are the first and second best links to be tolled.

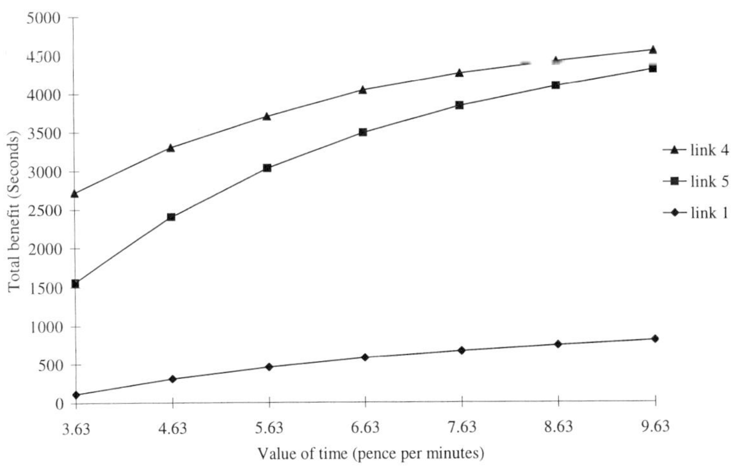

Figure 4 Results of sensitivity test of value of time with single tolled links

Benefits increase as value of time increases as expected. The difference between the benefits from tolling on link 4 and 5 decreases as the value of time increases. When the value of time increases the cost of travel also increases resulting in more suppressed trips. As the value of time is increased, the demand on path P2 decreases significantly. With the absence or very small flows on path P2, tolling on either link 4 or link 5 similarly controls the demand between A-C on path P3 which is the major route for the traffic.

Figure 5 shows the results of the sensitivity test with the value of time for pairs of tolled links. The results show the variation of the third, fourth, and fifth best pair whereas there is no change of the first and second best pair. Similar to the results of the test with elasticity, the pairs (2,4), and (1,4) are the first and second best pair respectively for all values of time tested.

Trip volume. The results of the sensitivity test with the trip volume are shown in figure 6. Link 4 is the first best link to be tolled within the range of trip volume tested. Link 5 appeared to be the less attractive tolled link with the high traffic volume, and in fact it even required the negative toll (subsidy) with the very high traffic volume.

As demand increases, the main approach to increase the benefit is to suppress the demand on each path. As mentioned earlier, link 4 has three relevant paths (P2, P3, and P5) that can capture more traffic in the network compared to other links. Therefore, the benefit generated from tolling on this link is high, since it can impose the toll on a high volume of traffic. On the other hand, the toll on link 5 diverted the traffic to other routes instead of suppressing them, which did not provide the significant benefit with the highly congested network.

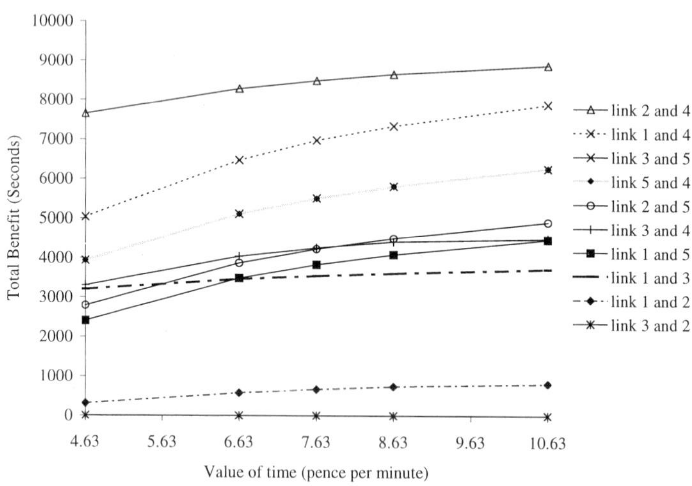

Figure 5 Results of sensitivity test of value of time with pairs of tolled links

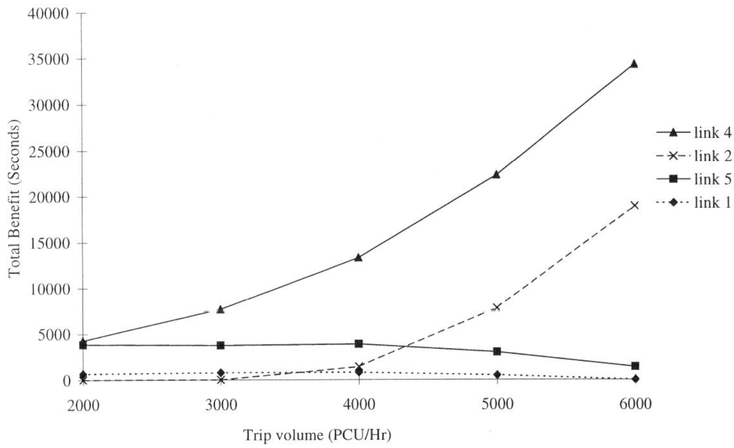

Figure 6 Results of sensitivity test of trip volumes with single tolled link

Tolling on link 2 can provide benefits under the more congested condition with the positive tolls whereas in other tests this link requires subsidy (negative toll) to produce the benefit. This is because as the demand increases congestion spreads over the paths in the network including path P4. Therefore, the toll on link 2 became positive in order to suppress the increasing traffic. On the other hand, as mentioned earlier when the congestion or traffic volume in the network is low, diverting traffic to the optimal pattern is the best approach. In this case, link 5, again, emerged to be the second best tolled link.

Figure 7 shows the results of the trip volume test with the pairs of tolled links. Again, the pair of link (1,4) is the best pair of tolled links. With low traffic volumes (2000 and 3000 PCU/hr), the second best pair of tolled links is (1,4) whereas with the high traffic volume, the second best pair of tolled links (1,3). There is also the variation of the order of the third, fourth, and fifth best pair to be charged under different trip volumes.

Trip distribution. The last test is the sensitivity test with the trip matrix distribution. Appendix B shows the detail of the alternative matrices. Figure 8 shows the results of the tests with single tolled links. The same set of links give the positive benefit, i.e. link 1, 5, and 4. Link 4 is the best link to be tolled under the base, first and second trip matrix. Link 5 is the best link to be tolled under the third and fourth trip matrix.

It is clear that link 4 is a better link to be tolled when the demand is equally spread between two O-D pairs or higher between A-C, see matrix base, 1, and 2 in Appendix B. On the other hand, when the demand is higher between B-C, link 5 is the best link to be tolled. The reason for this variation is as follows:-

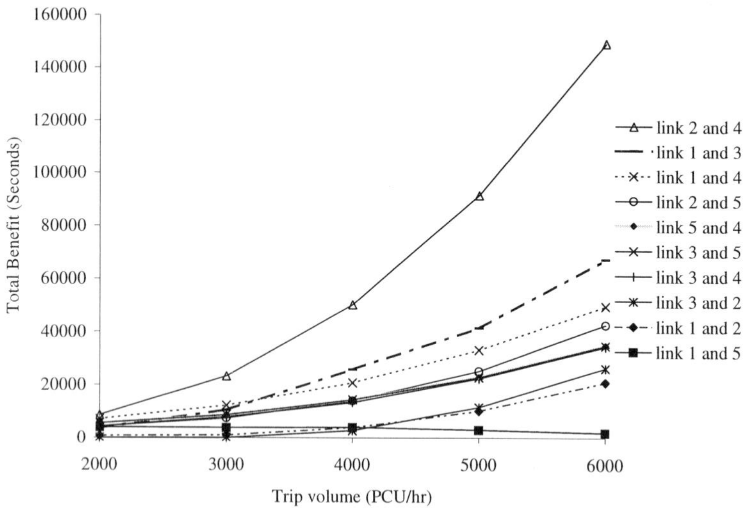

Figure 7 Results of sensitivity test of trip volumes with two tolled links

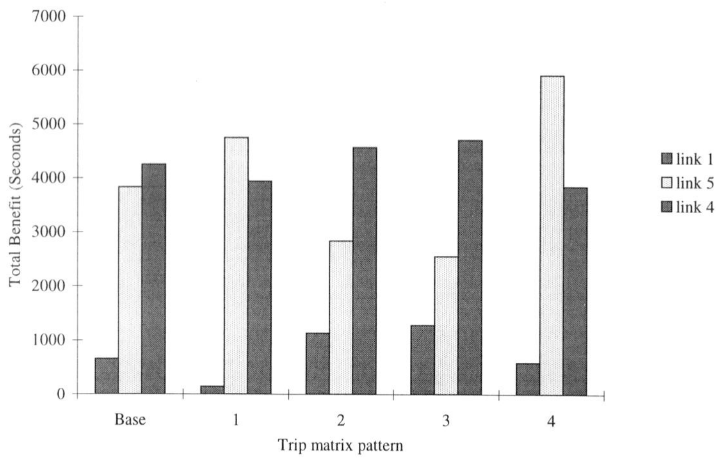

Figure 8 Results of sensitivity test of trip matrix pattern with single tolled link

- When the demand between A-C is high, all paths between A-C, i.e. path P1, P2 and P3 are highly used. Hence the congestion on all these paths increases and the best approach to increase the benefit for the network is to suppress the trips between A-C. In this case, the toll on link 4 can cover the demand on paths P2 and P3 of A-C, which are the paths with a higher traffic volume compared to path P1. On the other hand the toll on link 5 fails to capture the traffic on path P2.

- When the demand between B-C is high, the number of relevant paths with the low marginal cost of link 4 became the negative aspect, since the calculation of the optimal toll on link 4 for path P5 has to take into account the shadow price from the marginal cost of path P3 and P2 between A-C which distorted the toll to optimally control the demand on P5 between B-C. Basically there are different optimal tolls for each path and the best result is for the link which can impose the highest toll on demand between B-C.

Figure 9 presents the result of the test with the pairs of tolled links. Interestingly, the pair of links (2, 4) (which has generally been the first-best pair of tolled links) is not the best pair of tolled links with the fourth matrix. The fourth matrix has a very high traffic demand between B-C in and the pair (2, 5) became the best pair of tolled links.

The reasoning is similar to the discussion above for the case of single tolled link with the high demand between B-C. The tolls on links 2 and 5 can be adjusted to control the traffic between B-C better than the tolls on links 2 and 4 according to the shadow prices of other relevant paths. The variation of the best pair of tolled links is significant in this test. As we can see, apart from the first and second best pair of tolled links, there are changes in the optimal toll points when the trip matrix pattern is varied.

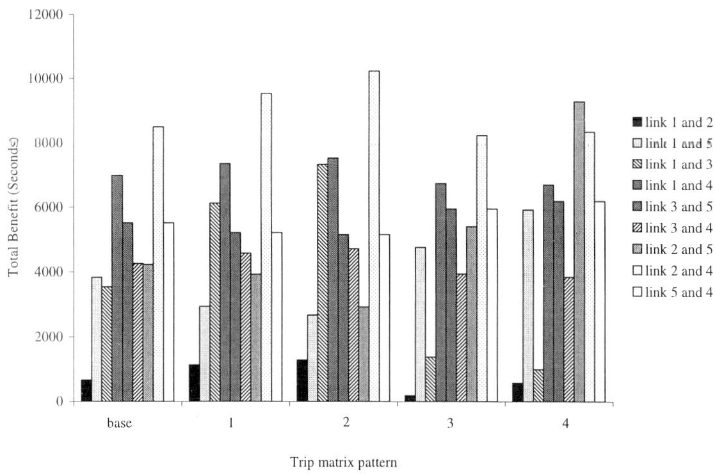

Figure 9 Results of sensitivity test of trip matrix pattern with pairs of tolled links

SENSITIVITY TEST WITH A LARGER-SCALE NETWORK

At the time of writing, the CORDON process could not be applied to a large network, partly due to run-time and partly due to the convergence of the traffic assignment algorithm. Thus, the test for a larger scale network was modified. A practical charging cordon usually adopts a uniform charge on all toll points. The assumption of the uniform charge, where we only need to find a single charge level for all toll points in the cordon, allows us to simply use the normal univariate optimisation method without the need to run the CORDON program i.e. a simple search is used.

Network Topology

Figure 10 shows the network used in this test. It should be noted that this is only a SATURN buffer network; links are assumed to be independent and there is no simulation of turning movements or junction control. There are 25 O-D pairs and 276 two-way links. Four different cordons were defined following the criteria for practical cordon design suggested by Shepherd et al (2001).

These four cordons are shown in figure 10. Two cordons are inner cordons concentrating on the city centre whereas the other two cordons are outer cordons covering the wider area of the city. The inner charging cordons are located inside the inner ring road, which is roughly formed by the local streets around the city centre. The outer charging cordons are located just inside the well-defined outer ring road. The outer cordon 1 also uses the coast as a natural boundary. The charges are imposed on inbound trips only. All cordons are closed which means travellers from outside the cordon cannot access the area inside the cordon without paying a charge.

The demand is treated as a single user class. Similar to the small network used in the previous section, the same demand function is adopted with the same PPM and PPK value (7.63 pence per minutes and 5.27 pence per km respectively). Three sensitivity tests are conducted consisting of variations in elasticity, value of time, and total trips. The trip pattern is not varied here.

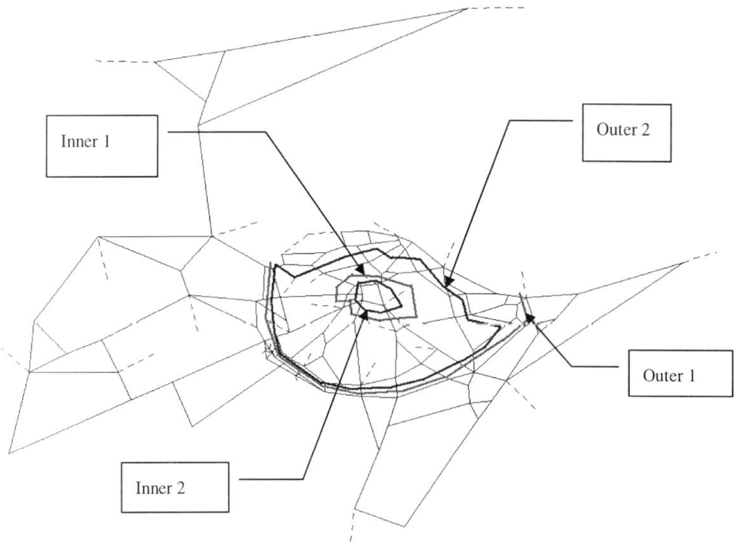

Figure 10 Larger scale network used in the test and the predefined charging cordons

Results for the Large-scale Network

Elasticity of demand. Table 2 shows the benefits for each cordon as the elasticity is varied. The value in the table is the optimal benefit for each cordon i.e. resulting from an optimal uniform charge. It shows that the second outer cordon performs best with the elasticity values of 0.38 and 0.58. In general as the elasticity is increased then the benefits increase. As the elasticity increases the difference of the benefit generated from both of the inner cordons and the second outer cordon decreases. Under the elasticity of 0.78 and 0.98, the first and second inner cordon performs better than the second outer cordon. Thus the optimal location for a cordon depends upon the elasticity of demand. The difference in benefit generated from the two inner cordons is small for all the elasticity values tested.

Table 2 Results of sensitivity test of elasticity with large-scale network

		Total benefit (thousands seconds)			
Cordon	toll points	Elasticity values			
		0.38	0.58	0.78	0.98
Inner 1	9	2,361.4	3,244.5	3,950.0	4,485.7
Inner 2	7	2,457.9	3,226.8	3,843.3	4,332.0
Outer 1	16	957.2	1,443.0	1,841.3	2,152.0
Outer 2	20	2,736.4	3,492.8	3,912.0	4,148.6

It is noticeable that as the elasticity value increases the benefit generated from each cordon also increases. The reason is that the higher the elasticity, the more trips will be suppressed rather then re-routed resulting in lower congestion levels in the network. The decrease of congestion will then decrease the cost of travel in the whole network.

The smaller cordons, inner1 and inner2, capture most of the traffic heading to city centre and through traffic. With low elasticity, these smaller cordons cannot suppress the through and city centre trips, instead the through traffic diverts to other areas and the traffic to city centre continues their journeys. The area with the relief of congestion is also very tight. The diversion to the other longer routes associated with the spread of congestion and the small effective area decrease the total benefit of the scheme. With low elasticity of demand, the wider cordon (outer2) can capture the long distance through traffic and divert them around the cordon while the traffic inside the cordon enjoys the reduction of the congestion in the area. The drawback of the outer1 cordon is the lack of alternative route for the through traffic and the cordon cannot capture the traffic from the north of the city.

When the elasticity is high, the smaller cordon becomes more effective, since the dispersion of the congestion to other areas and the diversion of the trips to a longer route is minimised due to the high sensitivity of trips to extra costs (either toll or re-routeing costs). This is because most of through traffic and traffic destined for city centre are suppressed. Despite the small area covered by the cordon, the improvement of the congestion in the city area is significant and this increases the benefit of the scheme.

Value of time. Table 3 presents the results of the sensitivity test with value of time. Interestingly, the total benefits from the outer1 and outer2 cordon increases as the value of time increases. On the other hand, the benefit from both of the inner cordons varied as the value of time increases. The second outer cordon performs best in most of the tested values, i.e. 5.63, 7.63, and 9.63 pence/minute. The gap between the benefit obtained from both of the inner cordons and the second outer cordon decreases as the value of time decreases. The first inner cordon gives a higher benefit under the value of time of 3.63 pence/minute.

Table 3 Results of sensitivity test of value of time with large-scale network

		Total benefit (thousands seconds)			
Cordon	toll points	Value of time			
		3.63	5.63	7.63	9.63
Inner 1	9	3,254.3	3,031.5	3,244.5	3,109.9
Inner 2	7	3,145.0	3,031.5	3,226.8	3,141.4
Outer 1	16	774.5	1,311.9	1,442.9	1,639.3
Outer 2	20	2,145.2	3,092.6	3,492.8	3,749.4

The changes of value of time change the ratio of PPM/PPK which affects the behavioural model of route-choice. The effect of the change of this proportion is not significant in the small network since all of the links have the same length. For the larger network, the length of each link varies. When the value of time is high the importance of the distance decreases and vice versa.

With high values of time, the travellers tend to avoid congested links (which imply longer travel time but shorter distances) and tend to use longer routes (which imply longer distance but less congestion). In this case, the outer cordon2 can re-route the long distance traffic around the city centre. On the other hand, when the value of time is low the longer routes are less acceptable and the inner1 cordon performed best, since drivers are prepared to pay for the shorter route through the city centre. Thus the value of time can also affect the optimal cordon location, but it can be difficult to interpret as value of time affects both route choice and demand responses.

Trip volume. The base matrix has a total of 121564 PCU/hr. Four matrices with different trip volumes, i.e. 0.4, 0.7,1.0, and 1.3 times base trips, but with the same trip distribution pattern were used in the following tests. Table 4 shows the results of the test with different trip volumes. The variation in total benefits and in terms of the best cordon is more significant compared to the other tests. Firstly the total benefits increase substantially as the volume is increased which is in-line with expectations as tolls are there to reduce congestion.

The first inner cordon performs best under a very low traffic volume (48625 PCU/hr) whereas the second inner cordon performs best when the trip volume increases to 85094 PCU/hr. With high traffic volumes, i.e. 151564 and 158033 PCU/hr, the second outer cordon performs best.

When the trip volume increases, the congestion spreads over the wider area of the network. The outer2 cordon can capture more traffic, since it has more tolled points. However this not the case for outer1 cordon. The drawback of outer1 cordon is the lack of alternative routes and it also included three major origins of the traffic inside the cordon. This results in a lower benefit compared to other cordons, despite the number of tolled points.

Table 4 Results of sensitivity test of trip volume with large-scale network

Cordon	toll points	Total benefit (thousands seconds)			
		Trip volumes			
		48,626	85,095	121,564	158,033
Inner 1	9	266.9	990.8	3,244.4	7,168.5
Inner 2	7	227.3	1,020.8	3,226.8	6,535.3
Outer 1	16	0	300.1	1,442.9	5,469.9
Outer 2	20	15.1	488.9	3,492.8	9,101.7

Analysis of Costs of the Scheme for the Large Scale Network

It should be noted that the benefit value shown in the previous section is the absolute value. In practice the equipment and operational cost of the system, which depends on the number of toll points, should be deducted from these benefits. This section reports on the effect of including of costs of the scheme, equipment and operational costs, on the analysis of the optimal cordon for the large scale network.

The equipment cost and operational cost are assumed to be £183,400 per toll point and £85,300 per toll point per year respectively (Oscar Faber, 2001). The next section describes the cash flow analysis where the discount rate is applied to spread the cost of equipment (implementation cost) over the life-time of the scheme.

Cash flow analysis of the costs of the scheme. If the implementation cost of the scheme paid in the initial year is X and we assume that the lifetime of the scheme (n) is 30 years, in order to compare with the benefit generated each year of the scheme, the implementation cost has to be spread over the next 30 years. Assuming that the discount rate (r) is 6% (DTLR, 2001), the calculation of the equivalent cost of implementation annually (C) in the next 30 years is as follows :-

$$C = \left[\frac{\left(1 - \frac{1}{(1+r)}\right)}{\frac{1}{(1+r)} \cdot \left(1 - \left(\frac{1}{(1+r)}\right)^n\right)} \right] \cdot X = \left[\frac{(1-0.943)}{0.943 \cdot \left(1 - 0.943^{30}\right)} \right] \cdot X = 0.0725 \cdot X$$

This means if the investment of the scheme in the initial year is £10,000,000, its equivalent investment cost over the next 30 years is £725,000 annually.

The equivalent investment cost of each cordon in the previous section is calculated along with its operational cost per year, and this is shown in table 5.

We allow both the operational cost per toll point per year and the equipment cost per toll point to change by ±10% of their current figure in the analysis of the total benefit of the scheme.

Table 5 Costs for the four cordons for the large scale network

Cordon	No of toll points	Total equipment cost (thousand pounds, £K)	Equivalent annual equipment cost (£K)	Operational cost per year (£K)	Total equivalent cost per year (£K)
Inner1	9	1,650	120	768	888
Inner2	7	1,283	93	597	690
Outer1	16	2,934	213	1,365	1,578
Outer2	20	3,668	266	1,706	1,972

Cost benefit analysis of the charging cordons. The benefit of the scheme in previous section has to be converted into the comparable format with the cost of the scheme in table 5. Firstly, the generalised travel time unit of the benefit has to be converted into the monetary value. This is done by simply using the value of time of 7.63 pence per minute to change the unit of seconds to pence. Secondly, the benefit is estimated for a one hour peak period. The benefit has to be calculated per year in order to be comparable to the cost in table 5. It is assumed that the system will be operated six hours per day (7am-10am and 4pm-7pm) and five days a week (Monday – Friday). This gives a total operating hours of 1,440 hours per year.

Elasticity of demand. Table 6 shows the results of the total benefit with the inclusion of the cost of the scheme. The top row is the elasticity value tested, i.e. 0.38, 0.58, 0.78, and 0.98. For each elasticity, there are three columns with different levels of the scheme costs, i.e. low, medium, and high (90%, 100% and 110% of estimated costs). The value in the table is the optimal benefit of each cordon; the unit of the benefit is in thousand pounds sterling.

Table 6 Cost benefit analysis of charging scheme as the elasticity value is varied

Cordon\EL	0.38			0.58		
Scheme costs	Low	Medium	High	Low	Medium	High
Inner1	3,525.4	3,436.6	3,347.9	5,142.4	5,053.7	4,964.9
Inner2	3,879.5	3,810.5	3,741.4	5,287.6	5,218.6	5,149.6
Outer1	332.6	174.8	17.0	1,222.1	1,064.3	906.5
Outer2	3,235.6	3,038.4	2,841.1	4,620.8	4,423.5	4,226.3
Cordon\EL	0.78			0.98		
Scheme costs	Low	Medium	High	Low	Medium	High
Inner1	6,434.4	6,345.7	6,256.9	7,415.3	7,326.5	7,237.8
Inner2	6,416.5	6,347.4	6,278.4	7,311.4	7,242.3	7,173.3
Outer1	1,951.6	1,793.8	1,636.0	2,520.4	2,362.6	2,204.8
Outer2	5,388.2	5,191.0	4,993.8	5,821.7	5,624.5	5,427.2

With no costs of implementation the optimal cordon was the outer2 cordon. When the cost of the scheme is included, the inner cordon 2 became the best location for the charging cordon instead of the outer2 cordon with the elasticity value of 0.38 – 0.58 regardless of the level of the scheme costs. It is obvious that the inclusion of the scheme cost reduces the benefit of the cordon outer 2 significantly given its higher number of toll points, see table 2. The variation of the elasticity did change the location of the optimal charging cordon where with the elasticity of 0.98 the inner 1 cordon is the best charging cordon regardless of the level of the scheme costs. With the elasticity of 0.78 the inner 1 cordon is the best cordon only with the low cost of the scheme whereas the inner2 cordon is the best charging cordon when the scheme cost is at medium and high level. This is because of the higher number of tolled points of inner1 compared to inner2 cordon.

Value of time. Table 7 shows the cost benefit analysis results of the charging cordons with the variation of value of time.

Once again when costs are included with the base value of time (7.63) the optimal cordon switches from the outer2 to inner2. However the limited variation of the scheme costs does not then affect the optimal location of the cordon with the value of time of 5.63, 7.63, and 9.63 i.e. it remains inner2. Only when the value of time decreases to 3.63, the inner1 cordon becomes the best cordon with the low and medium level of scheme costs.

Table 7 Cost benefit analysis of charging scheme as the value of time is varied

Cordon\VOT	3.63			5.63		
Scheme costs	Low	Medium	High	Low	Medium	High
Inner1	5,161.1	5,071.7	4,982.9	4,753.1	4,663.7	4,574.9
Inner2	5,138.3	5,068.8	4,999.7	4,930.4	4,860.9	4,791.9
Outer1	-0.8	-159.7	-317.5	983.3	824.4	666.6
Outer2	2,154.5	1,955.8	1,758.6	3,889.3	3,690.7	3,493.4
Cordon\VOT	7.63			9.63		
Scheme costs	High	Low	High	Low	High	High
Inner1	5,143.1	5,053.7	4,965.0	4,896.6	4,807.2	4,718.5
Inner2	5,288.1	5,218.5	5,149.5	5,131.7	5,062.2	4,993.1
Outer1	1,223.2	1,064.3	906.5	1,582.8	1,423.9	1,266.1
Outer2	4,622.2	4,423.5	4,226.3	5,092.1	4,893.4	4,696.2

Trip volumes. Once again the outrer2 cordon is replaced by inner2 for the base case as costs are taken into account. However results from Table 8 show that the outer2 cordon can become the optimal location when the traffic volume is very high regardless of the scheme costs. This is because when the traffic volume increases the congestion increases in the whole network and the outer2 cordon can cover and affect more traffic. This could generate a very high benefit despite the high cost of scheme.

But when the congestion is low the best approach to manage the network, as shown in the small scale network, is to re-route the traffic in the optimal pattern. For this smaller cordon concentrating on the central area of the city, inner2, can control the route choice more cost effectively than the wider cordon. The inner1 cordon also provides a relatively high level of benefit, but according to its higher number of the toll points compared to the inner2 cordon the benefit of the scheme is reduced by the costs.

Table 8 Cost benefit analysis of charging scheme as the trip volume is varied

Cordon\Trips	48,626			85,095		
Scheme costs	Low	Medium	High	Low	Medium	High
Inner1	-310.1	-398.9	-487.6	1,015.5	926.7	838.0
Inner2	-205.1	-274.1	-343.2	1,248.0	1,178.9	1,109.9
Outer1	-1,420.2	-1,578.0	-1,735.8	-870.6	-1,028.4	-1,186.2
Outer2	-1,747.6	-1,944.8	-2,142.1	-880.0	-1,077.2	-1,274.5
Cordon\Trips	121,564			158,033		
Scheme costs	Low	Medium	Low	Medium	Low	Medium
Inner1	5,142.3	5,053.5	5,142.3	12,328.1	12,239.3	12,150.6
Inner2	5,287.6	5,218.5	5,287.6	11,346.1	11,277.1	11,208.0
Outer1	1,222.1	1,064.3	1,222.1	8,596.3	8,438.5	8,280.7
Outer2	4,620.8	4,423.5	4,620.8	14,891.8	14,694.6	14,497.3

CONCLUSIONS

A charging cordon or toll point system is the most common charging regime normally used in introducing a road user charging scheme. The choice of the toll level and toll points depends on the demand characteristics and the objectives of the scheme. At the same time, the introduction of a toll also affects the resulting travel patterns and levels of demand. The problem to find the optimal cordon is categorised as a bi-level optimisation process. The CORDON process based on Verhoef (2000) and Shepherd et al (2001) associated with SATURN is developed to solve this optimal toll level problem.

Variation of the travel demand occurs across the day and it is widely accepted that this variation could cause a change in the optimal toll level of each tolled link. This paper presents the results of the tests of the effect of the variation of the demand on the location of the optimal cordon. The tests adopt two networks, a small artificial network and a large-scale network. For the small network, the tests are carried out using the CORDON process with a single tolled link and a pair of tolled links while changing elasticity, value of time, trip volume, and trip matrix distribution. For the larger scale network, the assumption of uniform charge on the charging cordon is applied and the univariate optimisation method is used to find the solution instead of the CORDON process.

The results for both networks show the effect of the demand characteristics on the location of the optimal cordon or tolled point(s). There are two distinct aspects to control the traffic in the network via charging, namely trip suppression and route choice or a combination of the two aspects. Whether to use trip suppression or route choice depends upon the demand characteristics. At the same time, the performance of each link in terms of these two aspects also varies according to the changes in demand characteristics.

Firstly, we discuss the optimal strategy for controlling traffic using tolls with the different demand conditions, and then we will discuss the characteristics of a link that could produce the optimal performance either by controlling demand level or route choice allocation with different demand conditions.

Different demand conditions form different traffic patterns and congestion levels in the network. The higher the congestion the more trips should be reduced. On the other hand, the use of route choice control can become the better strategy as well with the different demand conditions. The results of the small network show some interesting points as follows:

- The user with the low elasticity tends to continue using roads by changing the route to avoid the tolls rather than give up the trip. Therefore, the best approach to deal with the low elasticity of demand is to allocate the traffic in the efficient manner. In the extreme case of fixed demand all benefits are due to re-routeing.
- On the other hand, when the user is more sensitive to the toll (high elasticity) it is better to suppress the trip which can be best achieved by tolling links with few alternative routes or where the alternative route becomes heavily congested.
- The effect of changing value of time is complicated, since it affects both the route choice and demand level. The route choice decision is affected through the change of the importance of the distance of travel compared to travel time. The increase in value of time increases the generalised cost of travel and hence the number of suppressed trips is higher. Therefore, the best strategy for controlling traffic with the different value of time depends on the structure of the network and other demand characteristics.
- The change of trip volume causes the change in the congestion level of the network. The highly congested network has higher potential benefits and there is a need for the toll to reduce the traffic rather then divert the traffic.
- On the other hand, when the network congestion is light, it is better to allocate the traffic to other routes rather than suppress the trips.
- Finally, the change of trip distribution in the matrix causes the most significant effects on the optimal location of toll points. When the matrix changes, the concentration of congestion as well as route choice is altered. Given the same trip volume but different distribution between O-D pairs, the best strategy is to suppress the trips between the O-D pair that has a higher demand level.

Changing demand characteristics changes route choice and demand response in the network, and this can change the ability of each link to control either route choice or demand level. The results from the small network show a set of conclusion related to this issue as follows:-

- the link covering most of the congested paths and not providing the alternative route to the toll is the link that could perform best in terms of controlling demand level;
- the link with the relevant path with high marginal costs and providing alternative routes to the toll is the better link to be tolled if the main objective is to allocate the trips in the efficient pattern rather than suppressing them.

With the large network the cordons also have different abilities to affect the route choice and demand levels. The simple illustration in the small network is not sufficient for the results of the large scale network. The distribution of the congestion in the network varies by area and there are a high number of paths and O-D pairs in the network. However, the results still can show some interesting points as follows:-

- The performance of each cordon in terms of controlling demand level and route choice depends on the network structure and demand condition;
- The best strategy for each demand condition may be the mixture of the two aforementioned strategies, controlling demand level and route choice.

However, the results show the influences of the demand variation on the optimal cordon location.

- As elasticity of demand is increased then the optimal cordon moves from outer to inner.
- As the value of time is increased then the optimal cordon moves from inner to outer.
- As the trip volume increases then the optimal cordon moves from inner to outer.

The explanations for these changes in optimal cordon are complex and depend upon the trips which can be tolled and upon the alternative routes available. However one simple conclusion follows from the results :-

- When congestion has spread over a wide area a cordon which suppresses the most trips is preferable, in this case the outer cordon.

However the subsequent tests with the assumed scheme costs shows the significant effect of the cost of the scheme on the optimal location and hence optimal number of tolled points. The previously optimum outer cordon2 (in the base case) is replaced by a less expensive inner cordon.

It was also shown that the optimal cordon location was sensitive to the cost assumptions, as the optimal cordon could change for certain demand characteristics as the costs varied by only plus or minus 10%.

This paper has demonstrated that the optimal cordon location depends upon demand characteristics which can vary in reality throughout the day. Future research should consider how to design an optimal cordon which is robust to these changes in demand characteristics or which can vary the charges during its period of operation. The tests conducted in this paper are still based on several assumptions, single user class, uni-modal network (exclude public transport), single time period, and no departure time choices. Future research should look at the effect of relaxing these assumptions. There are also other objectives of road pricing schemes, e.g. improving environment and revenue generation. The location of the optimal cordon may be sensitive to these different objectives as well as to the change in demand characteristics.

ACKNOWLEDGEMENTS

This research is funded by the UK EPSRC. Thanks are due to Dr. Erik T. Verhoef for his collaboration in this research and also thanks to other members of the project, Dr. David Milne, Dr. David Watling, and Dr. Ronghui Liu.

REFERENCES

Arnott, R.J., de Palma, A. and Lindsey R. (1990)a. Departure time and route choice for the morning commuter. *Transportation Research* 42B.

Arnott, R.J., de Palma, A. and Lindsey R. (1990)b. Economic of a bottleneck. *Journal of Urban Economics* 27.

Button, K. K. (1992). *Transport Economics 2nd edition.* Edward Elgar Publishing, Cheltenham, UK.

DTLR (2001). Transport Economics Note URL: http://www.roads.detr.gov.uk/roadnetwork/heta/highway/03.htm

Goodwin, P.B. (1992). A review of new demand elasticities with special reference to short and long run effects of price changes. *Journal of Transport Economics and Policy* 26(2).

Hague Consulting Group and Accent Marketing and Research (1996). *The Value of Travel Time on UK Roads-1994*. Prepared for the Department of Transport.

Knight, F.H. (1924). Some fallacies in the interpretation of social cost. *Quaterly Journal of Economics* 38.

Marchand, M. (1968). A note on optimal tolls in an imperfect environment. *Econometrica* 36.

May, A.D., Bonsal, P.W. and Hills, P.J. (1998). The impacts of different road users charging systems. *Proc. of Transportation Into the Next Millennium*, Nanyang Technological University, Singapore.

Menon, A.P. (2000). ERP in Singapore – a perspective one year on. *Traffic Engineering and Control* 41(2).

Morden, L. (1996). Value pricing tested on 91 express lanes: California's first privately-financed tollroad implements a variable rate policy. *ITS international* 4/ 1996.

Oscar Faber (2001). Birmingham City Council Road User Charging Study-West Midlands Final Report (unpublished)

Pigou, A.C. (1920). *Wealth and Welfare*, Macmillan (London).

Press, W.H, Flannery, B.P., Teukolsky, S.A. and Vetterling, W.T. (1990) *Numerical Recipes : The art of scientific computing*. Cambridge University Press.

Sharp, C.H. (1966). Congestion and welfare, an examination of the case for congestion tax. *Economic Journal* 76.

Shepherd S., May A.D., Milne, D.S. (2001). Optimal Road Pricing Cordon, *Proc. of the World Conference of Transport*, forthcoming.

Swenson, C.R., Gilbertson, S.M., Byers, M. and Burris, M. (1999). Variable toll reduction as a practical transportation demand strategy. *Proc. Transport Frontiers for the Next Millennium: 69th Annual Meeting of the Institute of Transportation Engineers*, Las Vegas, USA.

Van Vliet, D. (1982). SATURN – a modern assignment model. *Traffic Engineering and Control* 23(12).

Verhoef, E.T. (1998) "Second-best congestion pricing in general static transportation networks with elastic demands" Discussion paper TI 98-086/3, Tinbergen Institute, Amsterdam-Rotterdam. *Regional Science and Urban Economics*, forthcoming.

Verhoef, E.T. (2000) "Second-best congestion pricing in general networks: algorithms for finding second-best optimal toll levels and toll points" Discussion paper TI 2000-084/3, Tinbergen Institute, Amsterdam-Rotterdam.

Vickrey, W. (1995). Some implications of marginal cost pricing for public utilities. *American Economic Review (paper)* 45, Nashville, Tennessee.

Walters, A.A. (1961). 'The theory and measurement of private and social cost of highway congestion'. *Econometrica* 29(4).

Wardman, M. (2000). The Value of Travel Time: A Review of British evidence. *Journal of Transport Economics and Policy* 32(3).

Wardrop, J. (1952). Some theoretical aspects of road traffic research. *Proc. of the Institute of Civil Engineers* 1.

Yang, H. and Huang, H.J. (1998). Principle of marginal-cost pricing: How does it work in a general road network?. *Transportation Research* 32A(1).

APPENDIX A: NOTATION

I	the set of OD-pairs, denoted i=1,…,I
T_i	the continuous number of users (or OD-flow) for OD pair i, with $T_i \geq 0$
$D_i(T_i)$	the inverse demand function for trips for OD-pair i, with $D'_i \leq 0$
J	the set of directed links in the network, denoted j=1,..,J
V_j	the continuous number of users (or link flow) on link j, with $V_j \geq 0$
$C_j(V_j)$	the average cost function for the use of link j, with $c_j' \geq 0$
i	index for OD pairs
j	index for links

APPENDIX B: TRIP MATRIX PATTERNS

Base trip matrix

Origin\Destination	A	B	C
A	0	0	1000
B	0	0	1000
C	0	0	0

Trip matrix pattern 1

Origin\Destination	A	B	C
A	0	0	1300
B	0	0	700
C	0	0	0

Trip matrix pattern 2

Origin\Destination	A	B	C
A	0	0	1600
B	0	0	400
C	0	0	0

Trip matrix pattern 3

Origin\Destination	A	B	C
A	0	0	700
B	0	0	1300
C	0	0	0

Trip matrix pattern 4

Origin\Destination	A	B	C
A	0	0	400
B	0	0	1600
C	0	0	0